21097487

£8·95₂.

PERGAMON INTERNATIONAL LIBRARY
of Science, Technology, Engineering and Social Studies
The 1000-volume original paperback library in aid of education,
industrial training and the enjoyment of leisure
Publisher: Robert Maxwell, M.C.

Related Pergamon Titles of Interest

Books

ALLUM
Photogeology and Regional Mapping

ANDERSON
The Structure of Western Europe

ANDERSON & OWEN
The Structure of the British Isles 2nd edition

GRAY *et al.*
Studies in the Scottish Lateglacial Environment

OWEN
The Geological Evolution of the British Isles

PRICE
Fault and Joint Development in Brittle and Semi-brittle Rock

ROBERTS
Introduction to Geological Maps and Structures

SIMPSON
Geological Maps

Journals
Journal of Structural Geology

Quaternary Science Reviews

The terms of our inspection copy service apply to all the above books.
Full details of all Pergamon books and journals and a free specimen copy of
any Pergamon journal available on request from your nearest Pergamon
office.

Field Geology
in the British Isles
A Guide to Regional Excursions

by

J. G. C. ANDERSON
Emeritus Professor of Geology, University College, Cardiff

PERGAMON PRESS

OXFORD · NEW YORK · TORONTO · SYDNEY · PARIS · FRANKFURT

F

U.K.	Pergamon Press Ltd., Headington Hill Hall, Oxford OX3 0BW, England
U.S.A.	Pergamon Press Inc., Maxwell House, Fairview Park, Elmsford, New York 10523, U.S.A.
CANADA	Pergamon Press Canada Ltd., Suite 104, 150 Consumers Road, Willowdale, Ontario M2J 1P9, Canada
AUSTRALIA	Pergamon Press (Aust.) Pty. Ltd., P.O. Box 544, Potts Point, N.S.W. 2011, Australia
FRANCE	Pergamon Press SARL, 24 rue des Ecoles, 75240 Paris, Cedex 05, France
FEDERAL REPUBLIC OF GERMANY	Pergamon Press GmbH, Hammerweg 6, D-6242 Kronberg-Taunus, Federal Republic of Germany

First edition 1983

Library of Congress Cataloging in Publication Data

Anderson, J. G. C. (John Graham Comrie)
Field geology in the British Isles.
(Pergamon international library of science, technology, engineering, and social studies)
Includes index.
1. Geology — Great Britain — Guide-books.
2. Great Britain — Description and travel —
1971-   — Guide-books. I. Title. II. Series.
QE261.A817   1983   554.1   83-6275

British Library Cataloguing in Publication Data

Anderson, J.G.C.
Field geology in the British Isles. —
(Pergamon international library)
1. Geology — Great Britain — Field work
I. Title
550'.7'23   QE45
ISBN 0-08-022054-1 (Hardcover)
ISBN 0-08-022055-X (Flexicover)

In order to make this volume available as economically and as rapidly as possible the author's typescript has been reproduced in its original form. This method unfortunately has its typographical limitations but it is hoped that they in no way distract the reader.

Printed and bound in Great Britain by Redwood Burn Limited, Trowbridge

Preface

The first three chapters provide information on the best means of studying geology in the field in the British Isles. In Chapters 4 to 9, 194 geological itineraries are described based on a number of centres. All can be done in a day. However, if detailed study is envisaged, many may require two or three days. Thus centres for which only two or three itineraries are suggested can readily be used as bases for a week's work.

No two geologists would agree which are the "best" 200 or so excursions in the British Isles; those chosen present as many facets as possible of British geology.

A novel feature is a final chapter on the geology evident on 31 journeys by road, rail and coastal boat.

Sketch maps indicate the routes of all the excursions but the reader should also have a good topographical map. Maps show the geology of each region, and more detailed geological maps are included for some excursions. To have provided such maps for all the itineraries would have greatly added to length and cost and is unnecessary for a number of districts for which excellent modern maps on the 1:50,000 or 1:25,000 scale are readily available. Topographical and geological map numbers are given for every excursion.

For descriptions of the structure of each region, reference should be made to an earlier volume in this series "The Structure of the British Isles", 2nd ed., 1980. Conversely the present book complements the earlier volume by providing local detail.

Acknowledgements

Valuable advice and help were given by Professor T. R. Owen, Swansea University College.

To several colleagues at University College, Cardiff, thanks are due for providing information and for reading the manuscript.

A number of the figures were drawn by Mrs. Margaret Millan, of the same College.

Contents

List of Illustrations

CHAPTER 1

Introduction

J. G. C. Anderson

Geology Department, University College, Cardiff CF1 1XL, UK

The British Isles rise to a maximum height of 4406 ft. (1344 m) above sea
level from shallow seas covering the North-West European continental shelf.
They range from $49^{\circ}52'$N to $60^{\circ}52'$N and from $1^{\circ}50'$E to $10^{\circ}50'$W. Their length
from Pednathise Head in the Isles of Scilly to Muckle Flugga in Shetland is
790 miles (1271 km). Yet this small part of the earth's land-surface
reveals nearly the whole geological succession and structures imposed by
Precambrian, Caledonian, Hercynian and Alpine orogenies.

From Hutton's time many basic geological principles were established in
these islands. Moreover, a number of the main stratigraphical divisions
were named after ancient (e.g. Cambrian) or more modern (e.g. Devonian)
British districts. For over 150 years therefore geologists from Europe
and later from all over the world have visited the British Isles to study
geology in the field. Most British Universities have large geology depart-
ments, and the subject is being increasingly taught in schools. More and
more people add to their interest in walking and climbing by making
geological observations.

This book is for all those who wish to see the varied and often beautifully
displayed geology of the British Isles for themselves and to use these
exposures to teach geology where it can best be taught, in the field.

Information about British geology is widely available from the maps and
memoirs of the Geological Survey (see beginning of Chapter 2), from papers
in scientific journals and from both popular and specialised text-books.
There are also excursion guides to a number of individual districts. These
Sources of Information and others are discussed in Chapter 2.

With some exceptions, existing guides give little or no information about
how to get to excursion centres, what type of accommodation is available
and how the centre can be reached and excursions undertaken using public
transport. Information is often lacking on weather conditions in different
regions at different times of the year and on difficulties or even dangers
which may arise, particularly in mountainous or coastal terrains. These
aspects of field-work in the British Isles are discussed in Chapter 3 - Ways
and Means.

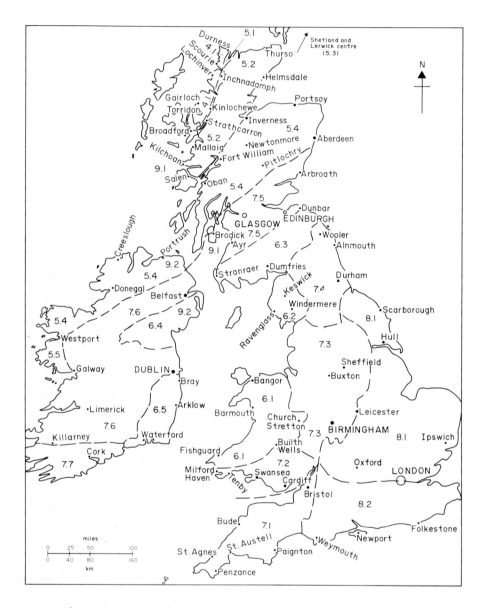

Fig. 1.1 Map of districts described and of excursion centres.

Chapters 4-9 consist of itineraries from a number of centres. The grouping
is based in the main structural regions as defined in an earlier book in
this series "The Structure of the British Isles" (Anderson and Owen 2nd ed.
1980), in which the stratigraphy and structural evolution of the regions
are described. Chapters 3-9 of the present book have the same numbers as
the corresponding chapters of the earlier book. In fact the present
volume can be regarded as complementary to the earlier volume as it
includes descriptions of sections and exposures which could not be
included in the earlier publication.

It is possible to complete each of the itineraries described in Chapters
3-9 in a day. Nevertheless, if it is intended to stay longer in one
centre, some of the excursions can be divided between two days, especially
if detailed study or systematic collecting is planned. For some excursions
how this division can best be made is indicated. Geologists may also find,
once they have become reasonably familiar with a district and with the
descriptive literature, that other excursions arise out of those described
in this book.

The geology, and its effects on the topography and economic development,
seen along some long-distance journeys in the British Isles, some by
private transport, some by rail or coach, and one or two by boat, are
described in Chapter 10.

Many formations show striking contrasts in the lithology of strata of the
same age, even within the relatively short distances of the British Isles.
Thus the Lower Palaeozoic geosynclinal turbidites of Wales are completely
different from the carbonate facies of N.W. Scotland, and the thick carbonates in the Lower Carboniferous of South Wales and much of England are
in marked contrast to the thin limestones, sandstones, shales, coals and
lavas of the Midland Valley of Scotland. These contrasts are emphasised
in a section on Comparison with other areas included in the introduction
to each region.

Some geologists, particularly from overseas, consequently may be interested
more in the development of a particular formation throughout the British
Isles. For them the regional groupings may be less convenient.
Accordingly, excursions on which significant exposures of each of the main
subdivisions of the stratigraphical column are seen are listed in an
appendix which also includes those on which glacial features are important.

Some of the itineraries given in Chapters 4-9 have been described in
existing guides; others are in districts for which descriptions of
excursions are sparse or not available, for example, parts of Central Wales,
the Southern Uplands and Highlands of Scotland, and Ireland. The various
authors of existing guides show wide variations in their approach, some
giving almost yard by yard detail, others a general description with some
suggestions on routes. A standard presentation has been adopted in
Chapters 4-9 which, it is hoped, falls between these extremes. It will be
realised that it is impossible, when dealing with the whole of the British
Isles, to go into too much detail, but all important outcrops, contacts,
structures and igneous intrusions are localised and means of access indicated.

It is possible that the omission of minor detail may have its positive
side. Some geologists may find that it adds to their interest to work
out details for themselves. Again, where sections are used for field

seminars, the students will have to work it out a priori having not had
the chance to read up all the details beforehand.

Fossil lists are not given, but fossil localities are indicated, and
stratigraphically significant fossils named. Restraint should be shown
in collecting.

Areas where photography or sketching is forbidden on security grounds are
very limited and usually clearly signposted.

The only land-frontier is that between Northern Ireland, part of the
United Kingdom, and Eire (Southern Ireland), and as long as the present
disturbances continue this will remain a sensitive area. Before undertaking excursions anywhere near this border, the local police should be
consulted and their advice accepted.

At certain frequently visited sites, geologists should collect sparingly
or not at all; this particularly applies to some classic fossil localities,
indicated in subsequent chapters.

It has been thought advisable to mention these restrictions, legally
enforceable or voluntary, but for most excursions they do not arise, and
they should certainly not discourage visitors. In the British Isles
fieldwork and reasonable collecting can be undertaken more easily than
in many other countries.

Maps for Excursions

In the itineraries which follow references are given to topographical and
geological maps. In the case of topographical maps, where a number only is
given this is the sheet-number of the 1:50,000 maps of Great Britain.

For geological maps E W 5, etc. is the sheet of the 1:63,360 or 1:50,000
map of England and Wales, S 5 etc., the sheet of the 1:63,360 or 1:50,000
map of Scotland and N I 5 etc. the sheet of the 1:50,000 map of Northern
Ireland. For all other maps a full reference is given.

Some references are given to maps which are out of print or being replaced,
as it is possible that these will become available while the book is in use.

A simplified geological map is included for every region but for reasons of
space and therefore of cost, it has not been possible to include a map for
every excursion. Those following the itineraries should have a good topographical map (easily obtained in the British Isles) and, as far as possible,
a geological map (see next Chapter and itinerary headings).

Only a selection of excursions from each centre can be given; from personal
visits and discussion these seem the most significant to the author but it
is realised that this is something on which no two geologists would completely agree. Other excursions will no doubt suggest themselves to readers,
either from experience when the centre is reached or from reading the
literature.

Reference

Anderson, J. G. C. and Owen, T. R.   1980.   The Structure of the British Isles, 2nd ed. Pergamon Press.

CHAPTER 2

Sources of Information

The most useful sources of field information in the British Isles for over one hundred years have been the maps of the Geological Survey of Great Britain and Ireland. Following the establishment (1922) of southern Ireland (Eire) as an independent state, a separate Geological Survey, based in Dublin, was set up for that country.

In 1965 the United Kingdom Geological Survey became a major part of the newly-established Institute of Geological Sciences (I.G.S.). The Geological Survey maps of Great Britain and Northern Ireland are prepared for publication by the United Kingdom Ordnance Survey.

Although on too small a scale for local field-studies, the Geological Survey map, in two sheets, of the whole of Great Britain (3rd ed., 1979) on the scale of 1:625,000 (approx. 10 miles to 1-inch) is well worth purchasing by any geologist who wishes to get an initial overall picture or who intends to travel through the country. The geology of the whole of Northern Ireland (and adjacent parts of Eire) is shown on a map on the 1:250,000 scale (approx. 4 miles to 1-inch), and there is a geological map of the whole of Ireland on the scale of 1:1,000,000 (approx. 16 miles to 1-inch).

Most of Great Britain is also covered by geological maps on the 4-miles to the inch scale. These have been published for the whole of England and Wales, and for Scotland (with a separate sheet grid from that for England and Wales) except for the Outer Hebrides and parts of the Highlands.

Some of these maps are out of print. It is understood that they will not be replaced; instead a new set of geological maps will gradually be published on the scale of 1:250,000 (approx. 4 miles to 1-inch).

For many years the most useful maps for field geology and other purposes have been the "New Series" maps on the scale of 1 mile to 1-inch (1-63,360) of England and Wales and the maps on the same scale (individually larger and based on a separate sheet grid) of Scotland. These have been largely replaced by maps on the 1:50,000 scale (approx. 1 mile to $1\frac{1}{4}$ inches) which are based on the same sheet grids as the 1-inch and have the same

sheet numbers. As the Scottish maps are larger, the 1:50,000 maps are
mostly published as separate East and West sheets.

For some areas of particular geological interest which overlap sheet
boundaries "Special" maps on the 1:63,360 or 1:50,000 scale have been
published, e.g. for the islands of Arran and Anglesey.

Maps on the 1 mile to 1-inch scale (1:63,360) were published for Ireland
as a whole. For Northern Ireland these are being replaced by 1:50,000
sheets.

Some areas in Great Britain of special geological importance are now shown
on sheets on the 1:25,000 scale (approx. 1 mile to $2\frac{1}{2}$ inches), e.g. Central
Snowdonia and Church Stretton. This is the largest scale likely to be
required for excursions (unless further research is planned) but it should
be mentioned that geological maps on the scale of 1 mile to 6 inches
(1:10,560) are published for the British Coalfields. Maps on the same
large scale for most of Great Britain and Northern Ireland are on the
files of the I.G.S. and can be consulted in the relevant I.G.S. offices
in London, Edinburgh, Leeds, Aberystwyth and Belfast. Maps on this scale
for parts of Eire can also be seen in Dublin.

A number of the maps mentioned are, at the time of writing, out of print.
For parts of North and Central Wales "New Series" 1 mile to 1-inch maps
were never published, and the only geological maps on this scale are hand-
coloured sheets of an "Old Series" which are now rare. Mapping of parts
of the central Scottish Highlands and of the Outer Hebrides is, to date,
not sufficiently complete to publish maps on the 4 miles to 1-inch or
1:50,000 scales.

In general the geological maps on scales of less than 1:63,360 do not show
superficial deposits such as glacial drift. On the 1 mile to the inch maps
and those on larger scales the approach varies. Some are published as
"Solid" editions only, some as "Drift" editions only, and an increasing
majority as separate "Solid" and "Drift" editions. On a few the superficial deposits are shown, not too successfully, by overprinted ornaments.

A great deal of information of help in the field is contained in the
memoirs, reports, etc. published by the I.G.S. Some are related to sheets
of the geological maps, others to districts, others to special topics.
Great Britain and Northern Ireland are also covered by a series of
Regional Guides. These are useful to obtain an overall picture of the
geology of a region but in general individual sections are not described.
References are given at the end of Chapters 4-9.

Most I.G.S. maps, memoirs etc. in print may be purchased at the bookshop
in the Geological Museum and Institute of Geological Sciences in London
(address below). Regional publications are available or at any rate can
be inspected at I.G.S. offices outside London.

The addresses of the Geological Survey Offices of the I.G.S. are as
follows:-

> Geological Museum and Institute of Geological Sciences,
> Exhibition Road, South Kensington, London SW7 2DE.

> Murchison House, West Mains Road, Edinburgh EH9 3LA.

20 College Gardens, Belfast BT9 6BS.

Ring Road, Halton, Leeds LS15 8TQ.

I.G.S. Office, Aberystwyth, Dyfed, Wales.

Geological Survey Memoirs are stocked in H.M. Stationery Offices in the main British towns and cities.

Local geological maps are also sometimes obtainable in stationers and bookshops in tourist areas and some cities.

In Eire the Geological Survey office is at 14 Hume Street, Dublin 2 and publications are sold at the Government Publications Office, G.P.O. Arcade, Dublin 1.

A vast amount of information and many maps relevant to field geology are available in papers published in geological and other journals which include:

The Quarterly Journal of the Geological Society of London;

Replaced in 1971 by the Journal of the Geological Society of London;

The Geological Magazine;

The Proceedings of the Geologists' Association;

The Scottish Journal of Geology;

The Transactions of the Royal Society of Edinburgh;

The Proceedings of the Yorkshire Geological Society;

The Geological Journal of the Liverpool and Manchester Geological Societies;

The Proceedings of the Royal Irish Academy.

Of excursion guides, the most comprehensive are the series of pamphlets published by the Geologists' Association from 1958 onwards. These deal with the geology around some of the University towns and with several classic areas but a number of geologically very interesting areas are not covered.

The "Selected References" at the ends of Chapters 3-10 include some of the publications which have been mentioned. Fuller lists of references will be found in the companion work to this book already referred to (Anderson and Owen 1980), in the I.G.S. Regional Guides, and in other works cited.

A variety of topographical maps is available, suitable for excursion planning and assessment of terrain. The most useful are the 204 sheets of the Landrover Series of Great Britain (1:50,000 or approx. 1 mile to 1¼ inches) published by the Ordnance Survey. Other maps include the Bartholomew's National Map Series on the 1:100,000 (1.6 miles to 1-inch) scale.

For Ireland Ordnance Maps on the 2 miles to 1-inch (1:126,720 scale) are available and there are Bartholomew maps on the 4 miles to 1-inch (1:253,440) scale.

Ordnance Survey maps may be bought at H.M. Stationery offices (usually 1:50,000 only) and at the Government office in Dublin.

Both Ordnance Survey maps and other maps of local areas are sold by many booksellers and stationers.

The importance of knowledge of tide times is stressed. Tide-tables are printed for some coastal towns, and information can be obtained from harbour offices. Details for the whole of the British Isles are given in the Admiralty Tide Tables, Vol. 1 published by the Hydrographer of the Navy and obtainable through booksellers.

CHAPTER 3

Ways and Means

Geology can be studied in the field in most parts of the British Isles which lie below about 1000 ft. (305m) throughout the year. The summer months have obvious advantages both as regards weather and the length of the day. However, many of the geologically most interesting districts are also attractive to tourists and holiday-makers, and accommodation problems may arise, especially for groups. Those who can should visit such districts in the spring and autumn when the weather is usually mild and the days fairly long. In some districts spring has the advantage as the vegetation is thinner. Winter excursions in the British Isles above about 1000 ft. (305m), even in good weather, may be disappointing as exposures may be covered in snow. In the N. snow may cause difficulties down to sea-level, and in some winters low-level snow lies in the S. Owing to the comparatively high latitude, winter days in the Scottish Highlands and Islands are short. At all times of the year the mountain areas of Wales, the Lake District, Scotland and Ireland have high rainfall and are often mist-covered. The weather can change very rapidly; more lives are lost on the mountains and moorlands of the British Isles through exposure than by climbing accidents.

The coasts of the British Isles provide many spectacular geological sections. Cliffs can often be examined from below by walking between tide-marks. Tidal times should always be ascertained, as the rise and fall is considerable; along the Severn Estuary the tidal range can reach $55\frac{1}{2}$ ft. (17m). High winds and powerful wave-action are common. Many cliffs are unstable, and rock-falls have to be guarded against. Along some coasts if a tight schedule makes it impossible to carry out an excursion while the tide is low, good exposures can often be found in raised beach cliffs. The hire of a small boat is possible in some districts.

Most of the excursions described in Chapters 3-9 can be carried out with the use of public transport although private transport is a convenience for some and almost essential for others. All the main towns in the British Isles can be reached by fast inter-city train services or by coach services. The smaller centres mentioned are also on rail or bus routes and the island centres can be reached by ferries. Once at the centres the visitor in most

cases will find local bus or train services which will take him to a point
from which the fieldwork can be carried out by walking a reasonable distance.
There are some exceptions in the more remote mountain districts. Public
transport can render it possible to start an excursion at one point and
make a cross-country traverse to another. In the itineraries, attention is
drawn to the existence of public transport services but it would be mis-
leading to give schedules. The fare-structure of public transport is complex, and it is worth making sure that the cheapest fare is obtained that
will suit the aim of the excursion. In particular, attention is drawn to
the favourable 'day-return' fares available for many routes. 'Circular
tour' and 'Rover' tickets may also be an advantage.

For those using private transport there are motorways (freeways in the
U.S.A. sense) between many of the major cities, and surfaced roads in nearly
every district. Unsurfaced roads or tracks are, however, the only access
for vehicles on some of the itineraries in Wales, Scotland and Ireland.
In the North of Scotland there are long stretches of single-track road with
passing places. Blockage of higher roads by snow in winter is common.
Some roads on which a car can be driven are unsuitable for larger vehicles,
and in a few areas, e.g. on some of the roads of the Lake District, coaches
are prohibited in the holiday season. The hire of a car, or of a coach
with driver, is easy to arrange in most centres, although some days' notice
may be necessary, particularly for coaches. When dealing with the larger
(and sometimes more expensive) firms it may be possible to hire a car in
one town and to return it in another.

The British Isles are well supplied with hotels, prices ranging from
moderate to expensive. In most tourist areas and along many roads 'Bed
and Breakfast' can be obtained relatively cheaply. Camp-sites, caravan-
sites and youth hostels are widely distributed. Advance booking should
always be made for parties. Most cities and holiday towns have tourist
offices from which lists of accommodation at all price levels can be
obtained. Specialist booklets detailing camp-sites, etc. are published
annually and can be obtained from stationers.

Off the roads the British Isles are criss-crossed by a network of tracks
and footpaths (often 'rights-of-way'). Although most of the land bordering
roads, 'rights-of-way' etc. is privately owned, in general there is not
likely to be much objection to walking over rough ground, open moorland and
mountain country. Permission should, however, be sought to enter enclosed
land and farmland generally.

Special laws against trespass apply to works and industrial properties;
these include working quarries, and even some apparently disused quarries
may be part of the reserves of quarry companies active elsewhere.
Permission to enter working quarries can, however, sometimes be obtained.

Railway lines and motorways are all fenced in the British Isles, and
special trespass regulations apply; apart from emergency, parking at the
side of motorways is forbidden.

Much of the coastline is followed by public shore or cliff paths. With the
exception of harbour and defence areas, there is always access along the
intertidal zone; the warning given above about tidal rise is repeated. The
borders of freshwater are, however, often private.

Areas where photography or sketching is forbidden on security grounds are
very limited and usually clearly signposted.

CHAPTER 4

Precambrian Terrains

The North-West Highlands and the Outer Hebrides of Scotland are the only parts of the British Isles affected merely to a minor extent by events which took place after the Precambrian. In this chapter excursions are described to the mainland part of this kratogen. The Thrust-zone or Caledonian Front, which borders the kratogen to the E., is dealt with in Chapter 5.1. Mention of the Outer Hebrides is made in E10.28.

The region is sparsely-populated and deeply-indented with fjord-like sea lochs. The Lewisian "basement" makes bare, hilly ground with small lochs and peaty hollows. Above, the spectacular relic mountains of late Precambrian sandstone rise to over 3000 ft. (915 m); some are capped by Cambrian quartzite, which in the N. rests directly on the Lewisian.

Roads are few and often narrow although improved in recent years. Public transport is limited to once-a-day services from railheads on the lines N. and W. of Inverness to several of the villages. Private transport is therefore a major advantage. However, most of the excursions can be done by longish walks from the centres suggested. There are a number of hotels, mostly expensive. Parties can be accommodated only outside the holiday season. There are, however, numerous houses providing "bed and breakfast". Camping may be a solution, and in summer several organised camp-sites are available. The centres suggested are Scourie, Lochinver, Gairloch and Torridon but there are several alternatives, particularly if private transport is used.

The succession is as follows:-

	Jurassic	
	Permo-Triassic	
	Ordovician	
	Cambrian	
Upper Proterozoic	Upper Torridonian	⎧ Aultbea Group ⎨ Applecross Group ⎩ Diabaig Group
	Lower Torridonian	Stoer Group

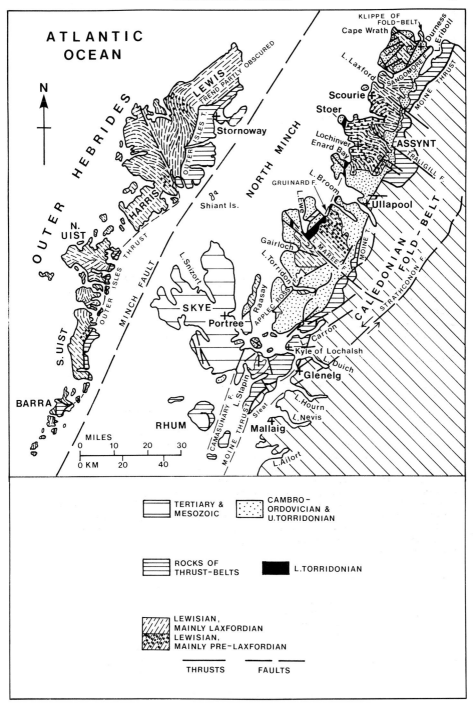

Fig. 4.1 Geological map of North-West Highlands and Outer Hebrides.

Lower Proterozoic		Laxfordian cycle
	Lewisian	Inverian cycle
Archean		Scourian cycle

For some itineraries not included in the present account the reader is referred to Barber and others (1978) and to Johnson and Parsons (1979).

Comparison with other areas:- The North-West Highlands and the Outer Hebrides are unique in the British Isles in that they are the only well exposed part of a kratogen and provide the only extensive outcrops of Archean and Lower Proterozoic rocks. These rocks are more closely related to gneisses etc. in North America than to other British Precambrian metamorphics, and the Torridonian sediments provide evidence of derivation from a source now across the North Atlantic. The shelf-sea Cambrian quartzites and Cambro-Ordovician thick carbonates are in complete contrast with the Lower Palaeozoic greywacke/shale sequences further S. (Chapter 6) apart from the limited Cambrian quartzites of the South-East Caledonian Front (E6.8).

However, the Precambrian and Cambro-Ordovician rocks of the kratogen form the autochthonous nappes of the Caledonian Front (Chapter 5.1), and the Lewisian, affected by Caledonian events, occurs in the Fold-Belt further E. (Chapter 5.2). Part of the Moinian metasedimentary succession (Chapter 5.2 and 5.4) is considered to be the stratigraphical equivalent in the Fold-Belt of the Upper Torridonian.

Scourie Centre (Fig. 4.2)

E4.1   South then North of Scourie

Lewisian.

Access:- Mainly coast walking. The geology can best be seen around low tide.

Maps:- * Topo:- 9   Geol:- ¼-inch to mile Scotland 5.

Walking distance:- 5½ miles (8.8 km).

Itinerary:- Walk S.W. from Scourie to Scourie More then a short distance across fields to coast. By following the low coastal cliffs about ¼-mile (0.4 km) to the S.S.E. a good section of granulite facies gneisses, developed during the main (Badcallian) phase of the Scourian cycle, can be seen. The rocks include metaperidites containing pyroxene, garnet and plagioclase, and intermediate and acid gneisses containing pyroxene, plagioclase and quartz. Apart from rare granite sheets and pegmatites, rocks containing potash felspar are absent. The granulites show coarse banding, while each lithological unit may form concordant sheets up to 40 m in thickness. The banding has a uniform shallow dip. Characteristically, mineral grains are equidimensional.

Return to Scourie then turn off the main road just W. of Scourie Hotel and follow the track to the pier. Pass through the gate above the old boat-house and follow the coast N.W. past Poll Eorna to the headland of Craig a' Mhail. This is the type locality for the Scourie dyke, made classic by

* For an explanation of references to maps here and in subsequent excursion descriptions, see latter part of Chapter 1.

Teall's (1885) description of the transformation of dolerite to hornblende-schist within a shear zone. The country rocks are granulite facies gneisses and a virtually planar contact with the N.E. margin of the dyke is well exposed. Within the shear zone the dolerite develops an LS fabric, and the shear zone is much wider in the gneiss outside the dyke than within the dyke.

Return to Scourie.

E4.2 North of Tarbet

Lewisian. Laxfordian front.

Access:- By private transport to Tarbet then by hill and coast walking. Can also be done by walking from Scourie to Tarbet and back. Some of the geology is best seen at low tide.

Maps:- Topo:- 9. Geol:- ¼-inch to mile Scotland 5.

Walking distances:- From Tarbet 3½ miles (5.6 km). From Scourie 8½ miles (13.6 km).

Itinerary:- Walk N. along the coast from Tarbet, examining a thick sequence of gneisses with bands of brown-weathering schists, characterised by a garnet-biotite-plagioclase-quartz assemblage. These have been developed within the Tarbet Laxfordian shear-zone and may be metasedimentary in origin. About ¼-mile (0.4 km) further N. there is a mass of strongly foliated but homogeneous rock of granitic composition. This contains potash-felspar, virtually unknown in the granulite facies area further S. (E4.1). Two thirds of a mile (1.1 km) further N.W. an interesting dyke, which thins down and terminates inland, is exposed on the intertidal platform. It has fairly planar margins and is unusual for the region in being a complex mixture of three rock types - a felsic, a mafic and a normal dolerite type.

The coast should be left here and a traverse made a short distance inland across Cnoc Gorm (Blue Hill) which takes its name from a strip of metagabbro running in a N.W. direction and which forms part of a Scourian sheet; large garnets occur. Continue N. to Rubha Ruadh (Red Point). A number of minor folds of the Scourian gneisses, defining a series of antiforms and synforms, are seen. Laxfordian deformation has resulted in the tightening up of these late Scourian folds, and the dykes that cut them are themselves folded. There are also an increasing number of pegmatites, often only slightly deformed.

The headland of Rubha Ruadh is largely made up of pink Laxfordian granite sheets. The S. limit of the granite is sharply defined, the boundary forming the Laxfordian front. Many ultramafic schistose lenses occur, a characteristic feature of the Laxfordian front.

From Rubha Ruadh walk S.E. along the ridge to Fanigmore thence by road back to Tarbet.

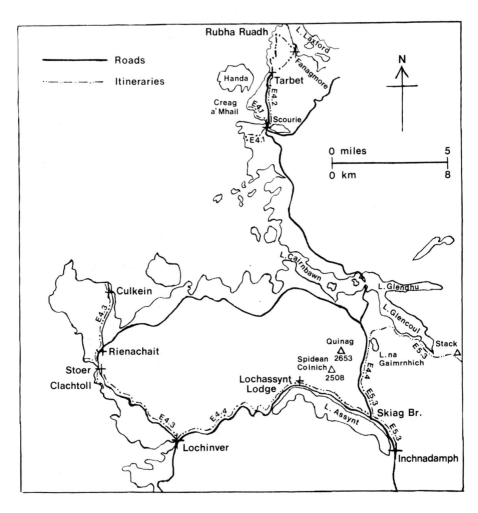

Fig. 4.2 Route map for Scourie, Lochinver and part of Inchnadamph centres.

Lochinver Centre (Fig. 4.2)

E4.3 Stoer Peninsula

Lewisian. Stoer group (Lower Torridonian). Upper Torridonian.

Access:- By private transport from Lochinver to Stoer then coastal and
road walking. Coastal sections best seen at low tide. Those without
private transport may be able to get accommodation at Stoer and to carry
out the excursion by walking.

Maps:- Topo:- 15. Geol:- $\frac{1}{4}$-inch to mile Scotland 5.

Walking distances:- From Stoer 8 miles (12.8 km). If private transport
is used 2.5 miles (4 km).

Itinerary:- The base of the Stoer group, a coarse, gneiss conglomerate,
should be studied near Clachtoll, about 1 mile (1.6 km) S. of Stoer. The
best exposure is $\frac{3}{4}$-mile (1.2 km) S. of Stoer cemetery and about 165 yds.
(150 m) E. of the road. The conglomerate passes upwards into dark red
muddy sandstone which in coastal outcrops E. of the great cleft rock
A'Chlach Thuill contains algal limestones. These show desiccation cracks
partially healed by subsequent accretion of limestone and other features
suggesting shallow water algal growth.

Further N. the base of the Stoer group is again seen 33 yds. (30 m) E. of
Stoer cemetery. This is the beginning of the type section which can be
followed to the W. along the N. shore of Stoer Bay. Where the coast
curves from a northerly to a westerly direction sandstone and siltstone
beds are seen, then at a rocky promontory trough cross-bedded red sandstones
with well rounded pebbles appear. Palaeocurrents flowed towards the E.
These sandstones are interrupted by three couplets each of muddy red sandstone followed by red shale and ripple-marked sandstone. Each is several
metres thick. One of the couplets contains small cupriferous nodules in
its upper part.

Continue along coast to the next promontory, Stac Fada, $\frac{1}{2}$-mile (0.8 km) W.
of Stoer. This gives its name to the Stac Fada Member of the Stoer Group,
a volcanic mudflow deposit 12 m thick, a massive muddy sandstone packed
with much-altered glass shards. The mudflow has injected and disrupted
underlying strata. Skirt a waterfall and descend a low cliff along a
slippery bedding plane. This leads to a limestone overlain by finely
laminated calcareous grey siltstone. Still further W. higher members of
the Stoer Group can be seen, although the shore is not everywhere
accessible. The strata include ripple-marked red sandstones, pebble-beds,
and siltstones.

Return to the road and continue N. to Rienachait, about 1 mile (1.6 km) N.
of Stoer. Here a conglomerate facies is exposed on crags above a gravel
pit.

Northwards along road to north coast of Stoer Peninsula, then follow cliff
and shore section westwards, starting at a bay 0.4 mile (0.6 km) E.S.E. of
Culkein. Here gneiss breccia at the base of the Stoer Group rests on a
highly irregular surface of Lewisian gneiss. The sedimentary features of
higher parts of the Stoer Group, mostly sandstones, can be readily studied
on the westwards traverse. The same conglomerate as that exposed at

Rienachait outcrops 0.2 mile (0.3 km) E. by N. of the mouth of a small
stream, and near the stream itself there are muddy sandstones and shales.

The most significant part of the section is seen at low tide on the floor
of a small bay 0.2 mile (0.3 km) N.W. of the stream. An irregular surface
of the Stoer Group is unconformably overlain by red Diabaig siltstones at
the base of the Upper Torridonian. The Diabaig siltstones are only about
10 m thick and are erosively overlain by very coarse Applecross sandstones
which to the N.W. cut out the Diabaig and rest directly on the Stoer Group.
The sandstones contain exotic pebbles.

Return to Stoer and Lochinver.

E4.4 North side of Loch Assynt

Lewisian. Torridonian. Cambrian. Glacial features.

Access:- This excursion is best done by private transport. However,
depending on seasonal schedules, it may be possible to travel by public
bus to Lochassynt Lodge and then to follow the itinerary by walking.

Maps:- Topo:- 15. Geol:- Special 1:63,360 sheet Assynt district.

Walking distances:- With private transport 4 miles (6.4 km). Walking from
Lochassynt Lodge to Loch na Gainmhich and back to Kylesku road junction
11 miles (17.6 km).

Itinerary:- The excursion should be started at Lochassynt Lodge and the
road followed E.S.E. to study Lewisian gneisses of the Scourie cycle which
can be seen to be cut by a number of W.N.W. Inverian basic dykes showing
varying degrees of alteration. A felsite sheet, possibly of Lower Devonian
age is also seen. Two and a half miles (4 km) E.S.E. of Lochassynt Lodge
the Lewisian is unconformably overlain by Torridonian sandstones with
gneiss debris; the irregular form of the old Lewisian surface can be
easily demonstrated. The lowest Torridonian rocks may belong to the Diabaig
Group but the sandstones higher up in the sequence are certainly Applecross.
Flat slabs of sandstone along the shore show good W.N.W. glacial striae.

The Torridonian is unconformably overlain by Cambrian Basal Quartzite.
½-mile (0.8 km) W.N.W. of the junction with the Kylesku road the Cambrian
dips S.E. at 20°, more steeply than the underlying Torridonian. The Basal
Quartzite is overlain by the Pipe-rock, the uppermost part of which is seen
at Skiag Bridge followed to the E. by the "Fucoid" Beds, the Salterella
Grit and a thin band of Ghrudaidh dolomite, the lowest division of the
Durness Limestone, above which comes the Sole Thrust (E5.3).

At a headland in Loch Assynt, close to the road 550 yds. (500 m) S.E. of
Skiag Bridge the massive lower portion of the Salterella Grit is underlain
by "Fucoid" Beds consisting of interleaved hard, dark-blue shales and
dolomitic, yellow-weathering gritty bands. A horizon 90 cm below the base
of the Salterella Grit has yielded fragments of Olenellus but these are
hard to find.

Return to Skiag Bridge and turn up the Kylesku road. To the W. Cambrian
quartzite forms a dip-slope running down from the summit of Spidean Coinich,
2508 ft. (762 m). A fine corrie separates Spidean Coinich and Quinag,
2653 ft. (808 m) which shows magnificent cliffs of Torridonian sandstone

with a cap of Cambrian quartzite. Loch na Gainmhich, about $\frac{1}{2}$-mile beyond
the summit of the road, is a glaciated rock-basin with an outflowing
waterfall, over Cambrian Pipe-rock, following a N.W. crush-zone. The
pipes (infilled worm-burrows) are up to $\frac{1}{2}$-inch (1.25 cm) in diameter.

Return to Lochinver.

Gairloch Centre (Fig. 4.4)

E4.5 vicinity of Gairloch

Lewisian. Torridonian.

Access:- Walking on roads, paths and shore. Private transport may be
used on the road section.

Maps:- Topo:- 19 Geol:- S 91.

Walking distances:- 10 miles (16 km). Reduced to 4 miles (6.4 km) by
using private transport on road.

Itinerary:- The Lewisian supra-crustal metasediments and metavolcanics are
the main objective of the excursion. From the back of the "Old Inn" at
Gairloch village follow a track, overlooked by crags of hornblende-schist
referred to as the Kerrysdale amphibolite - probably altered basic lavas
and tuffs. About 550 yds. (500 m) from the Inn turn left through a gate
then take a left fork where there are further exposures of the Kerrysdale
amphibolite. A short distance further on turn S.E. along a boundary wall.
A prominent ridge here consists of a banded iron-formation with alternations
of quartz and magnetite. A hundred metres S.E. along the ridge an F2 fold
(F2 and F3 folds are probably Laxfordian) cuts the schistosity of the iron-
formation and earlier F1 isoclinal folds (probably Inverian). Forty metres
S.W. of the folds a thin marble occurs. Walk Eastwards to an outcrop of
the brown-weathering Flowerdale schists, S. of a stream and S.E. of
Flowerdale Mains Farm. These quartz-plagioclase-biotite schists are the
typical metasediments of the Loch Maree Group.

Cross the stream by the footbridge and return to Gairloch by the track past
Flowerdale Mains Farm and Flowerdale House. The track follows the N.W.
Flowerdale Fault which has a dextral strike-slip displacement of over
1300 yds. (1200 m).

The shore section E. of the pier should next be examined. Metasedimentary
biotite-schists are in contact with acid gneisses of the Ard gneiss Group
which can be recognised by the presence of felspar augen which have partly
resisted cataclases. The junction, which is repeated several times due to
isoclinal folding or thrusting, is now regarded as a highly deformed unconformity between the supra-crustal cover and basement.

At the pier an amphibolite sheet is exposed, one of three cutting the Ard
gneisses belonging to the suite of pre-Badcallian metamorphism basic
intrusives.

The road should then be followed S.E. to Kerrysdale Farm. Go up the track
past the farmhouse for 300 m to a footbridge over a stream where another
track branches to the left. A small knoll on the right consists of garnet

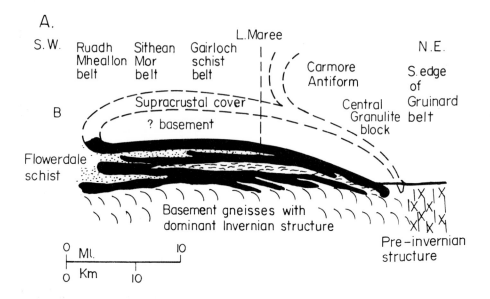

A.

S.W.   Ruadh   Sithean Gairloch   L.Maree   N.E.
       Mheallon Mor   schist   S.edge
       belt   belt   belt   Carmore   of
                                                     Antiform   Gruinard
B   Supracrustal cover   Central   belt
                        ? basement   Granulite
                                                                 block
Flowerdale
schist

                    Basement gneisses with
                    dominant Invernian structure
                                                                      Pre-invernian
                                                                      structure

       0   Ml.   10
       0   Km   10

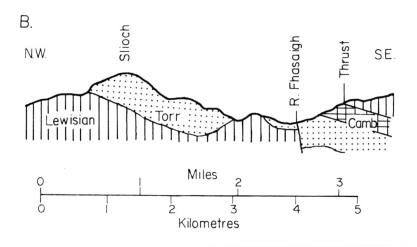

B.

N.W.   S.E.
                  Slioch   R. Fhasaigh   Thrust

              Lewisian   Torr   Camb

                              Miles   2   3
       0   1   3
       0   1   2   3   4   5
                              Kilometres

Fig. 4.3   Sections of North-West Highlands.
           A. Reconstruction of the Inverian Structure between Rudha
              Mheallon and Gruinard Bay (after Park);
           B. Cambrian/Torridonian/Lewisian relationships near Loch Maree.

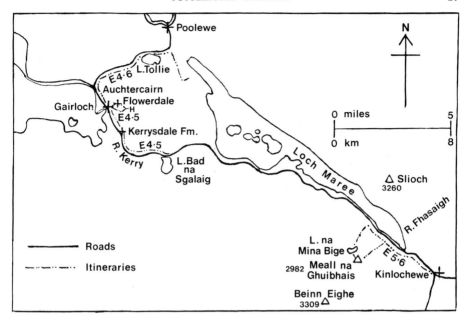

Fig. 4.4 Route map for Gairloch and Kinlochewe Centres.

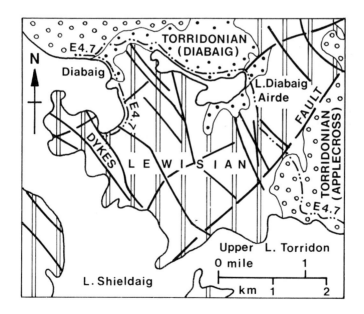

Fig. 4.5 Geological map and route map of Torridon Centre.

grunerite schist, representing a silicate facies of the banded iron-
formation seen earlier on the excursion. Cross the bridge and walk along
the grassy track to about 300 m from the bridge where yellow-weathering
graphitic quartz-mica-schists are exposed.

Return to the road and continue S.E. Torridonian on the low ground to the
S.E. is surrounded by Lewisian hills forming an exhumed Torridonian landscape. A short distance after the road enters a wood. Cross the R. Kerry
by a footbridge to a line of low crags where hornblende-schists of the
Mill na Claise amphibolite are overlain unconformably by basal Torridonian
breccia. Walk S.E. towards the prominent cliff on the skyline. The route
parallels the axial trace of the Mill na Claise N.W. closing F2 fold with
vertical plunge traced out by the amphibolite which contains three metasedimentary bands. Acid gneisses occur in the core of the fold.

Return N. to the road, crossing the R. Kerry at a power station and follow
the road E. to the top of the Kerry gorge. Siliceous biotite-schists are
seen in the river bank near the dam and on the N. side of the road
hornblende-schists of the Kerrysdale amphibolite show occasional tight F1
folds with schistosity refolded by abundant asymmetric F2 folds.

Return to Gairloch.

E4.6 Tollie district

Lewisian. Torridonian.

Access:- By private transport and hill walking or by road and hill walking.

Maps:- Topo:- 19 Geol:- S 91.

Walking distances:- With private transport 4 miles (6.4 km). Entirely by
walking 15 miles (24 km).

Itinerary:- From Gairloch N. then N.E. on Poolewe road. Above Auchtercairn
the Kerrysdale amphibolite is well-exposed separated by Flowerdale schist,
concealed by a wood, from the Aundrary amphibolite which forms crags further
up the road. The two amphibolites are probably the same repeated by an
early isoclinal fold. To the N. horizontal Torridonian sandstones form
prominent scarps. About $1\frac{1}{4}$ miles (2 km) N.E. of Auchtercairn the highly
sheared contact between the Aundrary amphibolite and the Tollie gneisses
is seen at the E. edge of a wood on the hillside; this tectonic contact
marks the Craig Bhan crush belt.

The object of much of the rest of the excursion is to study the Tollie
Antiform, a south-easterly plunging late fold which brings all previous
structures into near parallelism. Gneisses on the steep S.W. limb can be
examined in a quarry at the top of the hill 2.2 miles (3.5 km) N.E. of
Auchtercairn. The gneisses have been strongly sheared and flattened.
Veins of chocolate-coloured aphanitic pseudotachlite up to 2 cm thick are
seen; these are considered to be late Laxfordian and to have been formed
by frictional heating due to exceptionally rapid shearing movements.

Continue along the road past Loch Tollie to top of the hill leading down to
Loch Maree. Private transport should be parked here. Loch Maree marks a
major dextral transcurrent fault. From the road a track along the S.W. side
of Creag Mhor Thollaidh should be followed. The route coincides with a N.E.-

thrust which has brought the structurally lower rocks of Craig Mhor
Thollaidh S.W. over the gneisses of the Tollie Antiform. Crushed gneisses
with pseudotachlite veins are exposed on the E. of the track where it
crosses the stream.

Continue up the valley to the top of the steep section at a lochan. From
about 220 yds. (200 m) S. of the upper end of the loch cross the valley
and climb a grass gully through crags forming a hill. The summit consists
of a massive amphibolite dyke, with relic ophitic texture. The finer-
grained, schistose margins are generally concordant with the banding of
the country-rock gneisses but local discordances occur. When followed N.
the dyke bends through 90° and is cut off by the thrust. On the E. side
of the dyke an "early basic" body is exposed, interpreted as a pre-Badcallian
basic intrusion. The strong flattened fabric of the basic mass and of the
gneisses is considered to be Inverian. Dyke intrusion followed and then
Laxfordian folding and metamorphism.

Go down to the road and return to the E. end of Loch Tollie. Cross the
Tollie Burn where it leaves the loch and walk S.E. to a line of low crags
330 yds. (300 m) from the stream. Study of this exposure reveals the
following sequence of events:

1.   Formation of gneissose banding. S1

2.   Upright folding of gneissose banding. F2

3.   Sub-horizontal folding and foliation. F3, S3 (Inverian)

4.   Intrusion of basic dyke.

5.   Metamorphism and foliation of dyke. (Laxfordian)

Return to Gairloch.

Torridon Centre (Fig. 4.5)

E4. 7 Loch Torridon

Lewisian. Torridonian.

Access:- If staying at Torridon private transport is essential. If means
can be found to stay at Diabaig the whole excursion can be done on foot.
Walking on shore; low tide necessary.

Maps:- Topo:- 24 Geol:- S 81.

Walking distance:- 3 miles (4.8 km).

Itinerary:- From the pier at Diabaig walk S. along the shore. The first
exposures are of granite-gneiss which has undergone Inverian metamorphism.
A steep basic dyke of Scourie type is clearly later than the migmatisation.
Walk S. past a rocky promontory, to which it may be necessary to climb to a
path at the base of the cliffs, then descend again; the foliation of the
gneisses steepens abruptly in a shear zone. Several thin, well-foliated
dykes are seen.

Continue around the shore in a S.W. direction across the strike. The steep
foliation gradually becomes less steep until, at the S.W. corner of the bay,
the dip is 40°-50° N.E. This represents the N.E. limb of the Torridon

Antiform which, since it affects the dykes with their Laxfordian foliation,
may be a Laxfordian F2 structure analagous to the Tollie Antiform (E4.6).

Return by the path to the pier and walk N.W. then W. along the shore.
Coarse grey sandstone with gneiss fragments up to 10 cm at the base of the
Diabaig Group is exposed beneath trees between high water mark and the
road. The fragments soon become smaller as the succession is traced upwards
and the sandstone is followed by grey shales with thin ripple-laminated
sandstones; desiccation polygons and ripples are common in the shales.
Grey sandstone beds appear higher in the sequence and show signs of grading.
The top of the Diabaig Group is drawn at a coarse pink sandstone, with
trough cross-bedding, exposed in a wood just above high water mark 650 yds.
(600 m) N.W. of the pier. The overlying Applecross Group consists dominantly
of coarse, red sandstones.

Return to the pier. Along the road towards Torridon a stop should be made
near the N.E. corner of Loch Diabaigs Airde. Here the base of the Diabaig
and underlying gneiss can be seen; the succession is duplicated by a fault.
About half-a-mile (0.8 km) S. along the road there is a fine view of the
pre-Torridonian topography along the S. side of Upper Loch Torridon. Grey
gneiss can be clearly seen below the almost horizontal Applecross sandstones.
Above the N. side of the loch the same sandstones are magnificently exposed
on Beinn Alligin, 3232 ft. (985 m). Further E. the sandstones are capped
by Cambrian quartzite on Liathach, 3456 ft. (1054 m).

References

Barber, A. J. and others. 1978. The Lewisian and Torridonian Rocks of
    North-West Scotland. Guide No. 21. Geol. Assoc.
Johnson, M. R. W. and Parsons I. 1979. Geological Excursion Guide to the
    Assynt District of Sutherland. (revision of Macgregor and Phemister's
    Guide). Edin. Geol. Soc.
Phemister, J. 1960. British Regional Geology: Scotland: The Northern
    Highlands. 3rd ed. M.G.S.
Teall, J. J. H. 1885. The metamorphism of dolerite into hornblende schist.
    Q.J.G.S., 41, 133-144.

CHAPTER 5

Caledonian Terrains with Caledonian Metamorphism

5.1 North-West Caledonian Front (Figs. 5.1 and 5.2)

Thrust structures separating the north-west kratogen from the highly folded
and metamorphosed rocks of the Caledonian Orogen are spectacularly displayed
on the faces of the deeply dissected 3000 ft. (915 m) mountains near the
seaboard of North-west Scotland. Moreover, the district was one of the
first in the world where large-scale reversals of normal stratigraphical
order, consequent on thrusting, were demonstrated, and it has attracted
geologists from many countries for over 80 years. From Loch Eriboll on
the N. coast to Assynt, major, but individually discontinuous, thrusts
bring westwards autochthonous nappes consisting of Lewisian slices with
unconformable Cambrian. From Assynt southwards, Torridonian also appears,
and folding becomes significant. Beneath these nappes there is generally
a zone of imbricated Cambrian resting on a basal thrust of "sole". Above
the autochthonous nappes the Moine Thrust, certainly continuous for over
120 miles (192 km) brings forward an exotic nappe consisting of Moinian
metasediments of the fold-belt. Mylonite is developed on a large **scale**
along many of the thrusts. The undisturbed stratigraphical succession is
clearly seen in the foreland (Ch. 4).

The information in Ch. 4 about roads, public transport and accommodation
also applies to most of the present region. However, access by public
transport is somewhat easier in the S. as the railway from Inverness runs
along Loch Carron and on to Kyle of Lochalsh from which there is a frequent
ferry to Skye (does not cross on Sundays) connecting with bus services to
Broadford and other places on the island.

Five centres are suggested: Durness, Inchnadamph, Kinlochewe, Strathcarron
and Broadford.

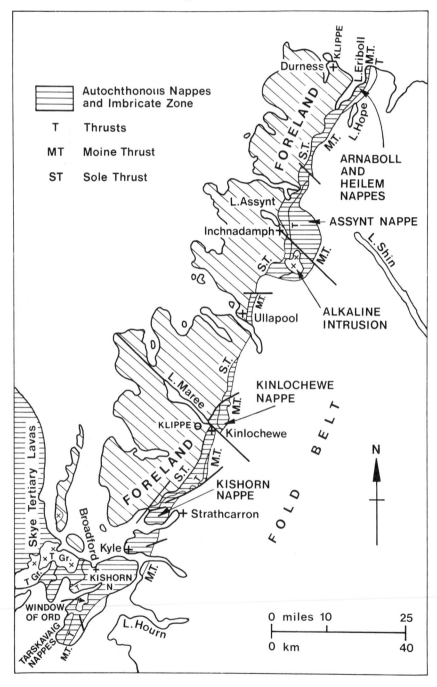

Fig. 5.1 Geological map of North-West Caledonian Front.

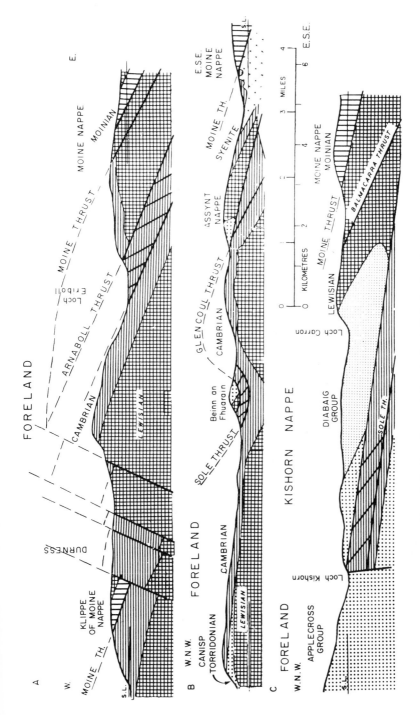

Fig. 5.2 Sections of North-West Caledonian Front.
A. Loch Eriboll–Durness;
B. Assynt; C. Loch Carron–Loch Kishorn.

The succession involved in the Thrust-zone is as follows:-

Durness Limestone	-	Cambro-Ordovician

Serpulite Grit		
Fucoid Beds	-	Cambrian
Pipe-rock		
Basal Quartzite		

Upper Torridonian	-	Upper Proterozoic
part of Moinian (metamorphic)	-	Upper Proterozoic
Lower Torridonian	-	Upper Proterozoic
Lewisian	-	Lower Proterozoic and Archean

The structural units are broadly:-

Moine Nappe - exotic
(continuous for 120 miles (192 km) in present region and
probably continuous for 250 miles (400 km) in British Isles)

Autochthonous Nappes
(individually discontinuous; various local names)

Imbricate Zone - autochthonous (not always present)

Kratogen - autochthonous

In the Assynt district there are extensive alkaline intrusions of Lower
Devonian age involved in the Thrust-zone.

Comparison with other areas:- The clarity with which the nappes and thrusts
are displayed makes the region unique in the British Isles. Historically
and in its own right it is scientifically important on a world-rating.

The formations in the autochthonous nappes are exposed in normal order in
the foreland (Ch. 4), and the metasediments of the exotic Maine Nappe are
the leading edge of a complex assemblage extending far to the E. (Ch. 5.2)
and S.E. (Ch. 5.4).

The alkaline intrusions are nearly unique in the British Isles, apart from
the Ben Laoghail syenite in the Northern Highlands.

Durness Centre (Figs. 5.3 and 5.4)

E5.1   Loch Eriboll

Lewisian. Moinian. Cambrian. Caledonian nappes. Underground river.

Access:- By private transport. Hill and coast walking.

Maps:- Topo:- 9 Geol:- S 114. ¼-inch Scotland 5.

Walking distance:- 6 miles (9.6 km).

Itinerary:- A stop should be made a mile (1.6 km) E.S.E. along the road
from Durness where the Allt Smoo plunges into a pothole in the Durness
Limestone. By walking N. to the coast the resurgence of the underground
river can be seen in the large Smoo Cave. A short distance to the E. a

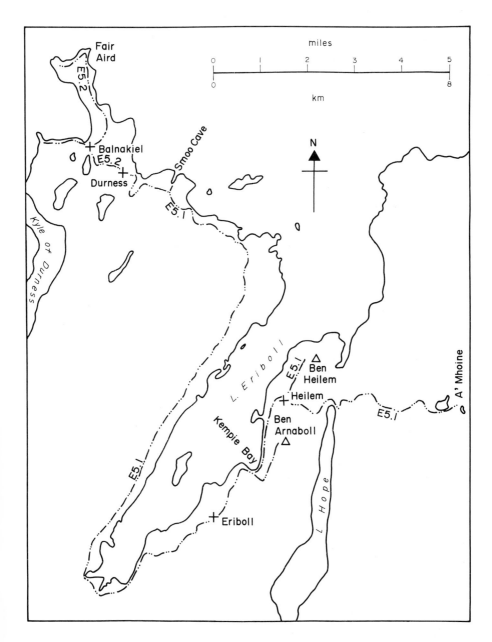

Fig. 5.3 Route map for Durness-Loch Eriboll area.

fault brings up the Lewisian, which further E. still is uncomfortably overlain by Cambrian followed by Pipe-rock. On to Loch Eriboll where Lapworth
carried out his classic work on the thrusts.

Round head of loch to the hamlet of Eriboll where the light-coloured Eilean
Dubh (II) Group of the Durness Limestone (Cambrian) is underlain S.E. of the
road by the darker Ghrudaidh (I) Group. Park at Kempie Bay. Just N. of
bend of the road steeply inclined beds of Pipe-rock within the Arnaboll
Nappe outcrop; the pipes are visible on the lower surfaces.

By leaving the road and ascending S.E., Lewisian (probably Laxfordian)
granitic gneiss and amphibolite of the same nappe can be seen to contain
an isoclinal syncline of Cambrian quartzite with basal conglomerate.
Walk N. a mile (1.6 km) to study the Arnaboll Thrust which carries
mylonitised Lewisian over imbricated Cambrian of the Heilem Nappe. Return
to transport and continue to furthest N. point of road. Walk N. to W.
slopes of Ben Heilem. Here two sets of imbricated beds within the Heilem
Nappe are seen, a W. set involving Fucoid Beds, Serpulite Grit and Durness
Limestone and an E. set in Pipe-rock and Fucoid Beds.

After returning to transport continue E. along road to the E. side of Loch
Hope. Here the Moine Thrust brings forward altered Lewisian over which a
higher thrust carries mylonitised Moinian. This is overlain, apparently
by inversion without thrusting, by a higher strip of Lewisian overridden
by Moinian. The flaggy granulites of this assemblage are well seen to the
E. on the moorland of A'Mhoine, the type locality.

Return to Durness.

E5.2   Durness and Fair-aird Head

Moinian. Cambrian. Ordovician. Caledonian klippe. Shell sand dunes.

Access:- Road and coast walking.

Maps:- Topo:- 9 Geol:- S 144. $\frac{1}{4}$-inch Scotland 5.

Walking distance:- 11 miles (17.6 km).

Itinerary:- Walk N.W. on Balnakiel road. The highest Ordovician Groups
(VI and VII) of the Durness Limestone are exposed on the outskirts of
Durness. They consist of limestone and dolomites dipping E.S.E. Fossils
may be found with difficulty in dark limestones somewhat above the base of
Group VI (Croisaphuill). Along or near the coast W. of Balnakiel a descending succession from Group V to the light grey Group II (Eilean Dubh)
can be seen to consist of both dolomites and limestones. Fossils occur in
some of the limestones but a great deal of searching is required. Algal
structures are present near the top of the Eilean Dubh Group.

Walk N.W. along the shore which is fringed by dunes of shell sand. The
higher ground of Fair-aird Head consists of Moinian and Lewisian rocks
overlying the Durness Limestone (the thrust is not exposed). This klippe
of the Moine Nappe (E5.1) proves a minimum displacement of 10 miles (16 km).
Moinian granulites form picturesque offshore stacks.

Return to Durness.

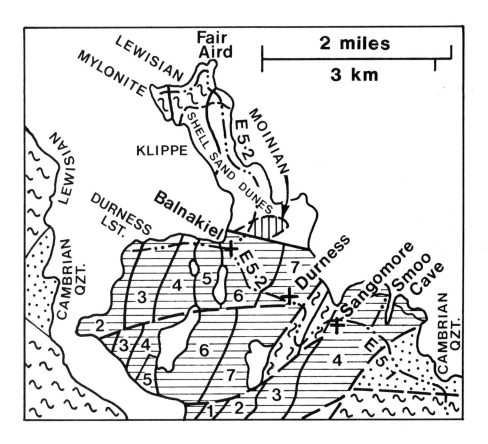

Fig. 5.4 Geological map with routes of Durness area.
Nos. 1 to 7 subdivisions of Durness Limestone.

Inchnadamph Centre (Fig. 5.5)

Three excursions, all involving full days and possibly divisible into two days each, are suggested. Other excursions are described by Johnson and Parsons 1979.

With private transport available, it may be easier, from the accommodation aspect, to stay in Lochinver (Ch. 4).

E5.3   N.E. side of Loch Assynt, Loch Glencoul and the Stack of Glencoul

(Fig. 4.2)

Lewisian. Moinian. Cambrian. Lamprophyre and felsite sheets of probable early Devonian age. Imbricate Zone, Glencoul Nappe and Moine Nappe.

Access:- By car to near Loch Gainmhich thence by rough walking. The first part of the excursion, to the Imbricate Zone, can, however, be done on foot from Inchnadamph.

Maps:- Topo:- 15 Geol:- Special 1:63,360 Sheet Assynt district.

Walking distances:- 9 miles (14.4 km). From Inchnadamph to Imbricate Zone and back on foot 5 miles (8 km).

Itinerary:- Follow the road to Loch Assynt then the Kylesku road to a point 380 yds. (350 m) N. of Shiag Bridge. On the side of the road Fucoid Beds are overlain by Salterella Grit. Walk N.E. for 275 yds. (250 m) then N. for about 200 yds. (180 m) to where the footpath to Achmore crosses a small burn. The Sole Thrust, sheared Salterella Grit and Fucoid Beds are crossed on the way. A traverse up the burn shows the Imbricate Zone with the succession from the Fucoid Beds to dark dolomite (Durness) repeated again and again.

Swallow holes occur in the dolomite. Near Achmore a major thrust brings forward Ghrudaidh and Eilean Dubh dolomites with lamprophyre sills.

Those without transport should return from here. With transport, after returning to the road travel N. The fine view of the foreland to the W. is described in E4.4.

Leave transport N. of the bridge over the stream draining Loch Gainmhich and walk N. over the Torridon sandstone onto the Lewisian gneiss before turning E. towards Loch Glencoul. The Torridonian can be seen to thin out, and half-way to Loch Glencoul Cambrian quartzite is seen within a few feet of Lewisian. Continue down the slope towards Loch Glencoul. About ¼-mile (0.4 km) from the loch the Pipe-rock is followed by undisturbed, yellowish-weathering Fucoid Beds. Near the top of this outcrop the Sole Thrust occurs, and between this thrust and the Glencoul Thrust which brings forward Lewisian there is an Imbricate Zone about 33 ft. (10 m) thick.

Continue N.E. along the foot of the gneiss crags to the loch near which a narrow ledge has to be negotiated. The Lewisian gneiss passes down into pink and green mylonite above the Glencoul Thrust beneath which there is Durness dolomite. The thrust cuts across the mylonite foliation and across N.-S. folding of this foliation.

Walk to the head of Loch Beag then climb towards the Stack of Glencoul passing from the gneiss onto Cambrian quartzite with felsite sheets. The Moine Thrust is not exposed on the Stack but a thin wedge of severely

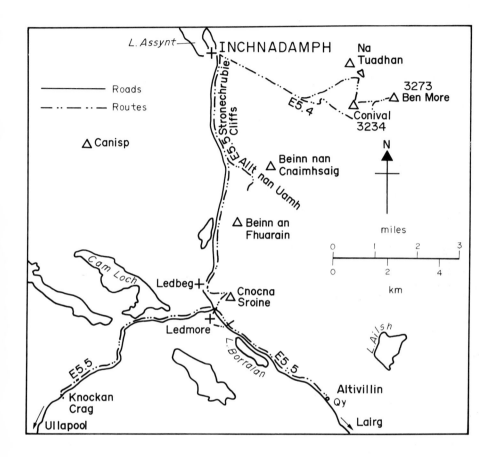

Fig. 5.5 Route map for part of Inchnadamph area.

deformed Pipe-rock has been dragged along it and shows flattened pipes. Immediately above some 65 ft. (60 m) of mylonite form precipitous cliffs. The mylonite displays at least two sets of folds. By rounding the Stack and walking S.E. it is possible to see Moinian metasediments in which the metamorphic grade increases rapidly away from the Moine Thrust.

Return to the road by the same route.

E5.4   Ben More Assynt and Conival

Lewisian. Torridonian. Cambrian. Minor intrusions of probable Lower Devonian age. Folds and thrusts. Underground drainage. Glacial features.

Access:- By rough walking and mountain scrambling. The higher parts of this excursion should be undertaken only in good weather.

Maps:- Topo:- 15 Geol:- Special 1:63,360 sheet Assynt district.

Walking distance:- 10 miles (16 km).

Itinerary:- From Inchnadamph Hotel walk E.S.E. along the N. side of the Traligill River; there is a rough track for about the first 1½ miles (2.4 km). The stream runs across Durness Limestone in the Imbricate Zone and flows partly underground. Immediately S.E. of a point where the river reappears after having followed an underground course for ¼-mile (0.4 km) there is a particularly good exposure of a thrust bringing dark Ghrudaidh dolomite over lighter Eilean Dub dolomite.

The route continues over the N. shoulder of Cnoc nan Uamh (the hill of the cave) and across the road to a large torrential cone at the W. foot of Conival. Ascend N. to a hollow on the hillside which marks the Glencoul Thrust bringing Cambrian quartzite with felsite sills over Durness Limestone.

Follow the hollow S.E., keeping to the N. side until the pool near the Bealach Traligill (pass) is well in view then ascend the hillface above the scree-covered slopes to a conspicuous greenish scar or break. Here the Ben More Thrust brings highly sheared Lewisian gneiss over reddened sheared Cambrian quartzite; a thin lens of dark sheared Torridonian grit is inter-posed between the quartzite and gneiss. It should be noted that the Moine Thrust has been reinterpreted as a comparatively minor thrust splitting what has been termed the Assynt Nappe.

Follow the gneiss S.E. round the hillside until the overlying Torridonian grits and conglomerates are met. Continue in the same direction until the gneiss is reached once more. The Torridonian lies in an overturned anti-cline.

Ascend to the col between Ben More and Conival; the Cambrian quartzite can be seen to cross the Torridonian into the Lewisian. Follow the ridge of Cambrian quartzite with felsite sill to the summit of Ben More, 3272 ft. (997 m). From the summit there is a view to the N.W. of Cambrian quartzite folded into an upright, slightly overturned anticline and syncline in the great precipice of Na Tuadhan.

Return to the col then ascend over Cambrian quartzite with felsite sills to Conival, 3234 ft. (986 m). Descend N. by E. to the lochan between the summit and Na Tuadhan then S.W. down a tributary to the R. Traligill and on to Inchnadamph.

E5.5   Beinn an Thuarain, Loch Borralan and Knochan

Lewisian. Moinian. Cambrian. Alkaline intrusions of probable Lower
Devonian age. Thrusts and klippes.

Access:- By private transport and some hill walking. Without private
transport and by a long road and hill walk it is possible to get as far as
the N.W. end of the Loch Borralan intrusion. At some seasons this might be
reduced by using the Lairg public bus service. Even with private transport
it may be worth dividing the excursion between two days.

Maps:- Topo:- 15 Geol:- Special 1:63,360 sheet Assynt district.

Walking distances:- With private transport 4 miles (6.4 km). Without,
14 miles (22½ km) to Ledmore River and back.

Itinerary:- E. of the road S. of Inchnadamph Durness Limestone forms the
Stronechrubie cliffs. Leave the road 2½ miles (4 km) S. of Inchnadamph and
follow the Allt nan Uamh (stream of the cave). About ¾-mile (1.2 km) upstream, on the S. side, four caves occur in folded and faulted Durness
Limestone. At the top of the drift slope nearby the Ben More Thrust carries
Torridonian over the limestone. Round the hillside to the S.W. Cambrian
quartzite comes in above the thrust in inverted sequence. Beinn an Thuarain
and Beinn nan Cnaimhsaig to the N. are klippes preserved in a synform.

Follow road S. to Ledbeg, at the N.W. end of the Loch Barralan intrusion.
In the road cutting 165 yds. (150 m) S. of the end of the Ledbeg Farm
track red augite-nepheline-syenite encloses xenoliths of marble with
pyroxene selvages. Walk across the moorland about ½-mile (0.8 km).
Low crags at the N. foot of Cnoc na Sroine consist of pink leucocratic
syenite, and in slopes N. of these outcrops melanite-syenite with fresh
nepheline can be found.

Walk S.W. to junction of Lairg and Ullapool roads then S. to footbridge over
Ledmore River near Ledmore where nepheline-syenite is again exposed; further
S., where the stream turns W., biotite-pyroxenite is seen.

Follow the Ledmore River upstream to where it comes close to the road. In
the river nepheline-syenite is exposed. Up the Allt a' Bhrishdish, a tributary from the N.E., the nepheline-syenite is succeeded rapidly by bright
pink parthosite, partly brecciated.

Those with transport should continue S.E. to the Altivillin Quarry, 1.5
miles (2.4 km) beyond the S.E. end of Loch Borralan. This is the type-
locality for Borralinite-pyroxene-nepheline-syenite. Pseudoleucite spots
are penetratively deformed but the rock is cut by undeformed pegmatites,
showing that igneous activity overlapped some of the thrust movements.

Return to road junction and travel S.W. to Knochan Crag, 1½ miles (2.4 km)
S.W. of Knochan village. This is part of a Nature Reserve. The sequence
of Pipe-rock, Fucoid Beds, Salterella Grit and a small thickness of
Ghrudaidh dolomite is easily seen. The last is truncated by a thrust above
which there is a thin zone of crushed white Eilean Dubh dolomite. This is
overridden, along the gently S.E. inclined Moine Thrust, by Moinian metasediments. There is evidence of repeated movements with mylonite formation
antedating the clean-cut thrust.

Return to Inchnadamph, noting the clearly displayed foreland succession to the W. particularly well seen on Conival.

Kinlochewe Centre

Kinlochewe can be reached by public bus from Achnasheen railway station.

E5.6   S.E. end of Loch Maree (Fig. 4.4)

Torridonian. Cambrian. Klippe of Kinlochewe (Kishorn) Nappe. N.E. and N.W. faults. Glacial features.

Access:- By road and hill walking or by private transport and hill walking.

Maps:- Topo:- 19, 25 Geol:- S 92.

Walking distances:- 9.5 miles (15.2 km). With private transport 4.5 miles (7.2 km).

Itinerary:- Along road following S.W. side of Loch Maree. The loch marks a major dextral transcurrent fault (see also E4.6). A number of N.E. faults are brought out by features on both sides of the valley. In the 2 miles (3.2 km) from the end of the loch the road crosses a syncline with Torridonian of the Applecross Group on its flanks and Cambrian quartzites in its centre; exposures are poor owing to moraine cover. All these rocks are part of the foreland.

Leave the road 2 miles (3.2 km) beyond the loch and follow a stream S.W. which flows out of Loch na Mina Bige. The succession of Torridonian (Applecross) Basal Cambrian quartzite/Pipe-rock/Fucoid Beds/thin, sheared Salterella Grit can be readily followed. Olenellus has been recorded in the Fucoid Beds. Then, 1.3 miles (2.8 km) S.W. of the road, the Cambrian is overridden along the Kinlochewe (Kishorn) Thrust by sheared Diabaig Torridonian. This is succeeded, S.E. of the loch, by Applecross Torridonian, itself cut by several minor thrusts. The rock surfaces are highly glaciated. In clear weather it is worth continuing to the top of Meall na Ghiubhais, 2982 ft. (909 m). The Torridonian of this mountain is a klippe of the Kinlochewe (Kishorn) Nappe resting in a synform. To the E. the main outcrop of the nappe and the thrust is seen in the cliffs on the S.E. side of the fault-determined, overdeepened valley of the R. Fhasaig (Fig. 4.3B). To the N.W. of this valley rises the great mass of Slioch, part of the foreland, on the W. slopes of which the old Lewisian landscape rises beneath the Torridonian cover.

Southwards from Meall na Ghiubhais the corries and cliffs of the Torridonian/Cambrian quartzite mountain Beinn Eighe are a spectacular feature.

For the return to Kinlochewe a more direct route may be taken E. to the road.

Strathcarron Centre (Fig. 5.6)

Strathcarron is on the Inverness—Kyle railway and there is limited accommodation near the station. There is further accommodation at Lochcarron
(Jeantown) $2\frac{1}{2}$ miles (4.0 km) to the S.W.

E5.7   Loch Kishorn and Loch Carron

Lewisian. Torridonian. Cambrian. Kishorn and Moine Thrusts and Nappes.

Access:- By private transport or public bus (when available) to head of
Loch Kishorn then by road walking.

Maps:- Topo:- 24 Geol:- S 81, 82.

Walking distances:- Head of Loch Kishorn to Lochcarron, 7 miles (11.2 km);
to Strathcarron 10 miles (16 km).

Itinerary:- A start should be made from the W. side of the head of Loch
Kishorn. Here sedimentary structures can be studied in the sandstones and
grits of the Applecross Group which form spectacular mountains to the W.

Cross bridge to Tornapress to outcrops of Eilean Dubh dolomite (Durness
Limestone) brought down by N.E. fault following valley. On the hillside
above Tornapress the Kishorn Nappe (thrust not well exposed) directly overlies the dolomite; the Imbricate Zone occurs some 4 miles (6.4 km) to the
N.E.

Along road towards Lochcarron. Half-a-mile (0.8 km) from where it bends
sharply E. a minor thrust brings Cambrian quartzite over the dolomite and
to the E. of this the Kishorn Thrust carries over Diabaig hard fine
sandstones and slightly altered shales. The Torridonian is inverted (in
the upper limb of the Lochalsh recumbent syncline, see also E5.9) and to
the S. the Applecross Group underlies the Diabaig. Two-and-a-half miles
(4.0 km) from the sharp bend Lewisian pegmatite overlies the Torridonian
sandstone along an irregular contact. Although there are signs of movement
at the junction this is an inverted unconformity. The Lewisian which follows
consists of both acid gneisses and metadolerites.   South of Loch Carron the
Lewisian is split by the Balmacara Thrust; this is difficult to identify N.
of the loch.

South of the road, just past the summit, mylonite associated with the Moine
Thrust is exposed in a small stream. Moinian flaggy granulites of the Moine
Nappe are seen further down the road, and both mylonite and flaggy granulites
outcrop on the shore S.W. of Lochcarron village. The Moinian shows signs
of retrograde metamorphism.

Return to Lochcarron or Strathcarron.

E5.8   North of Strathcarron

Lewisian. Torridonian. Cambrian. Thrusts. Raised beach.

Access:- Hill and road walking.

Maps:- Topo:- 25 Geol:- S 82.

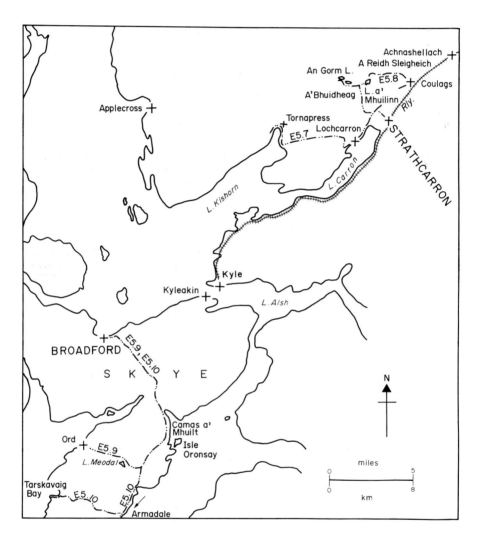

Fig. 5.6 Route map for Strathcarron and Broadford Centres.

Walking distance:- From Strathcarron 12 miles (19.2 km). From Lochcarron
13½ miles (21.6 km).

Itinerary:- Up Amhairn Bhuidhearn which crosses Lochcarron road ¼-mile
(0.4 km) S.W. of its junction with the Strathcarron station road. For 1¼
miles (2 km) upstream sheared Lewisian gneiss of the Kishorn Nappe is
exposed. The Kishorn Thrust dipping S. at about 30° makes a conspicuous
overhanging feature both in the main stream and in a tributary from the
West; 3 ft. (1 m) of mylonite occurs. Under the thrust is a downward
succession of Serpulite Grit, brown-weathering Fucoid Beds, good Pipe-rock
and Basal Quartzite. The Basal Quartzite is thrust over Serpulite Grit
and "Fucoid Beds" just upstream of where three streams join.

From here a short traverse N.W. leads to the E. end of a small loch where
the Cambrian unconformably overlies Applecross Torridonian. At the W. end
of the loch the latter is thrust over Cambrian which is again seen to rest
on Torridonian to the N.W., at An Gorm Loch. At the W. end of the loch
Torridonian is once more thrust over Cambrian.

Return to the small loch, and traverse E. to Loch a' Mhuilinn to the N. to
head of Allt Reidh Sleigheich.

The descent to the road at Coulags provides a fine section of Torridonian,
Cambrian Basal Quartzite and Pipe-rock. Across Strathcarron the mountains
are in the Moine Nappe, brought down by a vertical component of the north-
easterly Strathcarron Fault which dies out to the S.W.

At Coulags a klippe of Lewisian of the Kishorn Nappe is thrust over "Fucoid
Beds".

Return along road. The 50 ft. (15 m) raised beach is well seen. The
absence of the 100 ft. (30 m) raised beach in this and some other Western
Highland valleys is thought to be due to the presence of a glacier at that
time.

Broadford Centre (Fig. 5.6)

There is a fair amount of accommodation at Broadford, which is also a centre
for study of Tertiary igneous rocks (E9.1 and E9.2).

E5.9   Broadford and Ord

Lewisian. Torridonian. Cambrian. Liassic (Lower Jurassic). Tertiary
dykes. Kishorn Nappe. Moine Nappe. Window of Ord.

Access:- By private transport and some shore walking or by public bus to
junction of Ord and Armadale roads then road and shore walking.

Walking distances:- 8 miles (12.8 km). With private transport 1 mile
(1.6 km).

Itinerary:- The Broadford shore section (low tide required) shows Liassic
fossiliferous sandy limestones and shales with N.W. Tertiary basic dykes.

Those with private transport should stop on the road between 2.5 and 3.5
miles (4.0 and 5.6 km) S.E. of Broadford to examine Applecross sandstones
of the Kishorn Nappe. Current-bedding shows that N.W. dip is "right-way-up";

this is the upper limb of the Loch Alsh recumbent syncline (see also E5.7).
Five miles (8 km) S.E. of Broadford the Applecross sandstone overlies grits,
greyish, flaggy sandstones of the Sleat Group. The Diabaig Group is either
thin or absent and most if not all of the Sleat Group is probably older than
the Diabaig. As the coast is approached conglomerates of the Sleat Group
appear. At Camas a' Mhuilt these are overridden along the Moine Thrust by
Lewisian which forms this part of the Moine Thrust. The latter Thrust is not
exposed but on the S. side of the bay sheared acid Lewisian gneiss showing
retrograde metamorphism can be studied. Similar gneisses are also seen along
the Ord road as far as Loch Meodal where the Moine Thrust brings them over
the Sleat Group. Half-a-mile beyond the loch Applecross sandstones appear.
These are in the Window of Ord, part of the imbricated foreland brought up
by an antiform. 1.7 miles (2.7 km) S.E. of Ord a thrust (not seen at road)
is followed by Torridonian resting on which is Cambrian quartzite. This ends
at the Kishorn Thrust, forming the W. side of the Window. A narrow outcrop
of Torridonian above the thrust is followed by Cambrian quartzite, quarried
for high-grade silica rock. Thin and partly-exposed Serpulite Grit and
"Fucoid Beds" are followed by Durness Limestone of the Eilean Dubh and
Sailmhor formations best seen by walking along the coast N.E. of Ord.
Fossils may be found with difficulty. Tertiary dykes occur.

Return to Broadford.

E5.10   Armadale and Tarskavaig

Lewisian. Moinian including "Tarskavaig Moines". Torridonian. Tertiary
dykes.

Access:- By public bus to Armadale then road and shore walking. By private
transport and shore walking. It is also possible to cross by boat to
Armadale, and return the same day, from Mallaig (E5.18 and E5.19).

Maps:- Topo:- 32 Geol:- S 61, S 71.

Walking distances:- From Armadale to Tarskavaig Bay and back 13 miles
(20.8 km). With private transport 1 mile (1.6 km).

Itinerary:- At Armadale Pier flaggy Moinian granulites of the Moine Nappe
can be studied. They are probably separated by a thrust within the nappe
from Lewisian to the W.

Along Armadale-Broadford road then branch off on road which for $1\frac{1}{4}$ miles
(2.0 km) crosses Lewisian gneiss of the Moine Nappe. This overrides the
Sleat Group (Torridonian) of the Kishorn Nappe along the Moine Thrust (not
exposed). The Torridonian and numerous N.N.W. Tertiary basic dykes of the
Skye swarm are seen for the next $3\frac{1}{2}$ miles (5.6 km). The Torridonian occurs
in an antiform, and $\frac{1}{4}$-mile (0.4 km) N.W. of where the road bends sharply in
that direction the Tarskavaig Nappe comes down; this intervenes between the
Kishorn and Moine Nappes.

Continue to Tarskavaig Bay, where there are exposures of gritty granulites
of low metamorphic grade. These are the "Tarskavaig Moines" regarded as
evidence for correlating part of the Moinian with the Torridonian. On the
S. shore of the bay the Tarskavaig granulites can be seen thrust over contorted sandstones and black shales of the Sleat Group of the Kishorn Nappe.
Numerous Tertiary dykes occur.

The Taiskavaig Nappe appears to consist of three nappe-sheets.

Return to Broadford.

5.2   The Northern Highlands (Fig. 5.7)

The Northern Highlands, for present purposes, comprise the whole of mainland Scotland N.W. of the Great Glen Fault, apart from regions described in
Chs. 4, 5.1 and 9.1. Most of the region is mountainous, rising to 3877 ft.
(1180 m) at Carn Eige; the W. coast is deeply incised by sea-lochs. The
interior is the most sparsely populated region in Britain; accommodation
is limited. It is crossed by only a few roads and one railway line.
Further S. another line reaches the W. coast at Mallaig from Fort William.
A border of relatively low ground underlain mainly by Old Red Sandstone,
occurs along the E. coast and continues to Orkney. Here there are a number
of small towns, more roads and the railway from Inverness to Thurso.

Four centres are suggested: Thurso, Helmsdale, Inverness and Mallaig.

The stratigraphical succession of the Northern Highlands is:-

> Tertiary
> Cretaceous
> Jurassic
> Triassic
> Carboniferous
> Upper Old Red Sandstone
> Middle Old Red Sandstone
> Lower Old Red Sandstone
> Moinian - Upper Proterozoic (mainly)
> Lewisian - Lower Proterozoic and Archean

Outcrops E. of the Moine Thrust identified as Lewisian amount to more than
100 square miles. Acid and intermediate banded rocks comparable to the
Lewisian ortho-gneisses of the foreland (Ch. 4) predominate, together with
basic gneisses, serpentinites, talc-schists and eclogites and metasediments.

The Moinian consists cominantly of altered psammitic, felspathic sediments,
often termed granulites; quartzites are rare. The assemblage also contains
plentiful mica-schists, but few true limestones or marbles; thin, lenticular
beds of calc-silicate rock are, however, fairly common.

The following formations have been recognised, placed by Johnstone and
others (1969) in three divisions:-

Loch Eil Psammite	Loch Eil Division
Glenfinnan Striped Schist⎱ Lochailort Pelite ⎰	Glenfinnan Division
Upper Morar Psammite ⎫ Morar Striped and Pelitic ⎪ Schist ⎬ Lower Morar Psammite ⎪ Basal Pelite ⎭	Morar Division

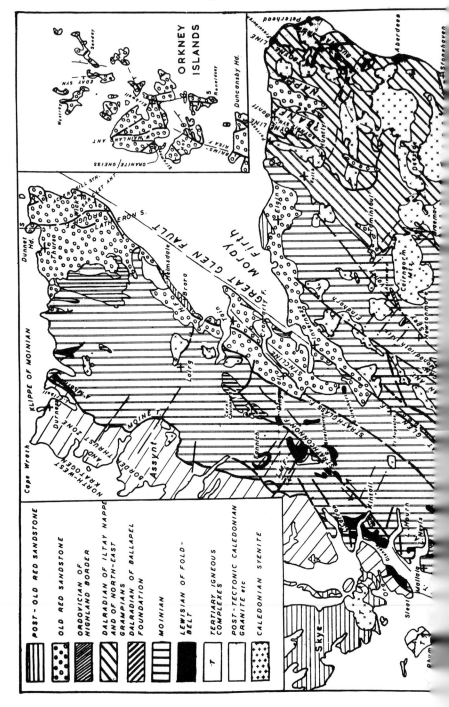

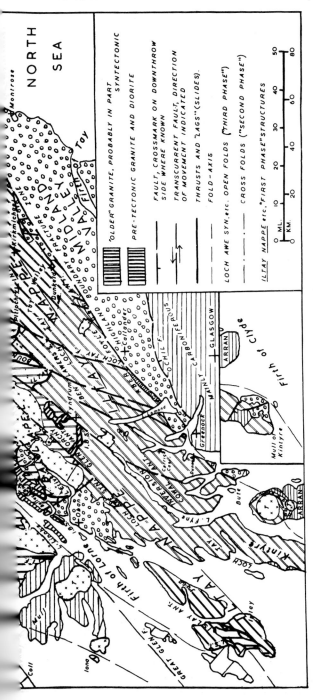

Fig. 5.7 Geological map of Norhern and Grampian Highlands.

F.W.S. = Fort William Slide
I.B.S. = Iltay Boundary Slide
I.F.   = Inninmore Fault
L.M.F. = Loch Maree Fault
L.S.T. = Loch Skerrols Thrust.

These divisions have been found to hold good throughout much of the south-western part of the Northern Highlands, but cannot be extended with certainty further E. and N. The Sutherland Moinian stratigraphy and structure are difficult to establish partly because of involvement with migmatite complexes.

The Moinian Assemblage may not consist of a single, continuous, strati-graphical succession but may contain an "Older Moinian", affected by Pro-terozoic events and a "Younger Moinian" of late Precambrian age in which all or nearly all the folding and metamorphism are Caledonian (5.6).
The Loch Eil Division almost certainly the youngest, would be part of the "Younger Moinian". For fuller discussion see, for example, Anderson and Owen 1980, 70-72.

The Carn Chuinneag-Inchbae granite gneiss in Rossshire pre-dates the main Moinian fold-phases (5.6) and has an aureole of hornsfelsed sediments which have largely escaped the regional metamorphism.

A number of post-tectonic Caledonian intrusions cut the Moinian. These are mainly calc-alkaline plutons except for the Ben Loyal Syenite in the far north which is similar to the alkaline intrusions of Assynt (Ch. 5.1).
The Old Red Sandstone rests unconformably on the Moinian metamorphics and the granites. In Caithness, and round the Dornoch and Cromarty Firths, Lower, Middle and Upper Old Red Sandstone have been recognised. In Orkney the Middle Old Red Sandstone rests unconformably on Moinian and is followed by the Upper, which contains basic volcanics (Island of Hoy).

At Innimore Bay, on the N.E. side of the Sound of Mull 300-600 ft. (92-183 m) of the Upper Carboniferous strata rest on the Moinian and underlie Trias and Tertiary lavas.

West of Loch Linnhe E. to S.E. dykes and elongated bosses of quartz-dolerite are probably of Permo-Carboniferous age. Camptonite and monchiquite dykes trending in the same direction in the N. of the region, including Orkney, and in the S.W., are also probably Permo-Carboniferous.

Mesozoic strata occur in a number of small outliers in the W. and in fairly extensive downfaulted strips on the E. coast. The Trias consists of sandy and conglomerate strata; the Jurassic ranges from Liassic to Kimmeridgian. The Estuarine series at Brora contains coal which has been worked. The Kimmeridgian near Helmsdale is of particular interest. It is downfaulted against Moinian, the Helmsdale granite and the Middle Old Red Sandstone. Bailey and Weir (1933) have shown that the fault was active in Kimmeridgian times and from the submarine fault-scarp were derived huge angular blocks of Old Red Sandstone now forming breccias interbedded with normal marine Kimmeridge sediments (E5.14 and E5.15).

All pre-Old Red Sandstone rocks have had a complex tectonic and poly-metamorphic history. The Old Red Sandstone has undergone considerable folding and faulting. Intra-Devonian movements preceded the Upper Old Red Sandstone. There was considerable intra-Jurassic (see above) and post-Jurassic and the Mesozoic strata have also undergone moderate folding. Considerable faulting postdated the Tertiary (Eocene) lavas.

Comparison with other areas:- The Lewisian rocks are similar to those of the foreland (Ch. 4) although affected by Caledonian events. As mentioned

above, the Moinian is a complex assemblage probably spanning considerable
Precambrian time. It also occurs in the Grampian Highlands (5.4). The
younger part of the Moinian is almost certainly the equivalent within the
Fold-Belt of the Upper Torridonian of the foreland (Ch. 4).

The Old Red Sandstone is a fluviotile/lacustrine formation generally
similar to that of Britain N. of the marine Devonian facies. It differs
from the nearest large Old Red deposit, that of the Midland Valley (Ch. 7.5)
in the presence of Middle Old Red Sandstone of great thickness but resembles
the Midland Valley development tectonically in the occurrence of late
Caledonian, pre-Upper Devonian movements.

Post-tectonic Caledonian granites closely resemble those of the Grampian
Highlands (5.4).

The Kimmeridge boulder beds of Helmsdale, derived from an active submarine
fault-scarp, (E5.14 and E5.15) are unique in the British Jurassic.

Thurso Centre (Fig. 5.8)

There is plentiful accommodation in Thurso which is the terminus of the
railway line from Inverness and has local bus services.

E5.11   John o' Groats, Duncansby Head and Dunnet

Middle and Upper Old Red Sandstone. Volcanic neck and basic dyke of
? Permo-Carboniferous age.

Access:- By public bus to John o' Groats or by private transport. Shore
and cliff-top walking. Low tide necessary.

Maps:- Topo:- 12 Geol:- S 116.

Walking distance:- 4 miles (6.4 km).

Itinerary:- The shore-section at John o' Groats shows thin-bedded, dull
red sandstones, the John o' Groats Sandstone, the highest subdivision of
the Middle Old Red Sandstone. A band of pale-weathering flagstone opposite
the hotel has yielded fossil fish of the genera Microbrachius, Dipterus,
Tristichopterus, Mesacathus and Glyptolapis.

Walk along coast to E. Near high tide mark and forming blown sand there is
a deposit of white shell sand. On the W. side of the Ness of Duncansby and
at its furthest out point there are small vents filled with agglomerate and
dykes of nepheline-basalt. Between the two necks and on the E. side of the
headland a fish bed contains the same genera as those in the bed at John o'
Groats and, in fact, this bed, having curved round a syncline pitching N.,
outcrops in the Bay of Sannick E. of Ness of Duncansby.

On the E. side of this bay a N.W. fault brings up the Caithness Flagstones,
stratigraphically below the John o' Groats Sandstone. These are almost
horizontal and form the spectacular Duncansby Head, the north-easterly tip
of mainland Scotland.

Return to John o' Groats.

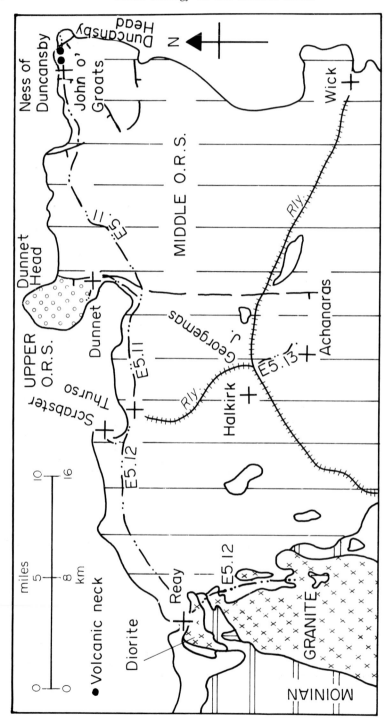

Fig. 5.8 Geological map with routes for Thurso Centre.

If private transport is available a diversion on the way back to Thurso should be made N. to Point of Ness, ½-mile (0.8 km) W. by S. of the village of Dunnat. Yellowish, current-bedded sandstone, dipping gently N.N.W. are cut by a N.E. basic dyke. The sandstones are of Upper Old Red Sandstone age, brought down against the Middle Old Red Sandstone to the E. by the powerful N.-S. Brough Fault.

E5.12   Thurso and Reay

Moinian. Middle Old Red Sandstone. Foliated Caledonian granite. Caledonian diorite. Boulder clay.

Access:- By private transport or public bus to Reay. Shore and moorland walking. Low tide necessary.

Maps:- Topo:- 10, 11, 12 Geol:- S 115, S 116. ¼-inch to mile Scotland 5.

Walking distances:- At Thurso 3 miles (4.8 km). At Reay 6 miles (9.6 km).

Itinerary:- Before leaving Thurso the shore section N.W. of the town should be visited. This shows a broad intertidal platform eroded in the Thurso Flagstones (part of Caithness Flagstones of Middle Old Red Sandstone age), dipping gently N.W. Near Scrabster Harbour the great thickness of boulder clay can be appreciated.

At Reay the section in Middle Old Red Sandstone on the E. side of Sandside Bay should be examined. From calcareous, flaggy beds near where the coast runs from N. to N.E. fossil fish have been obtained, including Homacathus, Mesacathus, Coccusteus and Osteolepis.

Return to road; from a bridge ¼-mile (0.4 km) E. of the church (just W. of road-fork) follow Achnarasdal Burn. From 400 yds. (366 m) upstream augite-diorite outcrops; this is clearly earlier than the Middle Old Red Sandstone but is probably post-tectonic. From ½-mile (0.8 km) upstream to where the stream bends sharply S. Moinian granulites are seen. Moinian granulites are present. A small outcrop of Middle Old Red Sandstone conglomerate comes next, resting to the S. on the N. end of the large Strath Halladale Granite which, from its structure, appears to have been intruded before the Caledonian movements had been completed.

Return to road, using a track on S.W. side of the burn.

A conspicuous feature on the N.E. side of Reay is the Dounreay Fast Breeder Reactor. Conducted tours are organised by the Central Electricity Generating Board in Thurso.

E5.13   Achanarras, near Halkirk

Devonian: Fish Bed in Middle Old Red Sandstone.

Access:- By private transport. By bus or train to Georgemas Junction then walking on road. Application to visit the quarry must be made, in advance, to the Geology and Physiography Section, Nature Conservancy, Foxhold House, Thornford Road, Crockham Common, Newbury, Berkshire RG15 8EL.

Maps:- Topo:- 11 Geol:- S 116.

Walking distance:- From Georgemas Junction and back 8 miles (12.8 km).

Itinerary:- Branch off on track to W. from the Lybster road 2 miles
(3.2 km) S. of Georgemas Junction. The fish bed, a limey flagstone,
10-12 ft. (3-3.7 m) thick, is at the top of Passage Beds between underlying
thick Flagstones and overlying Thurso Flagstones (both in Caithness Flagstone Series, Middle Old Red Sandstone). Fossil fish include Diplacanthus,
Pterichys, Dipterus, Homosteus and Osteolepis. (See Phemister 1960).

Helmsdale Centre (Fig. 5.9)

Helmsdale is on the Inverness-Thurso railway, and there are a few public
buses on the main road. A fair amount of accommodation is available but
it would probably be difficult to find room for parties.

E5.14   Ord of Caithness and coast N. of Helmsdale

Caledonian Helmsdale Granite. Upper Jurassic boulder beds of Kimmeridgian
age. Raised beach.

Access:- By public bus or private transport on Wick road to Ousdale then
road, path and shore walking. Low to fairly low tide necessary.

Maps:- Topo:- 17 Geol:- S 103, S 109.

Walking distance:- 6.5 miles (10.4 km).

Itinerary:- From Ousdale walk S.W. up road to the Ord of Caithness. The
pinkish Helmsdale Granite is first seen then flaggy siltstones and sandstones of Middle Old Red Sandstone age (in a disused quarry) then granite
again at the Ord and downhill towards Helmsdale. The contact, an irregular
unconformity, is not exposed on this traverse.

Leave the road about ¼-mile (0.4 km) N. of Navidale and follow a path S.E.
which ends in a steep drop to the shore. Walk along the shore to the N.E.
where the north-easterly Helmsdale Fault, marked by light-weathering breccia,
runs out to sea, separating the Helmsdale Granite to the N.W. from
Kimmeridgian. The latter is seen in almost continuous intertidal exposures
below the path which leads back to Helmsdale. The strata, which dip S.E.,
are black shales, sandstones and lenticular boulder beds with blocks of
Old Red Sandstone. Black shales on the shore below Navidale House, among
other localities, yield Anavirgatites. Near Helmsdale a "50-ft." (15 m)
raised beach is seen; this is the furthest N. raised beaches occur.

E5.15   Shore from Helmsdale to Brora.

Upper Jurassic (Oxfordian; Kimmeridgian with boulder beds). Raised beaches.

Access:- Road and shore walking. Rail, public bus or private transport
from Brora back to Helmsdale. Low to fairly low tide necessary. Those
with limited time can walk from Helmsdale to the Portgower "Fallen Stack"
and back.

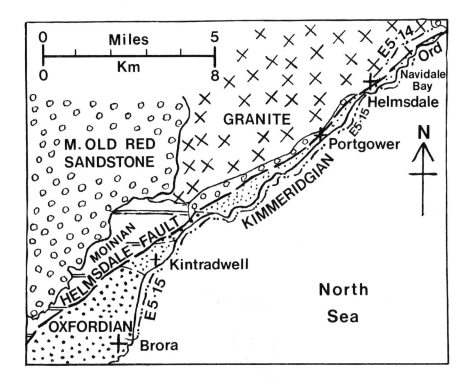

Fig. 5.9 Geological map with routes for Helmsdale Centre.

Maps:- Topo:- 17 Geol:- S 103.

Walking distances:- Helmsdale to Brora 13.5 miles (23.6 km). Helmsdale
to "Fallen Stack" and back 4 miles (6.4 km).

Itinerary:- Walk S.W. from Helmsdale for about a mile (1.6 km) then go onto
shore and continue in same direction. Kimmeridgian shales, sandstones and
lenticular boulder beds with large blocks of Old Red Sandstone are continuously exposed, dipping E. by S., ammonites, corals and lamellibranchs
occur. About ½-mile (0.8 km) S.S.W. of the hamlet of Portgower the "Fallen
Stack", 100 ft. (30 m) long, rises above the general level of the foreshore.
This can be regarded as a very big boulder among the others and like those
consists of Old Red Sandstone.

Exposures continue along the shore, but time may be saved by following the
road to Loth Burn and then returning to the shore. On the N. side of
Lothbeg Point Kimmeridgian strata occurs but to the S.W. a shallow anticline brings up dominantly sandy beds which are either low Kimmeridgian
or Upper Oxfordian (referred to as Corallian on some maps). The sandstone
surfaces show a few large ammonites. Near Kintradwell Farm the Kimmeridgian
reappears with a high proportion of shale and boulder beds with smaller
fragments than is the case further N.E. Ammonites, including Rasenia
uralensis, are fairly common.

Exposures cease here for a considerable distance owing to the cover of
intertidal sand, a storm beach and 25 ft. (7.5 m) and 50 ft. (15 m) raised
beaches. At Brora the Brora River cuts white Oxfordian sandstones.

Inverness Centre (Fig. 5.10)

Inverness can readily be reached by rail or road. There is plentiful accommodation of all kinds, and local public bus services are available.

E5.16   Inchbae and Garve

Inchbae (Carn Chuinneag) pre-tectonic Granite. Moinian. Glacial features.

Access:- By private transport. By rail to Garve then by connecting public
bus to Inchbae Lodge. Road walking.

Maps:- Topo:- 20 Geol:- S 93.

Walking distances:- 7.5 miles (12.0 km). With private transport 2 miles
(3.2 km).

Itinerary:- From Garve travel on Ullapool road and alight at Inchbae Lodge.
Walk on road about half-a-mile (0.8 km) upstream to examine outcrops of
coarse, granitic augen-gneiss.

Return to Lodge to study exposures of Moinian in river. Pelitic metasediments, interbedded with psammitic, are hornfelsed; this thermal metamorphism took place before the regional. (More extensive exposures of
hornfels occur in rather remote country 5 miles (8 km) to the N.E. on the
margin of the Carn Chuinneag Granite).

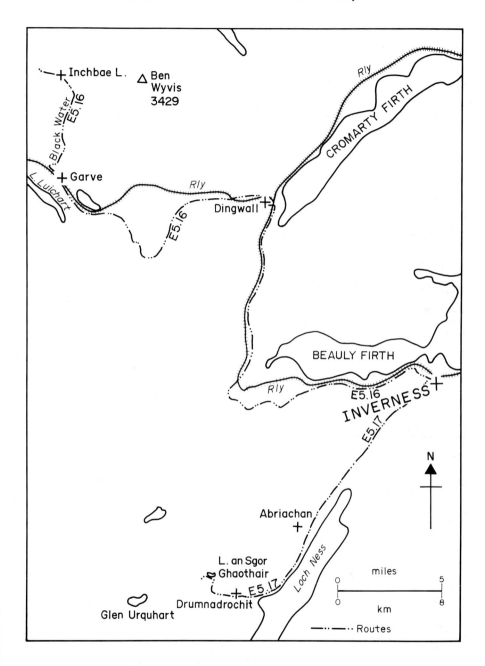

Fig. 5.10 Route map for Inverness Centre.

For those without private transport there should be plenty of time to walk
along the road back to Garve to catch the evening train to Inverness. The
glaciated nature of the landscape is evident, and there is a widespread
cover of morainic drift. Moinian metasediments (mainly psammitic) show
through in places, for example where the road first crosses from the left
to the right bank of the Black Water.

E5.17   Glen Urquhart and Loch Ness

Lewisian Inlier. Moinian. Caledonian Abriachan Granite. Great Glen Fault.
Glacial deposits.

Access:- By public bus to Drumnadrochit, returning from Abriachan road end.
Private transport reduces the walking distance but it should be noted that
the track to the Lewisian limestone quarry is unsuitable and that there are
few places to park along Loch Ness.

Maps:- Topo:- 26 Geol:- S 83.

Walking distance:- 9 miles (14.4 km).

Itinerary:- On the way to Drumnadrochit huge deposits of fluvioglacial
sand and gravel, some moundy some terraced, are seen between Inverness and
Loch Ness.

Walk along road up Glen Urquhart for $1\frac{1}{4}$ miles (2 km) from Drumnadrochit
then up rough track to Loch an Sgor Ghaothair near which there are several
exposures of Lewisian limestone. The most readily seen is in a disused
quarry 250 yds. (230 m) S.W. of W. end of loch. The limestone is a coarse,
white type with thin biotite-rich bands and flanked by green calc-silicate
rocks. It occurs in a sharp anticline pitching S.E. On both sides of the
limestone outcrop acid Lewisian gneiss occurs and to the N.W. there is
serpentine.

Return to Drumnadrochit then walk N.E. along Loch Ness road. The deep
trench eroded along the Great Glen Fault is readily appreciated. Diagonally
across the loch southwards, the hills are in the Foyers Granite Complex,
considered to be the transcurrently faulted continuation of the Strontian
Complex (E.10.24).

Road cuttings show Moinian flaggy granulites, some highly quartzose, others
with considerable biotite. There is marked shattering by the Great Glen
Fault but some parts are almost unaffected. These show steep isoclinal
folding with axial planes determining the flaggy partings. Thick metadolerite dykes and thinner metadolerite sheets cut the metasediments.
Near Milestone 12 from Inverness films of malachite occur on joints in the
granulite.

Near Milestone 9 the red Abriachan Granite is seen and outcrops of this
rock, which is also shattered, continue to the N.E.

Mallaig Centre (Fig. 5.11)

Mallaig can be reached by road or rail from Fort William and the south; there is a moderate amount of accommodation. The port is in the Morar district where a satisfactory Moinian sedimentary succession was first worked out, consisting of Basal Pelite/Lower Morar/Morar (Striped and Pelitic) Schists and Upper Morar Psammite (some subdivisions have been recognised within these).

This succession occurs on the flanks of the large Morar Antiform. Gneiss and amphibolite in the centre are now generally accepted as Lewisian but their relationship to the Morar succession is controversial (see below E5.18).

The boat service to Armadale makes it possible to use Mallaig as a base for excursion (E5.10) in southern Skye.

E5.18   Shore of Loch Nevis E. of Mallaig

Lewisian. Moinian. Morar Antiform. Tertiary dykes.

Access:- Walking on rough shore and steep hillsides. Low to fairly low tide needed.

Maps:- Topo:- 40 Geol:- 61.

Walking distance:- 6.5 miles (10.4 km).

Itinerary:- Examine shore section in vertical Upper Morar Psammite on W. side of line near Morar station. The beds strike N. and current-bedding indicates they become younger W. Tertiary dykes of the Skye swarm head N.

Follow the road E. then N. round Mallaig Harbour. Shore and cliff exposures of the Morar striped schists occur with a central pelitic band rich in garnets. In the upper part of the striped schists thin bands and lenses of a calc-silicate rock are plentiful.

Follow the road past a housing estate and down to the shore at Mallaig-vaig. Approximately 100 yds. (90 m) E. of this hamlet the Lower Morar Psammites are well exposed showing both current and graded bedding indicating younger beds to the W.

Follow a track to Mallaigmore House, 100 yds. (90 m) N.E. of which the shore section shows dominantly siliceous psammite which probably forms the lowest part of the Lower Morar Psammite. To the N.E. this is followed by acid gneisses rich in biotite interpreted as Lewisian. Bands of amphibolite may be pre-Moinian dykes intruded into the Lewisian. To the E. siliceous Moinian psammite again appears. The contact with the Lewisian runs along a gully beside a fence and may be seen below a huge boulder at high water mark. Due to metamorphic and structural events interpretation is difficult, and visiting geologists can try to decide whether the junction is a slide or an unconformity.

Beyond the Moinian a wider outcrop of Lewisian follows as far as the head-land of Srôn Raineach. On the E. side of the latter, crumpled semi-pelitic schists of the Basal Moinian Pelite occur (not present on the W. side of the Lewisian).

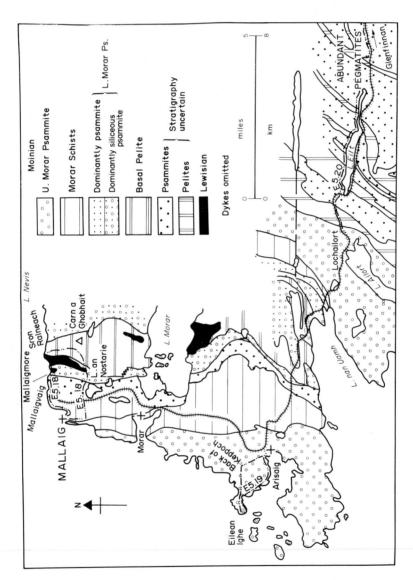

Fig. 5.11 Geological map with routes for Mallaig Centre.

If time is limited, a return may be made from here, but for the full excursion an ascent should be made, following a narrow gully S., up the steep
hillside on the N.W. flank of Carn a' Ghobhair. Further exposures are seen
of the Basal Pelite. Continue along the W. flank of the mountain, cross a
small stream which drains a lochan and climb a small hill some 300 yds.
(270 m) S. of the stream. Here the closure of quartzose Moinian psammite
can clearly be seen.

An interpretation is that first phase movements involved isoclinal inter-
leaving of Lewisian and Moinian (perhaps previously brought into contact
by sliding), a second the formation of a westerly-closing recumbent antiform and a third the formation of the Morar Antiform.

Walk W. across country round the N. side of Loch an Nostare to the road on
the S. side of Mallaig.

E5.19   Coast N.W. of Arisaig

Moinian (Upper Morar Psammite). Late Caledonian, Permo-Carboniferous and
Tertiary dykes.

Access:- Train or private transport to Arisaig. Coastal and road walking.
Low to fairly low tide necessary.

Maps:- Topo:- 40 Geol:- S 61.

Walking distance:- 6.5 miles (10.4 km).

Itinerary:- Walk N.W. along shore and if tide is sufficiently low to most
westerly point of Eilean Ighe. The low-grade of the psammites makes it
possible to study well-preserved sedimentary structures. Pebbly beds occur,
becoming more abundant towards the W. "Right-way-up" structures show that
the beds become consistently younger towards the W; the strike is northerly.

Tertiary basic dykes, trending N., of the Skye swarm are abundant. Several
late Caledonian dykes, striking in the same direction, can be seen, for
example at Uchain Fhada, the furthest S. point of the shore. At the W. tip
of Eilean Ighe two E.-W. Permo-Carboniferous camptonite dykes are cut by
Tertiary dykes.

Return to Arisaig along road from Back of Keppoch.

E5.20   Lochailort to Glenfinnan

Moinian. Migmatites. Pegmatite veining. Late Caledonian, Permo-
Carboniferous and Tertiary dykes. Glacial features.

Access:- By train or private transport to Lochailort, returning from Glenfinnan. The excursion can also be carried out, using rail or road, from
Fort William (Ch. 5.4).

Maps:- Topo:- 40 Geol:- S 61, 62.

Walking distance:- 8.5 miles (13.6 km).

<u>Itinerary</u>:- From Lochailort station walk E. along road. The granulites
near the station belong to the Upper Morar Psammite on the E. flank of the
Morar Antiform, but the N.E.-striking bands of pelite and psammite which
follow cannot be satisfactorily fitted into the Morar succession owing to
high-grade metamorphism and regional migmatisation. From about $3\frac{1}{4}$ miles
(5.2 km) E. of Lochailort station the metasediments are referred to the
Glenfinnan Division, thought to be separated further N.E. from the Morar
Division by the Sgurr Beag Slide although this cannot be demonstrated on
the present traverse.

From excellent roadside exposures it can be shown that the metasediments
were subject to an early, recumbent fold phase, accompanied or followed by
migmatisation, and succeeded by tight, upright folding; there is some
evidence of a third phase which may be connected with movements which formed
the Morar Antiform. Late, sharply-defined pegmatites cross-cut the folds,
these become abundant from the head of Loch Eilt eastwards. Several microdiorite dykes can be seen. About 2.5 miles (4 km) E. of Lochailort station
two N.N.W. Tertiary basic dykes occur. One mile (1.6 km) E. of the head of
Loch Eilt a Permo-Carboniferous camptonite dyke cuts the metasediments.

From the road as it climbs to a col E. of Loch Eilt corries in the mountain
faces and nearby roches moutonnées are evidence of intense glaciation. A
stream from the S. has built a delta which almost crosses Loch Eilt. The
road descends to Glenfinnan at the head of Loch Shiel in a glaciated valley
along a N.E. fault, the continuation of which is marked by a mountain gully
N.E. of Glenfinnan.

5.3   Shetland (Fig. 5.12)

With the development of North Sea oil, Shetland is now visited by far more
geologists than previously. The biggest town, Lerwick, on the "Mainland",
can be reached by plane to Sumburgh at the S. of the same island or by boat
from Aberdeen. Although there is considerable accommodation, booking is
advisable.

Four excursions are described; other interesting sections are mentioned by
Mykura (1976).

The formations present in Shetland are:-

> Upper Old Red Sandstone
> Middle Old Red Sandstone
> Lower Old Red Sandstone
> Shetland Metamorphic Rocks, including units of
>   Dalradian, Moinian and (?) Lewisian ages

There are marked contrasts in the geology E. and W. respectively of the
Walls Boundary Fault, which is probably the continuation of the Great Glen
Fault (5.2).

Acid and hornblendic gneisses of Lewisian aspect form the N.W. corner of the
Mainland. These could be part of the Caledonian foreland or else a Northern
Highland-type inlier (5.2). The rest of the metamorphic rocks W. of the
Walls Boundary Thrust consist of impure quartzite, hornblendic gneiss and
muscovite-schist followed by green schists and calcareous rocks. These
metasediments are unlike those E. of the Fault.

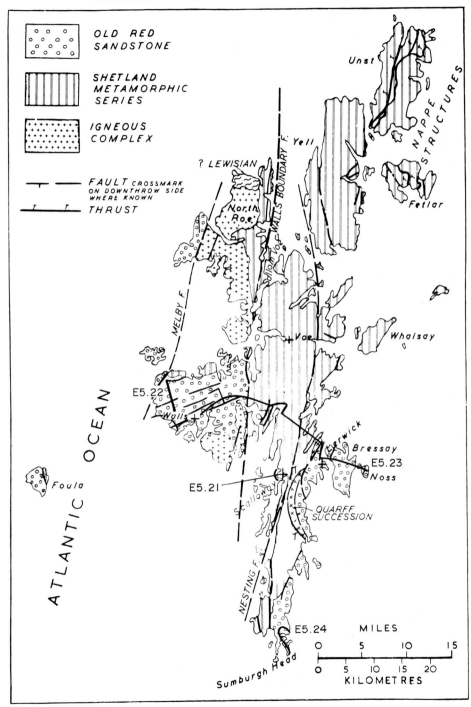

Fig. 5.12 Geological map, with routes, of Shetland.

The metamorphics of the Mainland E. of the Walls Boundary Fault form a long succession starting with psammites (granulites) like those of the Moinian; similar rocks make up Yell. On the mainland the psammites are followed by quartzites, pelites, metalimestones, grits and altered spilitic lavas. There are two belts of migmatites, granites and pegmatites. Along part of the E. coast of the Mainland gneisses, semipelites and gritty limestones, tectonically separated from the metasediments to the W., are termed the Quarff Succession (Fig. 5.5).

In Unst and Fetlar serpentinites, metagabbros, etc., are involved in a Nappe Pile. Post-tectonic Caledonian granites occur in both the E. and W. Mainland.

The Walls Sandstone, of Lower/Middle Old Red Sandstone age, outcrops in the W. and contains basalt, andesite and rhyolite lavas as well as ashes. Upper and high Middle Old Red Sandstone strata occur near Lerwick.

Pre-Caledonian events affected the gneisses of probable Lewisian age in the N.W. Mainland. The younger metamorphic rocks have undergone at least two phases of folding and metamorphism of Caledonian age. Late (or possibly partly Post) Devonian structures are major N. by E. faults including, from E. to W., the Nesting Fault, the Walls Boundary Fault and the Melby Fault. The Walls Fault is marked by intense shearing and shattering. Large dextral displacement, claimed for the Walls Fault, complicates its correlation with the Great Glen Fault.

The Lower/Middle Old Red Sandstone of the Walls Peninsula has undergone intense folding along early E.N.E. axes and later N.N.E. to N. axes. The Old Red Sandstone of the East Mainland is gently folded along mainly N. axes.

Comparison with other areas:- Most of the metamorphics correlate broadly with the Moinian/Dalradian of the Grampian Highlands (5.4). However, the metasediments W. of the Walls Boundary Fault are of a different character, and it is not possible to equate them with the Moinian/Dalradian sequence, although on radiometric evidence the Metamorphism is Caledonian. The Old Red Sandstone resembles that of the rest of Scotland.

Lerwick Centre

E5.21   Scalloway to Ness of Burwick

Shetland metamorphics (probably Dalradian).

Access:- By public bus or private transport to Scalloway. Coastal walking. Fairly low tide necessary.

Maps:- Topo:- 4 Geol:- S 126 with parts of S 123, S 124.

Walking distance:- 5.5 miles (8.8 km).

Itinerary:- On the E. side of Scalloway Harbour, below the Castle, the Laxfirth metalimestone is exposed. This is separated from the Girlsta metalimestone (believed to be older), exposed about the middle of the harbour, by micaceous psammites.

Walk W. past Westshore then N. to Ness of Burwick. There are numerous exposures of migmatised and non-migmatised rocks of the Colla Firth Group including the East Burra Pelite and also of foliated granite and pegmatite.

E5.22   Walls Peninsula

Shetland metamorphics. Old Red Sandstone. Granite.

Access:- By private transport. Some road and shore walking. Low to fairly low tide necessary for shore exposures.

Maps:- Topo:- 3 Geol:- S 127 with parts of S 125, S 126, S 128.
                    ¼-inch to mile S 1 and S 2.

Walking distance:- 10 miles (16 km).

Itinerary:- Travel N.W. from Lerwick and round head of Weisdale Voe where the Weisdale Limestone (believed to be the base of the metasedimentary succession) is seen. Continue W. along the N. side of the Firth. The Walls Boundary Fault is crossed ¼-mile (0.4 km) E. of Bixter; both the metamorphics to the E. and the Walls Formation (Middle Old Red Sandstone) to the W. show considerable brecciation.

At the head of Effirth Voe make a diversion for about 1½ miles (2.4 km) along a road to the S.E. to examine the N. end of a large granite mass intruded into the Walls Formation. Continue W. across peninsula through Walls; there are numerous exposures of the Walls Formation, with vertical or steep bedding striking E.N.E. to E.S.E.

Continue along road W. and leave transport at Burn of Setter. Walk down burn to Head of Voe of Footabrough and follow rocky coast along N.W. side of Voe. The Walls Formation is strongly folded on E.-W. axes and shows cleavage, lineation and small-scale plastic folds; there are rhythmic sedimentary sequences.

Return to transport, park at Netherdale and walk down to Voe of Dale and along its N. coast. Here sandstones and pyroclastics of the Sandness Formation (Lower/Middle Old Red Sandstone), striking N.E. are intruded by a felsite sill. From Mu Ness there is a view N. of spectacular cliffs.

Return to road and to Walls then travel N.W. to Melby. Walk along coast and examine the Melby Formation (Middle Old Red Sandstone) consisting of sandstones with two flows of silicified rhyolite or ignimbrite. The shore section W. of Huxter shows the Lower Melby Fish Bed, and ½-mile (0.8 km) further S.W., at Pobie Skeo, the Upper Melby Fish Bed. The fish remains include Coccosteus, Dipterus, Homosteus and Mesacanthus; plant remains also occur.

Return to Melby and travel E. to Bay of Garth. On the W. side of the bay hornblende-schists outcrop. Walk along coast for ¼-mile (0.4 km) to E.N.E. to study schists, quartzites and limestones with D2 lineation and boudinage and minor folding locally refolded by D4 conjugate folds; a N.-S. porphyrite dyke occurs.

Return to Lerwick.

E5.23   Lerwick, Bressay and Noss

Old Red Sandstone.

Access:- Ferry. Road and moorland walking. Fairly low tide necessary for some exposures.

Maps:- Topo:- 4 Geol:- S 126 with parts of S 123, S 124.

Walking distance:- 6 miles (9.6 km), including Noss 9.5 miles (15.2 km).

Itinerary:- Walk to South Ness, at the S.E. outskirts of Lerwick.
Exposures of Middle Old Red Sandstone show sedimentary structures and
rhythmic sequences. Large plant remains can be found.

Return to harbour and cross by ferry to Bressay. On the S. side of the
little bay at the ferry landing tuffisite breccia, considered to have been
formed by uprising volcanic gases, is seen.

Walk across island. Alongside the road and track and on the E. coast there
are exposures of the Bressay flaggy sandstones and siltstones (Middle/Upper
Old Red Sandstone).

It is sometimes possible to get a boat across the narrow Noss Sound to Noss.
At the landing tuffisitic breccia is again seen. Walk S.E. across island.
Between outcrops of Bressay flaggy sandstones and siltstones there is an
upfaulted outcrop of current-bedded sandstones (also Middle/Upper).

Noss Head on the E. coast is a nearly vertical cliff in thin-bedded sandstones 592 ft. (180 m) high.

Return to Lerwick.

E5.24   Sumburgh

Old Red Sandstone.

Access:- By private transport. By public bus to Sumburgh Airport. Fairly
low to low tide necessary.

Maps:- Topo:- 4 Geol:- S 126 with parts of S 123, S 124.

Walking distance:- 4 miles (6.4 km).

Itinerary:- From the Airport walk along road round Pool of Virkie. Along
the N.W. shore current-bedded Middle Old Red Sandstone with local pebble
beds is exposed. From the furthest N. part of the shore follow track N.E.
past Exnaboe and on to the E. coast Shingly Geo on the S. side of the
conspicuous Point of Blo-geo. Here the Exnaboe Fish Bed is exposed with
typical Middle Old Red fish remains. To the S.E., in the direction of dip
there are sandstones and siltstones.

Return to Sumburgh Airport.

5.4   The Grampian Highlands (Scotland and Ireland) (Fig. 5.7)

Apart from the area around and N. of Aberdeen the Grampian Highlands are
noted for their rugged mountain scenery. Ben Nevis 4406 ft. (1345 m),
within a Devonian ring-complex, is the highest point in the British Isles,
and there are numerous other mountains, mostly of metasediments, rising to
above 3000 ft. (915 m). Structurally, the Grampian Highlands extend into
Ireland where the mountains rise to nearly 2500 ft. (760 m) in the N.W.

The Great Glen Fault (5.2) forms the N.W. boundary. The Highland Boundary
Fault, marking the S.E. margin, extends for 380 miles (608 km) from Stonehaven on the North Sea to Clare Island off the Atlantic coast of Ireland.
This is a complex fracture-zone, along which movements, not all affecting
the total length, have taken place from Arenig (or perhaps earlier) times
to Tertiary times.

There is a reasonably good road system in the Scottish Grampians, particularly in the relatively low N.E. and along the raised beaches of the W.
coast. Four railway routes traverse part of the region. Apart from difficulties due to the popularity of the region at peak holiday times, accommodation can readily be gained in a number of towns and villages.

In N.E. Ireland communications and accommodation are not difficult but in
the N.W. roads are fewer and often narrow, public transport infrequent and
accommodation more of a problem.

For the Scottish Grampians the following centres are suggested: Pitlochry,
Newtonmore, Portsoy, Aberdeen, Fort William, Oban, Glasgow and Brodick.
For Ireland: Portrush, Donegal and Creeslough.

The formations present in the Grampian Highlands are:

        Pliocene
        Tertiary volcanics
        Liassic

        Triassic ⎫ New Red Sandstone
        Permian ⎭

        Old Red Sandstone
        Lower Ordovician
        Dalradian Metamorphic Assemblage or Supergroup (partly Cambrian)
        Moinian Metamorphic Assemblage ⎫ Precambrian
        ? Lewisian (Mayo only)   ⎭

It is only in the N.W. of County Mayo, Ireland, that any rocks are seen in
the Grampian Highlands which may be a basement to the Moinian (Fig. 5.8).
Here gneisses have been correlated with the Lewisian (Sutton 1972), although
radiometric dates (Max and Soret 1979) give a younger age.

The Moinian Metamorphic Assemblage forms a large part of the Grampian Highlands in Scotland and considerable outcrops in western Ireland. For general
description see that given for the Northern Highlands (5.2). As in the
latter region they are best regarded as making up an assemblage of metamorphic rocks for both in Scotland and Ireland it appears unlikely that the
sediments belong to one continuous succession.

The Dalradian metasediments are much more varied than those of the Moinian.
Consequently many district successions, sometimes interpreted in the wrong

stratigraphical order, were worked out at an early stage in research. The
use of "way-up" techniques has made it possible to establish a sequence
which is broadly accepted by most geologists. Correlation tables by
Anderson (1965, Table I) and by Harris and Pitcher (1975, Fig. 12) give
most older locality names. The two latter authors have selected certain
locality names for subdivisions of the broad sequence but, on the other
hand, lithological terms have the merit of providing descriptions. Both
are, therefore, set out in the summary table below, which also shows the
top of the Moinian.

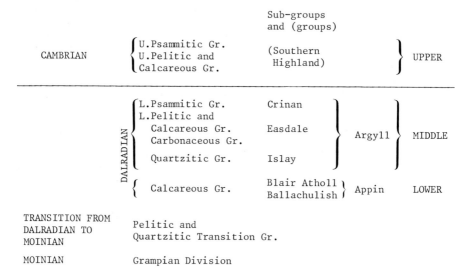

Where there is no tectonic contact between Moinian and Dalradian there is a
variable interbedded succession of pelites and quartzites between the
Psammitic Group and undoubtedly Lower Dalradian.

An important horizon at the base of the Quartzitic Group is a glacogene with
large boulders, often of nordmarkite, with interbedded upper and lower contacts, which has been traced from Aberdeenshire to the Atlantic coast of
Ireland (see also 5.6).

Stratigraphically, most of the Moinian and part of the Dalradian are late
Upper Proterozoic. The Upper Dalradian, on the other hand, is Cambrian as
pagetid trilobites of late Lower Cambrian age occur near Callander, Perthshire, in the Leny Limestone, which is part of the Upper Psammitic Group.
As acritarchs (Downie and others, 1971) of Lower Cambrian type occur down
to the Loch Tay/Tayvallich limestone horizon at the base of the Upper Dalradian the Precambrian-Cambrian boundary may be about here.

The structural and metamorphic history of the Moinian/Dalradian is complex
and has been much discussed (see, for example, Anderson and Owen, 1980,
Ch. 5.4).

Narrow strips of basic lavas, black shales, cherts and other sediments occur
along the Highland Border, mostly separated from Upper Dalradian by faults,
but in places, notably Arran, following on without structural or metamorphic

break. Fossils, some doubtfully diagnostic, indicate Upper Cambrian/Lower
Ordovician ages.

Many post-tectonic Caledonian intrusions of granitic and related rocks penetrate the schists. They include large masses, probably of batholithic form,
and ring-complexes such as that of Ben Nevis. Large numbers of hypabyssal
intrusions also occur, including north-easterly dyke swarms. Some thick,
easterly, quartz-dolerite dykes of presumed Permo-Carboniferous age are
present. In the W. of Scotland and in Ireland Tertiary basic dykes are
numerous (Ch. 9). Considerable outcrops of Lower Old Red Sandstone sediments
and volcanics unconformably overlie the metamorphic rocks. Middle and Upper
Old Red Sandstone conglomerates and sandstones cover a large area around
Inverness and S. of the Moray Firth. Upper Old Red Sandstone sediments rest
directly on the Dalradian in the Loch Lomond/Firth of Clyde district, where
they are overlain by a small outcrop of Lower Carboniferous strata. Both
Lower and Upper Carboniferous overlie the Dalradian at the S. end of Kintyre.
A patch of Upper Carboniferous occurs along the pass of Brander Fault E. of
Oban.

The Grampian Highlands in Ireland differ from those of Scotland in the
presence of extensive outcrops of Lower Carboniferous strata, including
thick limestones. In the extreme N.E., at Ballycastle, there is a Lower
Carboniferous succession, including basalt lavas, similar to that of the
Midland Valley of Scotland, followed by sandstones, shales and coals
belonging to the Upper Carboniferous.

On the W. coast of Kintyre, red sandstones, formerly assigned to the Upper
Old Red Sandstone, are now considered to be of Permian age. Triassic strata,
with reptiles, unconformably overlie the Old Red Sandstone S. of the Moray
Firth. Gravels in Aberdeenshire are believed to be Pliocene.

In N.E. Ireland, Triassic red conglomerates, sandstones and marls occur round
the rim of the Antrim Tertiary basalt lavas. These are followed in places by
Rhaetic and Lower Liassic shales and thin limestones, followed by Upper
Cretaceous white limestone, a hard chalk.

Comparison with other areas:- The Moinian is broadly comparable with that
of the Northern Highlands (5.2) but the size of the Great Glen Fault makes
detailed correlation difficult. Dalradian metasediments undoubtedly occur
in Shetland (5.3) and S. of the Highland Boundary Fracture-zone in Connemara
(5.5). The post-orogenic plutons are closely comparable with those elsewhere
in Britain (6.2, 6.3, 6.4 and 6.5) but the Caledonian ring-complexes are not
known elsewhere in these islands.

The Old Red Sandstone of the southern part of the Grampians is like that of
the Midland Valley (7.5) and that of the N.E. part similar to the Middle/
Upper sequence of the Northern Highlands (E5.11, E5.12). The sparse
Carboniferous outcrops of the Western Highlands are of the same facies as
Carboniferous of the Midland Valley (7.5) but those of North-West Ireland,
with their thick carbonates, are a continuation of the Carboniferous Limestone of Central Ireland (Ch. 7.6).

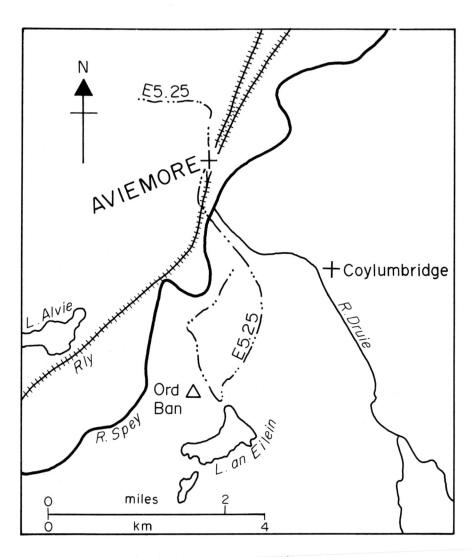

Fig. 5.13 Route map for Aviemore area.

Newtonmore Centre

What are probably some of the oldest rocks in the Grampians can be seen from
Newtonmore. It is on the main rail and road routes from Perth to Inverness,
and there is plentiful accommodation.

E5.25   Aviemore district (Fig. 5.13)

"Older" and "Newer" Moinian. Caledonian granite.

Access:- By train or private transport to Aviemore. Road and hill walking.

Maps:- Topo:- 35 Geol:- S 74.

Walking distances:- 8.5 miles (13.6 km). With private transport 2 miles
(3.2 km).

Itinerary:- From the main road $\frac{1}{2}$-mile (0.8 km) N. of Aviemore station walk
W. up hillside. Above the glacial drift there are exposures of Moinian
metasammites veined in a complex fashion by granite. This is the margin
of the Monadlaith Granite seen as a homogeneous, pink, quartzo-felspathic
granite further up the hillside.

Return to road, then on the S. side of Aviemore branch off S.E. and follow
the minor road to the N. end of Loch an Eilean. On the W. side of the road
(marked by an old kiln) there is a disused quarry in massive, coarse metalimestone with associated calc-silicate rocks. These lie at the base of
metapelites and striped schists in which delicate graded bedding (not easy
to detect in the field) indicates an upward succession towards the N.W.
Below the carbonate horizon there is foliated garnet-amphibolite.

Traverse the hillside N.N.W. along the N.E. slopes of Ord Ban to examine
exposures of pelitic gneiss, quartzite and gneissose psammite. These show
three generations of structures, all associated with high-grade metamorphism,
which are not present in the overlying pelites etc. They have been placed
by Piasecki (1980)in a Central Highland Division (or "Older" Moinian) and
are regarded as separated by a slide from a Grampian Division ("Younger"
Moinian) comprising the pelites etc. Three later structures can be distinguished in the latter and also a still younger deformation which formed
the antiform in which the Ord Ban outcrops lie.

From Aviemore there is a fine view to the S.E. of the large Cairngorm
Caledonian Granite, the largest area over 4000 ft. (1220 m) in the British
Isles; the corries which indent its N. face carry snow well into the summer.

E5.26   Dalwhinnie to Laggan Bridge and Kinlochlaggan (Fig. 5.14)

Moinian. Dalradian. Migmatite complex. Strathspey Granite. Glacial
features.

Access:- By train to Dalwhinnie. By private transport. Road and hill
walking.

Maps:- Topo:- 35, 42 Geol:- S 63, S 64.

Walking distances:- Dalwhinnie station to Laggan Bridge 7$\frac{1}{2}$ miles (12 km).
Dalwhinnie station - Laggan bridge - Newtonmore 14$\frac{1}{2}$ miles (23.2 km). The

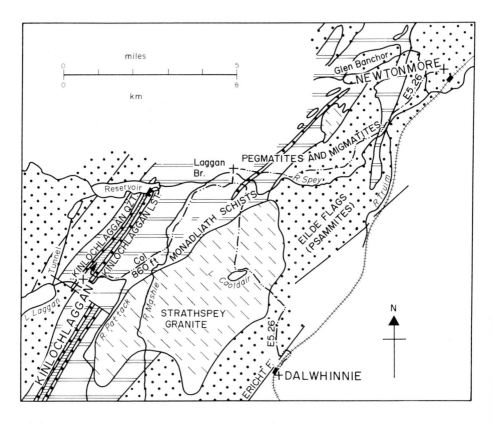

Fig. 5.14 Geological map with routes of Dalwhinnie-Kinlochlaggan-
Newtonmore area.

walk from Laggan Bridge to Newtonmore can be shortened by using a public
bus (infrequent).

Itinerary:- At the N. end of Dalwhinnie branch off on the Laggan Bridge
road. On the W. side of the road 0.3 mile (0.5 km) N.N.W. of the railway
a stream section shows pegmatite. Leave the road 1¼ miles (2 km) N. of the
railway and cross a small stream where this is an exposure of Moinian
striped schist with a calc-silicate band, cut by pegmatite.

Traverse N.W. across rough ground to the E. end of Loch Caoldair. Crags
above the N. shore of the loch provide good exposures of the coarse, sometimes pegmatitic, Strathspey Granite (Smith 1970). The loch occupies a
rock basin, and the topography is strongly glaciated. Traverse N. by E.
to the road across rocky ground which shows the presence of numerous metasedimentary inclusions in the granite, ranging up to large rafts. Continue
in transport or on foot N.W. to Laggan Bridge, where there are exposures of
migmatites on the N. bank of the R. Spey.

Return along the road to Newtonmore; further exposures occur of migmatised
pelites (Monadliath Schists), in which evidence can be seen for two, and in
places three, structural events. It is significant that the same structures
occur in the inclusions in the Strathspey Granite, showing that the latter
postdated the migmatisation.

If arrangements have been made to pick up private transport at Laggan Bridge,
a journey should be made S.W. to Kinlochlaggan. Immediately W. of the
bridge over the R. Mashie the migmatised Monadliath Schists are intruded by
metadolerite. At Kinlochlaggan white crystalline limestone with metadolerite, in the core of a N.E. synform with vertical axis, is enclosed on
both sides by quartzite and Monadliath Schists. Sparse boulders of soda-
granite occur in a bed between the limestone and the quartzite (cf. E5.28).

The flat col at 860 ft. (262 m) O.D. (on the main Highland watershed)
through which the road passes between the R. Mashie and the R. Pattack (a
river captured from the S.W.) was the overflow channel for the lowest of
the Glen Roy glacial lakes (E5.34).

Pitlochry Centre (Fig. 5.15)

From Pitlochry it is possible to see several Dalradian formations which were
recognised by Geikie when he first established the Dalradian "Series".
The town is on the main rail and road routes from Perth to Inverness.
There is plentiful accommodation but booking is advisable for groups as
Pitlochry attracts numerous visitors.

E5.27   Tummel and Garry valleys, Pitlochry to Blair Atholl

Dalradian. Glacial features.

Access:- Road and path walking. Return by rail from Blair Atholl.

Maps:- Topo:- 43, 52 Geol:- S 55.

Walking distance:- 7 miles (11.2 km).

Itinerary:- From Pitlochry cross the valley on the dam which encloses the
Pitlochry hydroelectric reservoir. The thick glacial deposits on the N.E.

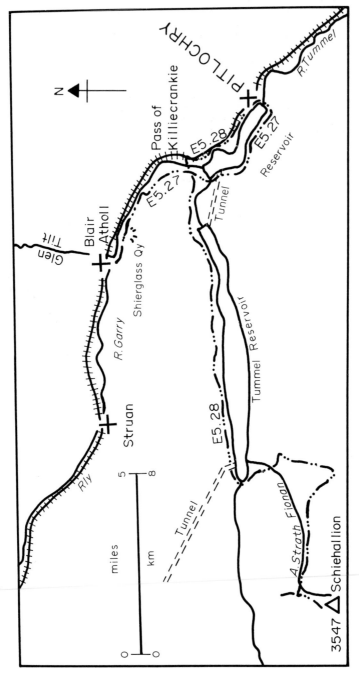

Fig. 5.15 Route map for Pitlochry Centre.

side hide a deep buried channel which had to be sealed by an impervious
"cut-off" to prevent leakage. The main part of the dam is on the Pitlochry
Schists which can be seen downstream along with some hornblendic schists
known as Green Beds which may originally have been ashy material.

Along narrow road on right bank to a short distance S. of where the valley
bends sharply W. then enter large disused quarry in a wood. This is in the
Loch Tay Limestone dipping N. 25° W. at 35°. The Limestone and the stratigraphically younger but structurally underlying Pitlochry Schists belong to
the Upper Pelitic and Calcareous Group (see beginning of this section), and
the Loch Tay Limestone may be the oldest part of the Dalradian Supergroup
which is of Cambrian age. The inverted sequence is within the Loch Tay
Inversion (F1) affecting a large part of the southern and mainly Upper
Dalradian belonging to the Iltay Nappe.

Continue along road upstream crossing a stratigraphically descending Middle
Dalradian succession of the Ben Lui Schists (more quartzitic than the
Pitlochry) Farragen Group (garnetiferous schists with hornblende schists),
Ben Lawers Schists (calcareous schists – Lower Pelitic and Calcareous Group)
and Ben Eagach Schists (graphitic). Near the confluence of the Tummel and
the Garry the last-named passes into the Carn Mairg Quartzite which sedimentary structures prove to be older.

Leave road and descend to footbridge over R. Tummel then follow path along
right bank of R. Garry to Loch Tummel road. Above the path there are exposures of the Killiecrankie Schists. It should be noted that the bedding
and F1 schistosity become steep as the inverted beds turn downwards. The
area is also one for studying the imposition on the F1 structures of N.N.W.
F2 structures and a few E.N.E. F3 structures.

At the end of the path turn W. for about 200 yds. along the Loch Tummel
road then on a narrow road which follows the hillside above the spectacular
Pass of Killiecrankie then descend to the R. Garry. The Killiecrankie
Schists separate the Carn Mairg Quartzite from the Schiehallion Quartzite
seen further N.; all three formations make up the Quartzitic Group. In the
Schiehallion Quartzite, pebbly in places, there are current bedding and
graded bedding which shows it to be younger than the Blair Atholl Limestone
(Lower Dalradian) outcropping on the hillside and worked in the Shierglass
Quarry. Across the Garry valley glaciated Glen Tilt marks the course of a
major N.E. fault.

Cross a footbridge over the river opposite Blair Atholl to a road leading
to the station.

E5.28   N. side of Schiehallion

Moinian. Dalradian.

Access:- Private transport. The road in Strath Fionan is not suitable for
larger coaches. Hill walking.

Maps:- Topo:- 42 Geol:- S 55.

Walking distance:- 3.5 miles (5.6 km).

Itinerary:- Follow the main Inverness road to Bridge of Garry then travel
W. to Tummel Bridge. Turn S. on Aberfeldy road then N.W. on narrow road

into Strath Fionan. Leave transport at bridge over stream.

Walk W. along road, N. of which are outcrops of the Meall Dubh Quartzite.
500 yds. (460 m) W. of the bridge current bedding shows that the steep S.
by W. dip is normal. 250 yds. (230 m) further on, and on the W. side of a
conspicuous acid dyke, make a traverse N. across rising moorland. Beneath
the quartzite there is a succession of graphite schists, limestones and
pelites, below which come thin beds of pure quartzite. Some 750 yds.
(690 m) N. of the road this succession passes down into flaggy psammites
which are the continuation of the Struan Flags of the Garry valley above
Blair Atholl and are of Moinian age.

The strike here is swinging round to the N.W. due to a large F2 fold
superimposed on the N.E. F1 structures (E5.27).

Return to the road then ascend the N. slopes of Schiehallion. Immediately
S. of the road the pale Strath Fionan limestone is seen. This is followed
by a thick succession of generally dark pelites and limestones which are
the continuation of the limestones etc. seen at Blair Atholl (E5.27). At
one time the strata N. from these limestones etc. to the Moinian were regarded as a repetition by folding of the strata to the S. Treagus and King
(1978) have, however, shown that the more northerly strata are stratigraphically older and have correlated them with the Ballachulish/Lochaber
succession (E5.34-E5.37). They also cast doubt, in this area at any rate,
on the existence of a Boundary Slide between the Dalradian and the Moinian.

200 yds. (183 m) S. by W. of the sharp bend in the most westerly headstream
of the Allt Strath Fionan the Schiehallion Boulder Bed is seen in an F1
fold. A wider outcrop can be studied by climbing to 0.6 mile (1 km) S.S.W.
of the road, i.e. to about the 1750 ft. (533 m) level. Carbonate and soda-
granite boulders occur in a sandy calcareous matrix. The Boulder Bed is a
widespread horizon in the Dalradian (cf. E5.45, E5.49) and in the late
Precambrian generally.

At about 1900 ft. (580 m) the Boulder Bed passes upwards into the demonstrably
younger Schiehallion Quartzite.

Return to the road.

Portsoy Centre (Fig. 5.16)

The coastal town of Portsoy has a fair amount of accommodation of various
kinds. It can be reached by public bus from Aberdeen and elsewhere, and
there are some local bus services.

E5.29   Cullen, Portsoy and Boyne Bay

Access:- By public bus or private transport to Cullen. Coast and road
walking. Fairly low tide necessary.

Maps:- Topo:- 29 Geol:- S 96.

Walking distance:- 10 miles (16 km).

Itinerary:- The coast section at Cullen is in the Cullen Quartzite
(Transition Group at top of Moinian) an impure quartzite with bands of
pelite containing garnets. Current-bedding shows that the northerly dip is

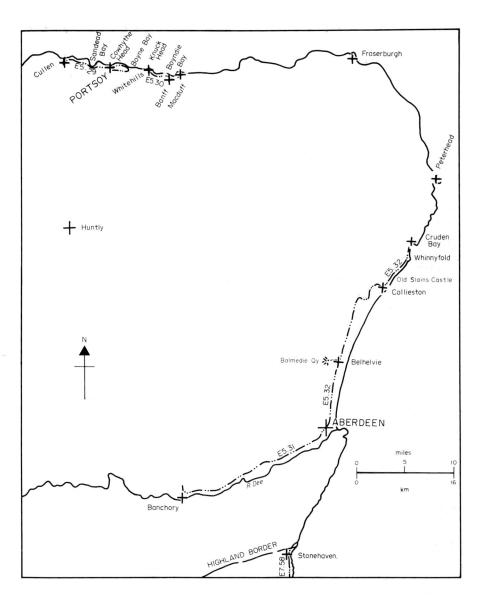

Fig. 5.16 Route map for Portsoy and Aberdeen Centres.

normal. A 30 ft. (9 m) raised beach is present.

Time can be saved by leaving the coast and walking E. on tracks and minor
roads to Sandend Bay, also backed by a 30 ft. (9 m) raised beach. On the
W. side of the bay the Garron Point Group consists of slightly calcareous
schists and flats with actinolite needles. In the bay itself the Sandend
Limestone in several beds is seen in natural exposures and in disused
quarries; there are associated black schists. The Sandend Limestone is
equivalent to the Blair Atholl (E5.27, E5.28) and is the local base of the
Dalradian. On the E. side of the bay two small patches of Old Red Sandstone conglomerate are seen. The more easterly covers a N.W. fault which
brings down the Durn Hill Quartzite which inland stratigraphically overlies
the limestone.

Follow a minor road E. to Portsoy. West of the swimming pool kyanite-
schists of the Portsoy Group occur. South-east of the pool metagabbro,
serpentine, mullioned quartzite and anorthosite outcrop in that order.

Follow cliff-path E. from Links Bay, E. of Portsoy. A migmatic tectonic
melange contains inclusions of siliceous rocks and limestones. On the
W. side of East Head an irregular sheet of red pegmatite has been worked
for rough-cast and for felspar. Schorl is a constituent.

Cowhythe Head is made of the felspathic, highly contorted Cowhythe Gneiss.
To the S., in Boyne Bay, the Boyne Limestone is worked in a large quarry;
it is probably equivalent to the Loch Tay Limestone (E5.27) and is the
base of the Upper Dalradian. On a smaller bay on the N.W. side of the main
bay the limestones abut against sheared phyllites derived from the Cowhythe
Gneiss. This is the Boyne Line but whether it marks the base of a Banff
Nappe, as has been claimed, is controversial.

Walk S. to road and back to Portsoy.

E5.30   Whitehills, Banff, Macduff

Dalradian. Old Red Sandstone.

Access:- By public bus or private transport to Whitehills returning from
Macduff. Coast and road walking. Fairly low tide necessary.

Maps:- Topo:- 29 Geol:- S 96.

Walking distance:- 5 miles (8 km).

Itinerary:- Walk N.E. from Whitehills Harbour to Knock Head. Here the upper
part of the Whitehills Group consists of vertical thick graded greywackes
(younger sides to E.). Pelitic beds contain andalusite. This is the Buchan
type of regional metamorphism developed at high temperature with low stress.
Continue to E. end of Boyndie Bay, passing two small patches of Old Red
Sandstone conglomerate, and examine shore sections below the W. end of Banff.
Graded bedding and other sedimentary structures can be examined, and it can
be shown that andalusite disappears eastwards. The axis of a major structure,
the Boyndie Bay Syncline, is crossed about 100 yds. from the W. end of the
section. E. of the axis the grits and slates form sharp folds with axial
planes dipping steeply W.

Continue through Banff and along the road E. from Macduff to Tarlair
swimming pool. Similar structures to those at Banff can be seen. By the
slaughter-house a number of beds of graded greywacke in an open syncline
contain boulders up to 3 ft. across of syenite, gneiss and granulite; they
may have been transported by floating ice.

Return from Macduff to Portsoy.

Aberdeen Centre (Fig. 5.16)

Aberdeen is on main rail and road routes, and there are good public bus
services. There are numerous hotels etc. but as Aberdeen is now "the oil
capital of Europe" accommodation may be difficult to get, especially for
parties.

The numerous outcrops of Caledonian granites have long been the basis of
the important Aberdeen granite industry, and the city is largely built of
this rock. Granite quarrying has diminished and there are numerous
abandoned workings as well as several still in operation. Rubislaw Quarry,
on the S.W. outskirts of the city, is 400 ft. (122 m) deep, well below
sea-level, and Kemnay Quarry, 15 miles to the N.W., is equally large (for
details of quarries see Anderson 1939).

The section of the Highland Boundary Fracture-zone (E7.58) on the North Sea
coast can readily be visited from Aberdeen by travelling to Stonehaven by
rail, bus or private transport.

E5.31   Banchory district

Dalradian. Caledonian granite. Glacial deposits.

Access:- By public bus on North Deeside road to near Banchory. By private
transport. Road and hillside walking.

Maps:- Topo:- 37, 38 Geol:- S 66, S 67, S 77.

Walking distance:- 3 miles (4.8 km).

Itinerary:- In the Dee valley moundy fluvioglacial deposits, partly eroded
to terraces, are conspicuous. Leave the bus about 2 miles (3.2 km) E. of
Banchory. The slopes N. of the road are formed of granite on the S. margin
of the large Hill of Fare Granite. About a mile (1.6 km) E. of Banchory
the Queen Hill Schists (Dalradian) appear. These are high-grade migmatites
with garnets, and sillimanite in some bands.

Walk to Hillhead of Arbeadie on the N. outskirts of Banchory, then along
track N. by W. for 300 yds. (275 m) to a disused quarry in the Deeside
Limestone (probably = Loch Tay Limestone, E5.27), folded along steep N.N.W.
axes. In addition to coarse metalimestone, calc-silicate rocks are present.

Return to Aberdeen.

E5.32   Belhelvie and coast from Collieston to Cruden Bay

Dalradian. Caledonian granite and gabbro. Glacial deposits.

Access:- By public bus or by private transport. Road and coast walking.
Fairly low tide necessary.

Maps:- Topo:- 30, 38 Geol:- S 77, S 87.

Walking distances:- 8.5 miles (13.6 km). Reduced if private transport
is used.

Itinerary:- N. of the R. Don, extensive deposits of moundy fluvioglacial
deposits are seen, some worked for sand and gravel. If travelling by bus
alight at Belhelvie. Walk N. along main road for ¼-mile (0.4 km) then W.
along road for 1½ miles (2.4 km) to Balmedie Quarry (Aberdeen County
Council). This is in Caledonian gabbro with veins of white, muscovite-
pegmatite.

Return to main road and travel by bus or private transport to Collieston.
The cliff section N. to Whinnyfold is in grits, greywackes, siltstones
and andalusite-schists, correlated with the Boyndie Bay Beds (E5.30).
They can be studied either by following the cliff top all the way by foot
or by using private transport on the road back from the coast and making
selected visits to the cliffs. Access to the cliff top is gained by
ascending a steep part at the N. end of a quarry at the N. of Collieston.
Localities of particular interest include: on the landward wall of the
isthmus connecting Aver Hill (a projection of the cliff ¼-mile (0.4 km)
N. of Collieston) to the mainland massive, "right-way-up" grits show
pebbles elongated in C. and boudins in C. On the S. wall of the promontory
on which Old Slains Castle is built, small-scale recumbent folds in anda-
lusite-schist pitching gently N. are displayed; the andalusite is later
than the folding. Further N. Coch Craig can be identified by a considerable
thickness of bright red boulder clay beneath which the south facing cliffs
show a beautiful recumbent fold with axial-plane cleavage; from graded
bedding it can be shown that the sense of movement is to the E. At the
Devil's Study, just S. of the cliff-waterfall known as Pissing Yad, a
spectacular recumbent syncline closing W., with flat axial-plane cleavage,
is seen.

At Whinnyfold the Dalradian metasediments are cut across by the pink acid
Caledonian Peterhead granite.

Continue to Cruden Bay and return to Aberdeen.

Fort William Centre

Fort William can be readily reached by rail or road. Accommodation is
plentiful but booking for parties at any rate is necessary as the town is a
very busy tourist centre, particularly as it lies at the foot of the highest
mountain in the British Isles (E5.35).

E5.33   Tulloch and Loch Laggan

Moinian. Pegmatite complex. Caledonian quartz-diorite. Glacial features.

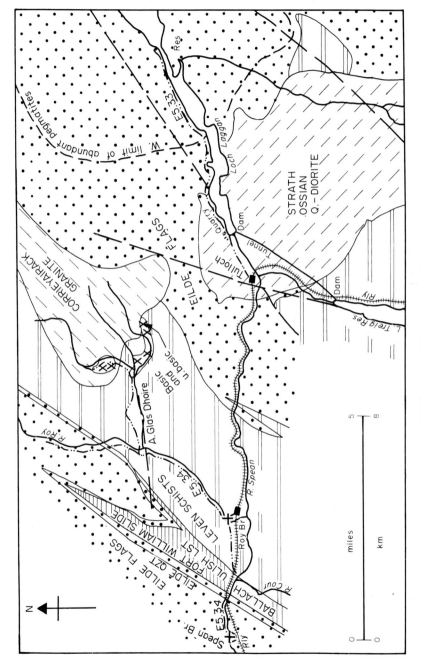

Fig. 5.17 Geological map, with routes, of Loch Laggan-Tulloch-Glen Roy area.

Access:- By train to Tulloch or by private transport. Road walking.

Maps:- Topo:- 41, 42 Geol:- S 63.

Walking distances:- 11 miles (17.6 km). With private transport 1 mile (1.6 km).

Itinerary;- On the sides of Glen Spean, below Tulloch, a "Parallel road", at 857 ft. O.D. (261 m) marks the shore of the lowest of the Glen Roy lakes (E5.34). Near Tulloch thick sand and gravel deposits, rising in places to a terrace not far below the same level, are part of the fill of the lake.

From Tulloch N.E. to the N. end of the Laggan Dam which extended Loch Laggan by 4.5 miles (7.2 km) as part of the Lochaber Power Scheme. The Strath Ossian quartz-diorite, with numerous orientated-basic xenoliths, is seen in a disused quarry. The intrusion has the unusual alignment, for a Caledonian pluton, of N.W., and the metasediments on its margins have been swung into the same direction. One mile (1.6 km) further N.E. the contact with meta-sediments is seen on the N. side of the road.

Continue along the road, noting outcrops of Moinian flaggy psammites con-taining thin calc-silicate lenses. Where the reservoir becomes narrow, thick, cross-cutting, pink pegmatites are abundant, these continue for 8.5 miles (5.6 km) to the N.E.

Return to Tulloch and Fort William.

E5.34   Glen Spean and Glen Roy (Fig. 5.17)

Moinian. Dalradian. Caledonian granite, basic and ultrabasic intrusions and dyke swarm. "Parallel Roads" of Glen Roy.

Access:- By private transport; the Glen Roy road is not suitable for large coaches. By train to Roy Bridge, returning from either the same station, or from Spean Bridge by train or bus. Road and hill walking.

Maps:- Topo:- 41 Geol:- S 62, S 63.

Walking distances:- With private transport 5.5 miles (8.8 km). Walking: Roy Bridge back to Roy Bridge 13 miles (208 km), to Spean Bridge 16 miles (25.6 km).

Itinerary:- From Roy Bridge up Glen Roy road for 1.4 miles (1.8 km) to a bridge over a tributary where garnetiferous Leven Schists with psammitic ribs containing calc-silicate lenses are cut by a N.E. porphyrite dyke. Continue to 4½ miles (7.2 km) from Roy Bridge where there is a classic view of all three "Parallel Roads". These shores of glacial lakes are at 1149 ft. (350 m), 1068 ft. (326 m) and 857 ft. (261 m) O.D. The highest lake drained through the col between the R. Roy and the R. Spey, the second through a col near the source of a tributary of the Allt Glas Dhoire (see below) and the lowest through a col E. of Loch Laggan (E5.26); the heights of the cols correspond to within a few feet with those of the "Roads".

Return down valley to almost opposite a major tributary from the W., the Allt Glas Dhoire. Cross the R. Roy by a footbridge ¼-mile (0.4 km) upstream of the confluence and traverse hillside to the N. of the tributary, taking the opportunity to examine the two lowest roads close at hand. Descend to

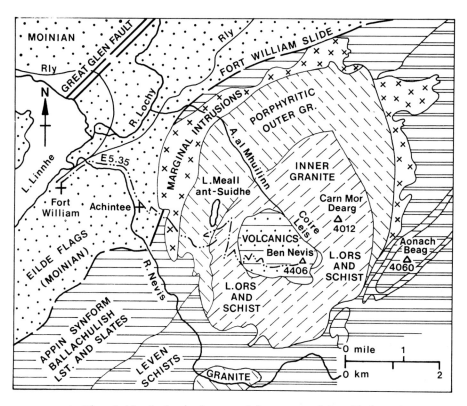

Fig. 5.18 Geological map, with route, of Ben Nevis.

the Allt Glas Dhoire ¼-mile (0.4 km) below where three headstreams come together. Mica-schists (Leven Schists, with vertical cleavage striking N.E.) are cut by N.E. porphyrite dykes of the Ben Nevis swarm (E5.35) and are followed upstream by a variety of coarse basic and ultrabasic rocks, mainly hornblendic. Pink granite and aplite veins, related to the S.W. end of the large N.E.-orientated Corrieyairack Granite, are abundant.

Return to road and continue down Glen Roy and Glen Spean to opposite the mouth of the R. Cour, a large tributary from the S. Examine section along right bank of R. Spean to Spean Bridge. Leven Schists pass downstream into metalimestone with dark schist - the Ballachulish Limestone in the Appin Synform (E5.35, E5.36). This is abruptly followed by the Eilde Quartzite across the Fort William Slide. The quartzite merges into the Eilde Flags (Moinian). Porphyrite dykes of the Ben Nevis Swarm (E5.35) are abundant.

Return from Spean Bridge to Fort William.

E5.35   Ben Nevis (Fig. 5.18)

Moinian. Dalradian. Caledonian Ring Complex and dyke swarm. Tertiary dyke. Glacial features.

Access:- By private transport or walking to foot of Ben Nevis path, at Achintee. Wide, but rough and steep, path to summit. The weather on the higher parts of the mountain can be severe, even in summer.

Maps:- Topo :- 41 Geol:- S 53.

Walking distances:- From Achintee 9.5 miles (15.7 km). From Fort William 13.5 miles (21.6 km).

Itinerary:- On the way to the foot of the path note should be taken of good exposures of flaggy Moinian granulite in the R. Nevis and of the strongly glaciated topography of the glen. The penstock of the Lochaber Power Scheme (E5.33) is well seen.

From the lower part of the path the Appin Synform (E5.34, E5.36) is seen across Glen Nevis. Calc-silicate-hornfels (Ballachulish Limestone with thermal metamorphism superimposed on the original) is followed by coarse quartz-diorite, one of the outermost (and oldest) annular intrusions. This merges into pinkish granite with large orthoclase phenocrysts (the Porphyritic Outer Granite). (On the N. side of the complex sharp, veining contacts are seen in places.) Numerous E.N.E. porphrite dykes are seen and one N.W. Tertiary basic dyke. The sharp junction between the Outer Granite and the Inner (later) is crossed three times in zig-zags of the path S. of Lochan Meall at-Suidhe. Dykes are nearly absent in the Inner Granite.

The path continues to rise across the Inner Granite to a height of 3000 ft. (915 m) where it passes over the Lower Devonian volcanic rocks forming the core of the mountain. Make a diversion to the S. of the path here to examine the chilled margin of the Inner Granite against the downthrown volcanics and to see the base of the latter resting on a few feet of Lower Devonian grits over Dalradian schists. From the 3000-ft. level (915 m) the path crosses frost-shattered volcanics (baked andesite and agglomerates). By walking a short way N. of the path as the summit is approached it is possible to see the inwardly-dipping volcanics in the high cliffs which form the backdrop of

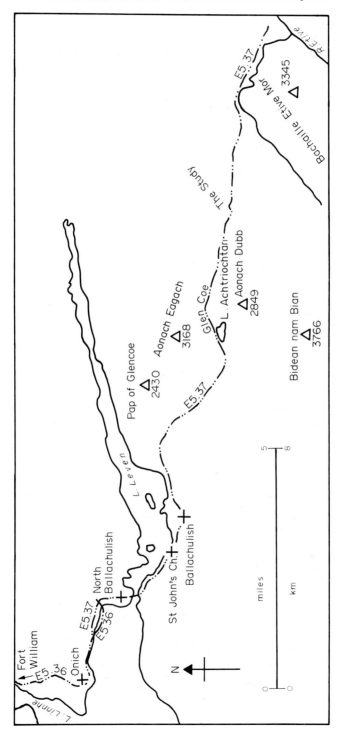

Fig. 5.19 Route map for Ballachulish-Glen Coe area.

spectacular Coire Leis. If there is snow on the ground great care should be taken approaching the edge of the cliffs as large, overhanging cornices form.

From the summit, 4406 ft. (1345 m) on a clear day (not frequent) there is a superb view, ranging from the Outer and Inner Hebrides across the Northern Highlands and the Great Glen to the Grampian Highlands as far E. as the Cairngorms.

Return by the same route.

E5.36   Loch Leven shore section, Onich to North Ballachulish (Fig. 5.19 and Fig. 5.20)

Moinian. Dalradian. Caledonian dykes. Raised beach.

Access:- By public bus to Onich, returning from North Ballachulish. By private transport. Road and shore walking. Fairly low tide helpful for some exposures.

Maps:- Topo:- 41 Geol:- S 53.

Walking distance:- 3.5 miles (5.6 km).

Itinerary:- The traverse shows most of the members of the Ballachulish Subgroup forming part of the nappe-complex of the Ballapel Foundation (cf. much thinner development in Perthshire, E5.28). Interpretation of structure has been controversial; it seems likely that not only the Appin Synform but also the Fort William Slide are D2 structures superimposed on D1 recumbent structures.

From Fort William to Onich the road follows Loch Linnhe in a deeply eroded hollow along the Great Glen Fault.

Start the traverse on the Loch Linnhe side of the point W. of Onich where Eilde Flags (Moinian), dipping at about 45° S.E. are exposed. These are brought into contact by the Fort William Slide with Ballachulish Limestone exposed at the point and followed to the E. by the Ballachulish Slates. Several N.E. porphyrite dykes of the Ben Nevis Swarm (E5.35).

Leave the shore for the road here on the N.W. side of which, at Onich, across a fault, the Appin Quartzite appears. To the E. there is a wide outcrop of folded Appin Limestone and Phyllite. Rejoin the shore section a mile (1.6 km) E. of Onich where the Appin Quartzite reappears on the S.E. of the Appin Synform. This can be proved by current bedding to be younger than the Ballachulish Slates, seen N. of North Ballachulish.

Both here and at Onich there is a well-developed raised beach at 35 ft. (11 m) O.D.

Return to Fort William.

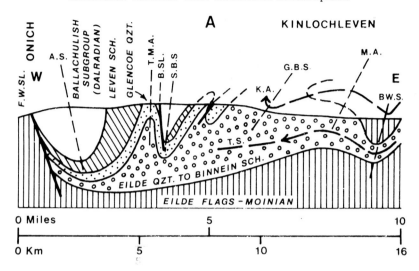

Fig. 5.20   Section of Onich area. Lochaber (after Hickman). Axial traces
of recumbent folds - long dashes. Axial traces of secondary
(upright) folds - short dashes. T.S. Treig Syncline. K.A.
Kinlochleven Anticline. F.W.Sl. Fort William Slide. A.S.
Appin Syncline. Bw.S. Blackwater Synform. G.B.S. Garbh Bheinn
Synform. M.A. Mamore Antiform. S.B.S. Stob Ban Synform.
T.M.A. Tom Meadhoin Anticline.

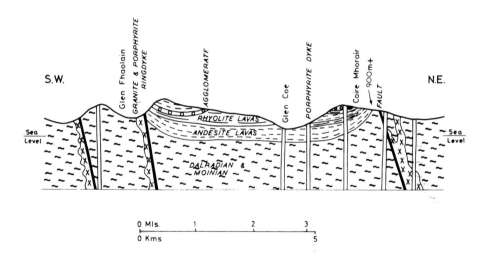

Fig. 5.21   Section of Glen Coe subsidence.

E5.37   Ballachulish and Glen Coe (Figs. 5.19 and 5.21)

Moinian. Dalradian. Lower Old Red Sandstone volcanics and granite. Lower
Old Red Sandstone ring dyke (Cauldron Subsidence of Glen Coe). Lower Old
Red Sandstone dyke swarm. Glacial features.

Access:- Private transport is almost essential to do the whole excursion
in a day. Parts may be done by using the rather infrequent public bus
service. Accommodation is available in Ballachulish. Road and some rough
mountain walking.

Maps:- Topo:- 41 Geol:- S 53.

Walking distance:- 3 miles (4.8 km).

Itinerary:- At the S. end of the bridge over Loch Leven glaciated exposures
of the Ballachulish Granite are seen. Along road to S.E. the Leven Schists
outcrop, with regional garnets contact-altered to cordierite. Further E.
the schists contain quartzitic beds, and then, at St. John's Church, the
Ballachulish Limestone, only 12 yds. (11 m) wide on the shore lies between
two branches of the Ballachulish Slide (a D1 structure). The S.E. branch
separates the limestone from the Appin Quartzite which dips off the Ballachulish Slates in a D2 antiform and is extensively quarried at Ballachulish.

Along Glen Coe road to E. end of Loch Achtriochtan. Looking S. the ring
dyke of granite, inclined steeply E.S.E., is seen in contact with Dalradian
phyllites (Leven Schists) overlain by Lower Old Red Sandstone volcanics
(andesites, rhyolites and agglomerates), the bedding of which is displayed
on the N.W. cliff of Aonach Dubh. By climbing (rough path) some 1250 ft.
(381 m) up the E. side of the stream which flows into the E. end of the
loch the base of the volcanics (andesite) can be reached, and the chilled
margin seen of the granite against the sunken block.

If private transport is available, the road should be followed for 4.5 miles
(7.2 km) to the E. where Moinian psammites appear from underneath the
psammites. Alongside the road at the Study rhyolites with contorted flow-
banding are seen. The effects of glacial erosion are spectacularly displayed. Near where the R. Etive bends S.W. the N.E. sector of the granite
ring is reached.

Return to Ballachulish and Fort William.

Oban Centre (Fig. 5.22)

Oban can be easily reached by road or rail. Accommodation is plentiful but
the town is a busy tourist centre. The Mull centre for the study of Tertiary
igneous rocks (Ch. 9.1) is reached by boat from Oban, and the town could be
an alternative starting point for two of the excursions (E9.7, E9.8).

E5.38   Vicinity of Oban

Dalradian. Lower Old Red Sandstone sediments and lavas. Tertiary dykes.
Raised beach.

Access:- Road and shore walking. Fairly low tide necessary.

Maps:- Topo:- 49 Geol:- S 44, S 45.

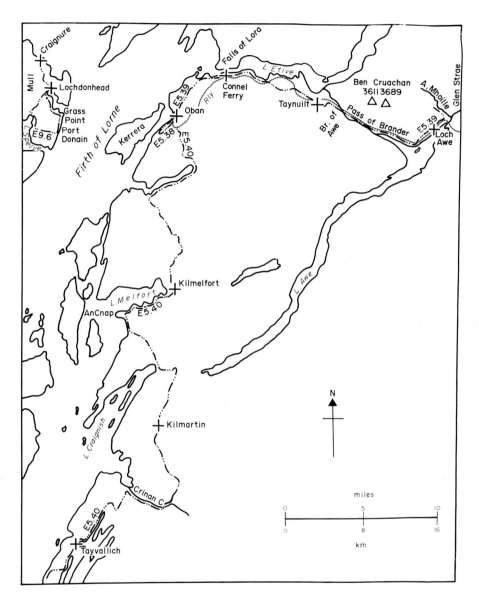

Fig. 5.22 Route map for Oban centre and Eastern Mull.

<u>Walking distance</u>:- 4.5 miles (7.2 km).

<u>Itinerary</u>:- The shore exposures round Oban Bay are mostly in the Easdale Slates (Dalradian Carbonaceous Group). These are unconformably overlain by Lower Old Red Sandstone conglomerate, although there is faulting at most of the contacts. Shales in the cliff behind Selma Cottage yielded Lower Old Red Sandstone fish and plants.

Walk S.W. from Oban on coastal road. This follows a raised beach at 30 ft. (9 m) O.D. backed by a high cliff of conglomerate with large boulders. There are a number of N.W. basic dykes of the Tertiary Mull Swarm. One mile (1.6 km) S.W. of Oban the conglomerate unconformably overlies Easdale Slates exposed on the shore. 1½ miles (2.4 km) S.W. of Oban turn off on a track up the conglomerate cliff and return on a path and track above the road. By walking a short distance S.E. of this route, Lower Old Red Sandstone lavas (basalt and augite-andesite) can be examined.

<u>E5.39   Pass of Brander and S.E. part of Etive Ring Complex</u>

Moinian. Dalradian. Lower Old Red Sandstone lavas. Carboniferous. Caledonian Ring Complex. Raised beaches.

<u>Access</u>:- By public bus to Loch Awe village. By private transport. Hill walking.

<u>Maps</u>:- <u>Topo</u>:- 49, 50 <u>Geol</u>:- S 45.

<u>Walking distance</u>:- 4.5 miles (7.2 km).

<u>Itinerary</u>:- If private transport is used, a stop can be made between Oban and Connel Ferry to examine Lower Old Red Sandstone lavas. At Connel Ferry there is a view of the Falls of Lora, a salt-water cataract which flows during certain states of the tide as the sea runs into, or out of, Loch Etive over a lava barrier.

Along Loch Etive raised beaches at several levels are evident. Both road and rail pass through the spectacular Pass of Brander eroded along a N.W. fault with a downthrow of over 3000 ft. (915 m) to the S.W. where the lavas overlie Dalradian. To the N.E. Dalradian pelites have thermal metamorphism superimposed on the region. There are a large number of N.E. dykes of the Etive Swarm.

A stop should be made, if private transport is used, at Bridge of Awe to examine sandstones and shales in the river section in which Upper Carboniferous plants have been found.

From Loch Awe village walk 1 mile (1.6 km) N.E. to the Allt Mhoille (private transport can be parked nearer the stream). To the N.E. Glen Strae marks a N.E. fault which to the S.W. cuts off the Pass of Brander Fault and continues down Loch Awe. In the first part of the Allt Mhoille section impure quartzite is seen, probably part of the Moinian Transition Group and separated by a slide from Dalradian to the S.E. ¾-mile (1.2 km) upstream from the road quartz-diorite of the Quarry Intrusion, the outermost and oldest partial ring dyke, appears. Upstream the intrusion gradually becomes darker and finer-grained and merges into a basic porphyrite against the Beinn a'Bhuiridh Screen. Here the Screen is a strongly schistose rock which upstream passes very gradually into baked but otherwise normal andesite.

1¼ miles (2 km) from the road the andesite is cut across by the Cruachan Granite.

The Screen therefore consists of lavas which have been partly converted to schists by stress during subsidence and by magmatic heat.

Return to road and Oban.

E5.40   Kilmartin and Tayvallich

Dalradian. Lower Old Red Sandstone lavas. Caledonian quartz-diorite. Tertiary dykes. Raised beaches.

Access:- This excursion is best done with private transport, although, depending on seasonal schedules, part at least could be carried out using an infrequent bus service.

Maps:- Topo:- 49, 55, 62 Geol:- S 25, S 36, S 44, S 45.

Walking distance:- 5 miles (8 km).

Itinerary:- From Oban S. nearly as far as Kilmelfort the route crosses Lower Old Red Sandstone volcanics of the Lorne Plateau. Along the S. shore of Loch Melfort N.W. tertiary basalt dykes of the Mull Swarm are plentiful, and raised beaches conspicuous. Leave the road and visit An Cnap, where Lower Devonian quartz-diorite has caused thermal metamorphism in the Dalradian Ardrishaig Phyllites. At the head of Loch Craignish the Ardrishaig Phyllites are overlain by the Crinan Grits, and between there and Kilmartin the grits are involved with the Tayvallich Limestone and slates in sharp folds. Graded bedding shows the grits to be older. Metadolerite intrusions are abundant.

Turn W. just before crossing the Crinan Canal and pass onto the Tayvallich Peninsula. About 2 miles (3.2 km) S.S.W. of Tayvallich village leave the road and walk W. to the coast where the Crinan Grits can be shown by graded bedding to be older than the Tayvallich Limestone. To the S.S.W. the limestone is succeeded by pillow lavas which, at An Aird, show "right-way-up" structure.

The Dalradian succession in normal order just described, occurs in the Loch Awe Syncline (probably D3) and in the upper limb of the recumbent fold which has brought about the Loch Tay Inversion (E5.27) within the Iltay Nappe.

Glasgow Centre

Glasgow is one of the centres for excursions in the Midland valley (Ch. 7.5) but it is also a good starting point for an excursion to a southern part of the Grampian Highlands.

E5.41   Gare Loch, Loch Long, Arrochar and Loch Lomond (Fig. 5.23)

Dalradian. Caledonian diorite. Old Red Sandstone. Glacial features.

Access:- Private transport is needed to do the whole excursion in a day. However, by travelling by train to Helensburgh and returning from

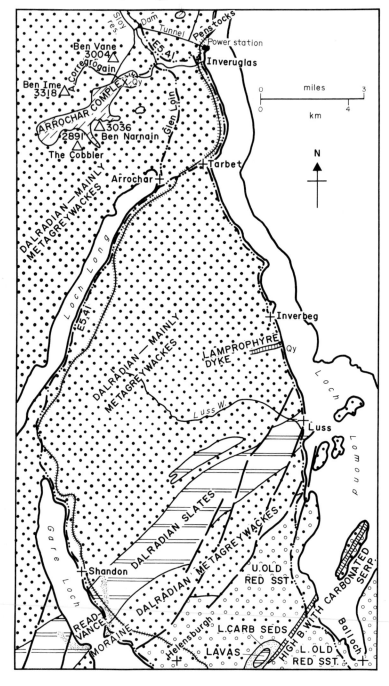

Fig. 5.23   Geological map with route of Gare Loch-Loch Long-Loch Lomond-
Loch Sloy area.

Garelochhead (or vice versa) it is possible to follow the Gare Loch section
on foot. Similarly, the N. part can be visited by taking the train to
Arrochar and back. Low tide makes it possible to see somewhat more of the
Gare Loch section but is not essential.

Maps:- Topo:- 56, 63 Geol:- S 30, S 38.

Walking distances:- With private transport 7 miles (11.2 km). Without,
Helensburgh to Garelochhead 8.5 miles (13.6 km). From Arrochar station
and back by Glen Loin 13 miles (20.8 km).

Itinerary:- The route from Glasgow to Cardross is described in E7.46.
Between Cardross and Helensburgh the Highland Boundary Fracture-zone is
crossed but it is not exposed. On higher ground to the N.E. carbonated
ultra-basics outcrop. The shore section for $1\frac{1}{4}$ miles (2.0 km) N.W. from
Helensburgh is in an outlier of Upper Old Red Sandstone conglomerates with
Highland fragments. To the N.W. there are a few exposures of low-grade
metagreywackes of the Dalradian Upper Psammitic Group within the Iltay Nappe.
More exposures occur on the hillside above the railway with graded bedding
showing that the S.E. dip is normal. To the N.W. the Luss Slates (Upper
Pelitic Group) outcrop.

The Gare Loch narrows are due to a late Readvance terminal moraine which
rests on 100-ft (30 m) raised beach clays. Metagreywackes, dipping S.E.,
are again seen near Shandon. These are known from other localities
(eg. E5.42) to be younger than the slates, which therefore occur in an
anticline or synform. The shore of the loch in this part is a security
area. Further N., however, exposures of metagreywackes and mica-schists
can be readily studied on the way to Arrochar.

From Arrochar (or Arrochar station if private transport is not being used)
continue to Tarbet and along Loch Lomond to Inveruglas where the penstocks
of the Loch Sloy Hydroelectric Scheme can be seen. Walk up construction
road to the Sloy Dam. Excellent sections are seen of metagreywackes
(referred to on older maps as schistose grits) and mica-schists; the
schistosity has been affected by two further deformations. Sparse garnets
occur. The dam extended and deepened the existing loch in a glacial basin.

Return to a point about $1\frac{1}{2}$ miles (2.4 km) from the main road then ascend
(to the W.) the Allt Coiregrogain valley for $\frac{3}{4}$-mile (1.2 km) to a large
quarry (worked to construct the dam) in pyroxene-mica-diorite. This forms
the N.E. end of the Caledonian Arrochar Complex made up of a wide variety
of igneous rocks.

If returning to Arrochar station it is shorter to turn S.E. before recrossing
the Sloy valley and then go over a low col and descend Glen Loin to the head
of Loch Long. Otherwise go back to the main road to return via Balloch to
Glasgow.

The metasediments seen on the outward journey are again traversed. Stop one
mile (1.6 km) S. of Inverbeg to examine an E.N.E. lamprophyre dyke (probably
Caledonian) in a disused quarry. Of particular interest are the xenoliths
of hornblendite and schists; the schists contain abundant garnet and some
stauriolite and are therefore of a higher grade than those of the country
rock.

The deeply glaciated hollow of the northerly part of Loch Lomond is a

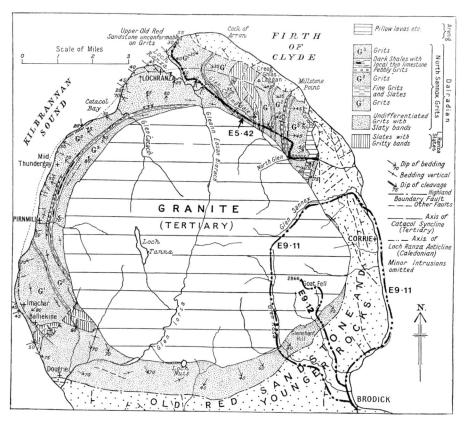

Fig. 5.24 Geological map, with routes, of North Arran.

striking feature. The Upper Old Red Sandstone outlier (see above) is hidden
under fluvioglacial sand and gravel. The Highland Boundary Fracture-Zone
with carbonated serpentinite forms a N.E. chain of islands.

Brodick Centre

Brodick is the best centre for studying the Tertiary igneous rocks of Arran
(Ch. 9) but it is also a base for an excursion to examine the Dalradian
rocks of the N. Arran which structurally is part of the Grampian Highlands.

E5.42   North Glen Sannox and Loch Ranza (Fig. 5.24)

Dalradian. Cambrian/Arenig. Hutton's unconformity. Old Red Sandstone.
Tertiary granite. Tertiary dykes. Glacial features.

Access:- The excursion is best done by private transport. By using the
Brodick-Loch Ranza public service it is possible to carry out at least part
of the excursion, depending on seasonal schedules. Alternatively, some
accommodation is available in Loch Ranza village.

Maps:- Topo:- 69 Geol:- Special sheet 1:50,000 Arran.

Walking distances:- With private transport 5 miles (8 km). By bus to
bridge over North Sannox Burn and back from Loch Ranza 9 miles (14.4 km).

Itinerary:- Park transport, or leave service bus, at bridge over North
Sannox Burn and examine section downstream. Fine grits and slatey beds of
the Dalradian Upper Psammitic Group within the Iltay Nappe dipping steeply
E. by S., are overlain, 150 yds. (140 m) below the bridge by black cherts
and basic lavas of late Cambrian/l. Arenig age. These show no discordance
with the Dalradian and share the same structures. Poorly preserved shells
of Acrotreta were found in shales 200 yds. (183 m) up a tributary coming in
from the N. Further up the same stream orange-weathering breccia occurs
between crushed spilite and crushed sandstone. The breccia is like the
carbonated serpentinite seen further N.E. along the Highland Boundary
Fracture-zone and is evidence that the fault to the W. of the Lower Old Red
Sandstone in North Glen Sannox is a component of that Fracture-zone. In
the main stream fault-bounded Dalradian grits intervene between the shales
and the Lower Old Red Sandstone conglomerates.

Return to bridge, just above which spectacular graded bedding in coarse
quartzose grits shows younger sides to the E. Upstream finer grits interbedded with slates are in contact with, and hornfelsed by, Tertiary granite
(E9.11).

Ascend hillside to road.

By private transport or by walking go on to summit of road. The route
crosses grits with mappable pelites. In the vicinity of the summit a wider
pelite (Loch Ranza Slates = Upper Pelitic Group) outcrops. The slates are
older than the grits on both sides. Graded bedding in the grits to the W.
is less frequently preserved owing to increasing cleavage. Several N.N.W.
Tertiary basic dykes are seen.

Descend to E. side of Loch Ranza village noting striking glacial features.
Walk along minor road and track on N.W. side of loch. Near the head of the
loch Dalradian grits dipping N.N.W. show inverted grading. Further N.W. the

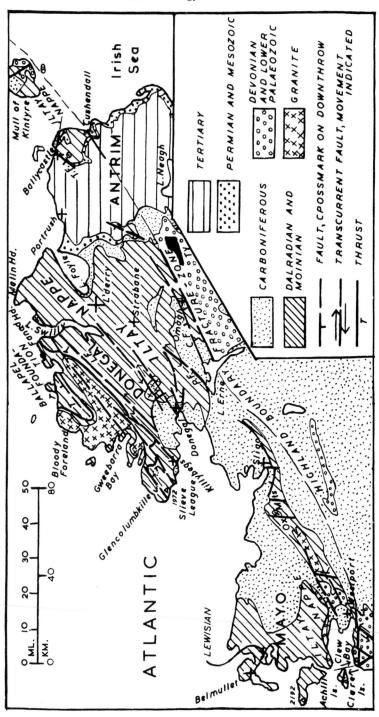

Fig. 5.25   Geological map of northern part of Ireland.
C.F. Cool Fault. L.F. Leannan Fault. P.F. Pettigo Fault.
T.P. Tow Fault. Ty. Tyrone Inlier.

grits, still inverted, dip S.E. The synform has been imposed on beds inverted by Caledonian movements by uplift at the margin of the Tertiary granite. Quarter of a mile (0.4 km) N.E. of Newton Point, at an unconformity made classic by Hutton, the Dalradian grits are overlain by Upper Old Red Sandstone. The base of the latter is cornstone, and the carbonate alteration has penetrated 4 ft. (1 m) into the Dalradian.

Return to Loch Ranza and Brodick.

Portrush Centre (Figs. 5.25 and 5.26)

For information about Portrush see E9.13.

E5.43   Antrim coast at and near Torr Head

Dalradian. Metadolerite. Cretaceous. Tertiary lavas.

Access:- By private transport on A2 through Ballycastle to Ballyvoy then by secondary road E. to Torr. By public bus to Ballyvoy then on foot.

Walking on road, rough coast, and hillside.

Fairly low tide necessary.

As an alternative to Portrush it may be possible to stay in Ballycastle.

With considerably longer travelling the excursion can be done from Belfast.

Maps:- Topo:- ½-inch Ireland 2   Geol:- N 8.

Walking distances:- With private transport 4.5 miles (7.2 km). From Ballyvoy 11.5 miles (18.4 km).

Itinerary:- From Torr descend a steep coastal slope. Exposures occur of quartz-schists and grits, sometimes calcareous, with some chloritic and epidote grits (green beds). Along coast to Torr Head where grey, gritty metalimestone with bands of coarse black marble is associated with intrusive metadolerite, in places transgressive. To the S., at Portaleen Bay where there is a large N.N.E. fault, felspathic grits and quartz-schists structurally underlie the limestone. These Dalradian rocks have a general N.W. dip but this is superimposed on one or two earlier deformations. In the metadolerite on the W. side of Torr Head, boudins are aligned towards the S.S.W. The Torr Head limestone is accepted as a continuation of the Loch Tay (E5.27) within the Iltay Nappe. However, it is doubtful whether the Loch Tay Inversion is present. From Portaleen Bay follow the stream to the road at Altmore Bridge where the schists are crowded with phenocrysts of albite. Ascend hill diagonally S.W. for 0.6 mile (1 km) to an exposure of horizontal Upper Chalk. N.W. over Carnmore, 1354 ft. (380 m), across Tertiary basalts (Lower Basalts) (E9.13).

Descend N.W. to road and return to Portrush.

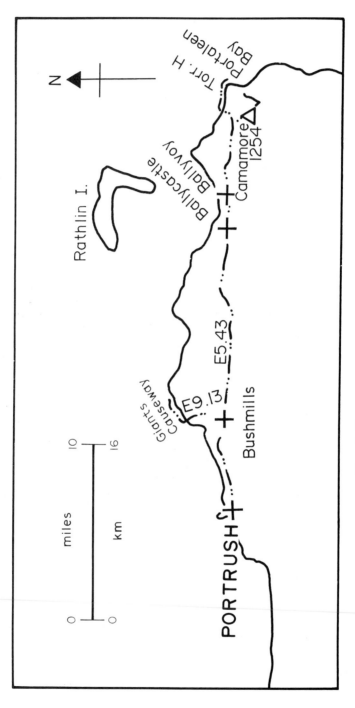

Fig. 5.26 Route map for Portrush Centre.

Donegal Centre (Figs. 5.25 and 5.27)

The town of Donegal can be reached by public bus or by private transport.
There is a fair amount of accommodation.

The region covered by this and the next centre is particularly sensitive
from the security aspect. While the present unrest continues it may not
be possible to carry out the whole or parts of the excursions described.
Police advice should be sought.

E5.44   S.W. of Lough Derg

Moinian. Pegmatite complex. Metadolerite.

Access:- By private transport. By walking, excluding S.E.-part along
Pettigo road. Road and moorland walking.

Maps:- Topo:- $\frac{1}{2}$-inch Ireland 3 Geol:- 1-inch Ireland 32,
                                            1:250,000 Ireland 1.

Walking distances:- 16 miles (25$\frac{1}{2}$ km). With private transport 6 miles
(9.6 km).

Itinerary:- Southwards from Donegal, then turn off E. on Pettigo road on
S. outskirts of Laghy. This part of the route is over Carboniferous
Limestone nearly completely covered with boulder clay. The ecological
change is marked as the road passes onto the metamorphics. Good exposures
of the striped siliceous granulites which make up most of the Moinian Lough
Derg Psammitic Group are seen near the Pettigo road 1$\frac{3}{4}$ miles (2.8 km) from
the main road. These dip N.N.W. at 35° on the N.W. flank of the Lough Derg
Antiform. On the N.W. side of these a narrow road leads N.E. This should
be followed on foot (it is suitable for cars only with difficulty) for 3
miles (4.8 km) to Ballykillowen Hill. On the way further exposures of the
psammites are seen, then, on the E. side of the hill, garnet-kyanite-
starmolite schist with tourmaline. It is doubtful whether these pelites
have an interbedded or tectonic junction with the underlying psammites.

A return can then be made to Donegal by walking along a minor road W. to the
main road on the N. side of Laghy. Those with transport should follow the
main road for 6$\frac{1}{4}$ miles (10 km) towards Pettigo. Numerous exposures of the
psammites are seen; the dip swings round to N.E. Metadolerites and garnet-
amphibolites occur, some showing striking boudinage. Both the metasediments
and altered igneous rocks are cut by pegmatites.

Return to Donegal.

E5.45   Slieve League

Dalradian. Metadolerite. Carboniferous.

Access:- By public bus through Killybegs to Carrick. By private transport
to Carrick and onwards by road on W. side of Glen River to Teelin Post
Office. Mountain walking.

Maps:- Topo:- $\frac{1}{2}$-inch Ireland 3 Geol:- 1-inch Ireland 22, 30,
                                            1:250,000 Ireland 1.

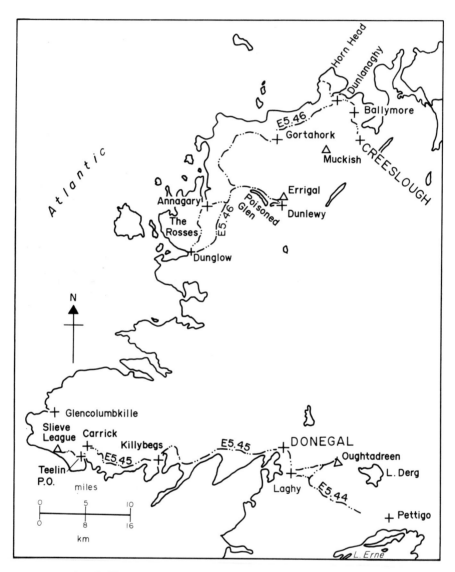

Fig. 5.27   Route map for Donegal and Creeslough Centres.

Walking distances:- Without private transport: 10 miles (16 km).
Excluding climb to summit of Slieve League 7¼ miles (11.6 km). With
private transport: 6 miles (9.6 km). Excluding climb to summit 3¼ miles
(5.2 km).

Itinerary:- On the outskirts of Carrick metadolerite outcrops, one of a
number of such outcrops in the district. Along the W. side of the Glen
River there are exposures of phyllites; their position in the Dalradian
sequence is problematic.

From Teelin Post Office follow track W. then W.N.W. on the N.E. side of a
small, marshy lough. A brown scar marks a boulder bed interbedded with
overlying quartzite (which can be shown to be younger) dipping steeply E.
by N. To the N.W. of the lough there are outcrops of metalimestone with
some schist - the Glencolumkille Limestone, to the S.E. of which the boulder
bed appears again with the quartzite dipping steeply under it. The
quartzite/boulder bed/metalimestone sequence correlates with the Dalradian
succession in Perthshire (E5.28) and elsewhere and is part of the continuation of the Iltay Nappe into Ireland. The boulder bed contains soda-granite
boulders as in Scotland. The metalimestone is in a steep syncline, and the
N.W. strike of this and flanking structures is due to F2 folding probably
analagous to that of Schiehallion (E5.28) and elsewhere.

If the weather is bad a return should be made from this point but in clear
weather the steep hillside should be climbed W.S.W. to the One Man's Path
which follows an arête between the steep landward slope and cliffs falling
up to 1900 ft. (580 m) into the Atlantic. Follow the path N.W. over an
unconformable cap of conglomerate with quartzite boulders (probably
Carboniferous) to the summit of Slieve League, 1972 ft. (601 m) where there
is another quartzite cap.

Return to Teelin Post Office and Donegal.

Creeslough Centre

Creeslough is best reached by private transport. Accommodation is limited.

E5.46   Creeslough to the Rosses

Dalradian. Metadolerite. "Older" Caledonian granite. "Newer" Caledonian
Ring-complex. Glacial features.

Access:- By private transport; route suitable for coaches of moderate size.
Some moorland and coastal walking.

Maps:- Topo:- ½-inch Ireland 1, 3, 4, 9, 10, 15   Geol:-   1:250,000
                                                            Ireland 1.

Walking distance:- 5.5 miles (8.8 km).

Itinerary:- At the road junction on the S.E. side of Creeslough banded
semi-pelitic schists show sporadic biotite flakes marking an outer zone of
the thermal aureole of the Donegal Granite. They belong to the Creeslough
Formation and are near the base of the Creeslough succession, correlated
with that of the Ballapel Foundation (E5.34-E5.37).

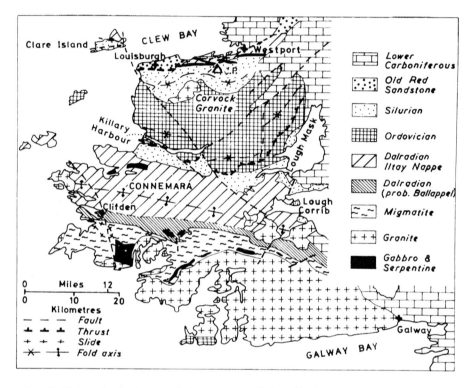

Fig. 5.28 Geological map of Connemara and its flanks. C.P. Croagh Patrick.

N. from Creeslough to beyond Ballymore where the Sessiagh Banded Quartzite
is seen in road side quarries; nearer Dunfanaghy the Marble Hill Dolomite
is exposed in old quarries. Through Dunfanaghy to N. end of Horn Head
Bridge. Walk N.W. to Claggan then W. to the coast at Micky's Hole, where
a cliff section shows a thrust bringing schists of the Falcarragh Group
with metadolerites thrust over the Ards Quartzite.

Return to bridge and follow the road S.W. to Gortahork then turn off N.W.
on Bloody Foreland road, which crosses the "Older" Granite intruded into
the Ards Quartzite. The granite contains many xenoliths and rafts of
metasediments. Continue to Gweedore and Crolly then to Arnagavy, $\frac{1}{2}$-mile
(0.8 km) beyond which the outer boundary of the Rosses ring-complex is
seen; medium-grained granite (Gc^1) is in contact with "Older" Granite.
Along road to cross-roads near Meenbannad School.

Examine the coarse granite (Gc^2) then walk E. along old railway track to
the first cutting where a porphyry dyke intrudes Gc^2 but does not cross
the contact into an inner ring, Gc^3. A swarm of porphyry dykes can be seen
by turning N. for 100 yds. (91 m) along a minor road at the end of the
cutting.

Return to crossroads then continue S. into the townland of Sheskinarone and
to a point $\frac{1}{4}$-mile (0.4 km) S. of the secondary road to Meenbannad. Near the
first cottage follow a rough track for about 100 yds. (91 m) to the E.
where Gc^3 is chilled against Gc^2. To the S. of this point, at the top of
the rise, and adjacent to a prominent mass of vein quartz, beryl can be
collected from a greisen.

On to Dunglow, Gweedore and Dunleavy. West of the last mentioned the great
corrie of the Poisoned Glen with its moraine-dammed lakes is a spectacular
feature, and to the N. of the road the quartzite (Ards) mountain of Errigal
rises to 2466 ft.(751 m).

Return to Creeslough.

5.5   Connemara and its flanks (Fig. 5.28)

The region between the Highland Boundary Fault just S. of Clew Bay and
Galway Bay with Connemara in its centre is the only part of the British
Isles where undoubtedly Dalradian rocks come to the surface S. of the Highland
Boundary and also the only part of the British Caledonides where strata as
young as the Silurian (Wenlock) have undergone regional metamorphism. It
is a sparsely inhabited and mountainous district rising to over 2600 ft.
(792 m).

Formations present are as follows:

> Silurian
> Ordovician
> Dalradian Metamorphic Assemblage or
> Supergroup (partly Precambrian, partly Cambrian).

The Dalradian rocks have long been known as the Connemara Schists. In addition to mica-schists and migmatites, they comprise quartzites and metalimestones. They include a boulder bed of glacogene type with a carbonate
horizon (containing the well-known Connemara Marble) below and a quartzite
above (E5.49).

There are only a few main roads, and public transport is scarce. Accommodation is limited to a few centres; those suggested are Westport and Galway.

Comparison with other areas:- The Dalradian rocks resemble those of the
Lower/Middle divisions of Donegal and Scotland, a conclusion supported by
the presence of the glacogene (E5.49). The Ordovician, with its Arenig
volcanics, is broadly like that of some other Lower Palaeozoic areas. The
Silurian is unique in the British Isles in having undergone regional metamorphism (E5.47).

Westport Centre (Fig. 5.29)

Westport can be reached by rail or road from Dublin and elsewhere. There
are local public bus services along the coast road to Louisburgh and a fair
amount of accommodation.

E5.47   West and S.W. sides of Croagh Patrick

Dalradian. Ordovician. Silurian. Ophiolites. Caledonian granite and its
aureole. Glacial features.

Access:- By private transport or public bus to Kilsallagh Post Office
(Louisburgh road). Walking on rough roads and hillsides.

Maps:- Topo:- $\frac{1}{2}$-inch Ireland 11 Geol:- 1-inch Ireland 74, 84.

Walking distance:- 8 miles (12.8 km).

Itinerary:- Examine shore section to E. of Kilsallagh School. Dalradian
(probably Upper Psammitic Group) metagrits show graded beds up to 8 ft.
(2.5 m) indicating inversion. Both bedding and cleavage show minor folds
with axes parallel to the N.W. strike.

Return to main road then walk S. on minor road for $\frac{1}{2}$-mile (0.8 km). Make
a diversion into fields to W. to study exposures of serpentinite and fissile,
fine-grained amphibolite. These occur along a major branch of the Highland
Boundary Fracture-zone, which here separates the Dalradian from Silurian to
the S. - the Croagh Patrick Quartzite. This can be seen in fields to the
S. of the serpentinites etc., dipping N.E. at 45° (an inverted dip from
evidence elsewhere). Return to the road then look at exposures to E., about
$\frac{3}{4}$-mile (1.2 km) from the main road, of greenish phyllites dipping N.
(Cregganbaun Group). Continue S. for another 1170 yds. (1070 m) to junction
with rough road to E.S.E. Look at field exposures of greenish impure
quartzites dipping N.N.E. at 50°. Turn S. on track 230 yds. (210 m) W.N.W.
of Bouris School. This track skirts a small mass of granite then, 835 yds.
(764 m) S.S.W. of Bouris School, reaches an outcrop of grey, in places rusty,
impure quartzite and phyllite dipping N. at 55°. These yield sheared
specimens of Hesperorthis, Streptelasmid corals and bryozoans, indicating
a Wenlock age. These beds, which are succeeded to the S. by quartzite,
belong to the Cregganbaun Group. The overall structure is therefore an
E.-W. Fl recumbent syncline (imposed on which are N.W. F2 folds).

Walk $\frac{1}{2}$-mile (0.8 km) S. by W. into moorland country to exposures of the
post-Wenlock Corvock Granite. On an embayment in the granite to the W.
Arenig pelites (E5.48) have been altered to beautiful spotted hornfels.

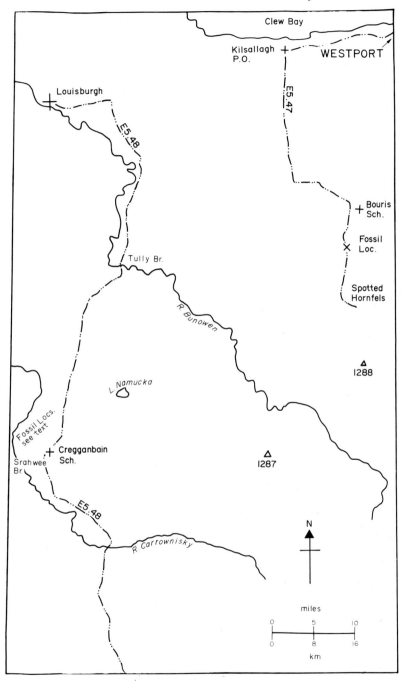

Fig. 5.29   Route map for Westport Centre.

Return to main road. From above Clew Bay there is a spectacular view of half-drowned drumlins forming numerous islands.

E5.48   Murrisk

Ordovician. Silurian. Caledonian intrusions.

Access:- By private transport. However, it is possible to see the N. part by taking the public bus to Louisburgh and walking to a mile S. of Cregganbaun and back.

Maps:- Topo:- ½-inch Ireland 11 Geol:- 1-inch Ireland 74, 84.

Walking distances:- With private transport 1 mile (1.6 km). By walking from Louisburgh (see above) 10 miles (16 km).

Itinerary:- For 4 miles (6.4 km) S. from Louisburgh towards Cregganbaun School the road crosses an expanse of peat with only a few exposures of impure quartzite. S.E. of the school an intrusion of granite occurs. In small roadside quarries 200 yds. (183 m) S.W. of the school and 250 yds. (229 m) S.S.W. of the school rusty-weathering, calcareous, impure quartzite yields distorted Wenlock fossils similar to those found near Bouris School (E5.47).

To the S. these beds are underlain by the Croagh Patrick Quartzite, with conglomerate at the base, on the S. limb of the Croagh Patrick Syncline (E5.47). The sediments are intruded by porphyry sills, seen mainly W. of the road.

Beneath the conglomerate a thin sequence of siltstones have yielded Llandovery fossils.

Outcrops occur further S. near the road of the Sheefry Group (Arenig), consisting of black, cleaved shales with cherts, spilites and basic tuffs followed by greywackes, shales and lenticular conglomerates. The Doo Lough hollow marks N.W. faults with total dextral displacement of 1.5 miles (2.4 km). At the Glenummera River the overlying Glenummera Slates (?) Llanvirn) are seen, made up of slates with subsidiary graded sandstone, and conglomerate lenses. S. of Doo Lough a short climb should be made up the W. slopes of Bengorm to examine the Mureelrea Grit (Llandeilo/Caradoc) which contains beds of ignimbrite.

Return to Westport or continue to Galway.

Galway Centre (Fig. 5.30)

Galway can be reached by rail or road from Dublin. There is plentiful accommodation.

E5.49   Glendalough, Lissoughter and Clifden

Dalradian. Glacial features.

Access:- By private transport or by Galway-Clifden public bus to Glendalough. On to Streamstown, N. of Clifden, by private transport or by public bus to Clifden on a second day.

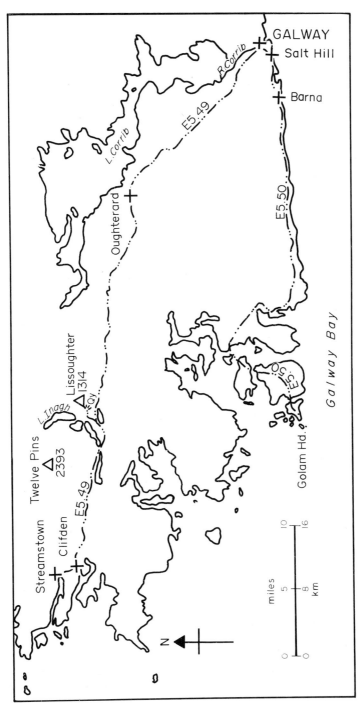

Fig. 5.30 Route map for Galway Centre.

Maps:- <u>Topo</u>:- ½-inch Ireland 10, 11, 14 <u>Geol</u>:- 1-inch Ireland 93, 94.

<u>Walking distances</u>:- At Glendalough 4 miles (6.4 km). Clifden to Streamstown and back 4 miles (6.4 km).

<u>Itinerary</u>:- The road N.W. from Galway to Oughterard approximately follows the contact between granite and metasediments with Carboniferous Limestone overlapping from the N.E. but boulder clay hides most of the solid rock. From Oughterard continue along road to W. end of Glendalough.

Leave transport and walk along minor road N.N.E. before branching off E. on a rough road which leads to a quarry in the Connemara marble (Lower Dalradian), a beautiful ophicalcite. Walk N. to W. slopes of Lissoughter where a boulder bed with granite clasts is exposed. Its presence proves the Lower/Middle Dalradian age of the Connemara metasediments. To the W. the overlying quartzite forms the strongly glaciated Twelve Pins.

Return to Glendalough and with private transport continue to Clifden then two miles N. to Streamstown. If a public bus is taken to Clifden it is necessary to walk to Streamstown.

On the E. side of the road just N. of Streamstown there is a working quarry in the Connemara marble. The beauty of the rock is enhanced by the plastic folding.

Return to Galway.

E5.50   N. shore of Galway Bay

Caledonian granite and dykes. South Connemara Series (? Ordovician).

<u>Access</u>:- Car transport is needed to see the ? Ordovician rocks but the granite can be studied by using the public bus to Salt Hill and back.

Maps:- <u>Topo</u>:- ½-inch Ireland 14 <u>Geol</u>:- 1-inch Ireland 104, 105, 113.

<u>Walking distances</u>:- From Salt Hill to Barna and back 8 miles (12.8 km). With private transport 3 miles (4.8 km).

<u>Itinerary</u>:- The migmatitic country rocks of the Galway Granite can be seen at the most northerly bridge over the R. Corrib in Galway.

At Salt Hill there are shore exposures of pink and grey granite cut by northerly porphyry dykes; quartzitic inclusions are present. N. of the road, 1½ miles (2.4 km) further on, coarse porphyritic granite occurs. Further W. still, beyond a W.N.W. fault, around Barna, the porphyritic granite shows N.W. foliation.

Those walking should return from here, but those with a car should continue to the end of the road near Lettermullen. Grey granodiorite can be seen at Cashla Bridge and other localities.

Walk to coast near Golam Head, where highly hornfelsed pelitic and silty sediments with visible andalusite are exposed along with tuffs and basic pillow lavas. The sediments contain clasts of Connemara Dalradian rocks (E5.49). This South Connemara Series, as it is termed, is tentatively referred to the Ordovician.

References

Anderson, J. G. C. 1939. The Granites of Scotland. Spec. Rep. on the
    Mineral Resources of Great Britain. Vol.32. M.G.S.
Anderson, J. G. C. 1965. The Precambrian of the British Isles. The Pre-
    cambrian, Vol.2. (ed. by K. Rankama). Interscience Publishers,
    New York.
Anderson, J. G. C. and Owen, T. R. 1980. The Structure of the British
    Isles. 2nd ed. Pergamon.
Bailey, E. B. and Weir, I. 1933. Submarine faulting in Kimmeridgian times,
    east Sutherland. T.R.S.E., 57, 429.
Downie, C. and others. 1971. A palynological investigation of the
    Dalradian rocks of Scotland. I.G.S. Rep. 71/9.
Harris, A. L. and Pitcher, W. S. 1975. The Dalradian Supergroup. Geol.
    Soc.Spec.Rep. No.6, 52.
Johnson, M. R. W. and Parsons, I. 1979. Geological Excursion Guide to the
    Assynt District of Sutherland. (revision of Macgregor and Phemister's
    Guide). Edin.Geol.Soc.
Johnstone, G. S. 1966. British Regional Geology. The Grampian Highlands.
    3rd ed. (revised from Read and MacGregor). M.G.S.
Mykura, W. 1976. British Regional Geology: Orkney and Shetland. M.G.S.
Phemister, I. 1960. British Regional Geology: Scotland: The Northern
    Highlands. 3rd ed. M.G.S.
Piasecki, M. A. J. 1980. New light on the Moine rocks of the Central
    Highlands of Scotland. J.G.S., 137, 41-59.
Smith, T. E. 1970. The structural characteristics of the Strathspey
    Complex, Inverness-shire. G.M., 107, 201-215.
Treagus, J. E. and King, G. 1978. A complete Lower Dalradian succession
    in the Schiehallion district, Central Perthshire. S.J.G., 14, 157-166.
Wilson, H. E. 1972. Regional Geology of Northern Ireland. Ministry of
    Commerce. Geol. Surv. of Northern Ireland.

CHAPTER 6

Caledonian Terrains without Caledonian Metamorphism

6.1   The Welsh Block and the Welsh Borderland (Fig. 6.1)

This is the classic region for the study of the Lower Palaeozoic, the three
systems of which are named after Cambria (Wales) and after two tribes (the
Ordovices and Silures) which formerly lived in the region. There is a long
northern and western coastline; most of the interior is hilly or mountainous.
In the N.W., Ordovician igneous rocks make spectacular, glaciated scenery
rising to 3560 ft. (1085 m) at Snowdon, the highest point in the British
Isles outside of Scotland. Exposures are excellent in many coastal and
mountain areas but some of the lower ground is mantled in glacial drift.

Private transport is an advantage for a number of excursions but many can
be done by using public transport and by walking. Six centres are suggested
- Bangor, Blaenau Ffestiniog, Barmouth, Church Stretton, Builth Wells and
Fishguard. All can be reached by rail or public bus.

The geological succession is as follows:

> Permo-Triassic
> Upper Carboniferous
> Lower Carboniferous
> Old Red Sandstone
> Silurian
> Ordovician
> Cambrian
> Precambrian

Precambrian subdivisions are given in the relevant sections (below).

Comparison with other areas:- Most of the Lower Palaeozoic rocks have long
been described as "geosynclinal" and in fact many of the sediments are of
the greywacke/shale/turbidite facies. In a broad sense therefore they
resemble those of the Lake District, the Southern Uplands and Ireland
described later in this chapter. The Cambro-Ordovician rocks, however,
are in complete contrast to the thick carbonates of N.W. Scotland
(Chapters 4 and 5).

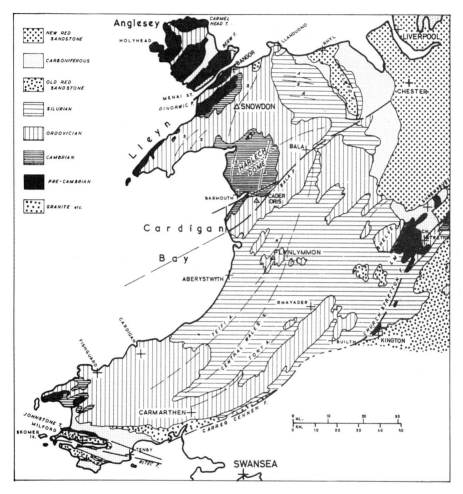

Fig. 6.1 Geological map of the Welsh Block and the Welsh Borderland.

Even within the present region, however, there is considerable variation. In the Welsh Borderland the Cambrian is of an epicontinental facies (E6.8) and the same applies to the Cambrian in S.W. Wales (E6.12). Both these districts are near the Caledonian "front".

Volcanic rocks are very widespread in the Ordovician, activity reaching its peak at different times in different districts. In some areas major activity continued until the Caradoc, whereas in the Lake District there was only relatively minor activity after the Llandeilo, and in the Southern Uplands post-Arenig volcanic rocks are uncommon.

The Upper Palaeozoic rocks resemble those of neighbouring regions.

Bangor Centre

Bangor is on the London-Chester-Holyhead railway and on main roads; there is plentiful accommodation. For student parties it may be possible to make arrangements at the University College hostel, Neuedd Reichel.

For very detailed descriptions of some of the excursions in the district reference should be made to Roberts (1979).

E6.1   Llanberis and Snowdon (Fig. 6.2)

Precambrian volcanics. Cambrian sediments. Ordovician sediments, volcanics and intrusions. Glacial features.

Access:- By private transport or public bus to Llanberis and on to Pen-y-Pass, returning from Llanberis. Walking on steep mountain tracks. The higher slopes should not be visited in bad weather. Part of the excursion may be followed by taking the rack railway (runs only part of the year) to the summit of Snowdon and descending by the path.

Maps:- Topo:- 115 Geol:- 1:25,000 sheet, Central Snowdonia.

Walking distance:- From Pen-y-Pass over summit of Snowdon to Llanberis 7 miles (11.2 km).

Itinerary:- With private transport a stop should be made at the N.W. end of the Llyn Padarn to examine glaciated crags of Precambrian rhyolite lava (Arvonian). The Llanberis lakes lie in a N.W. glaciated valley which is strongly overdeepened along a fault; in fact a bore showed that the rock-floor is at least 174 ft. (53 m) below the valley bottom.

A by-pass road along the S.W. shore of Llyn Padarn cuts the Cambrian Bronllwyd Grits forming a striking roche moutonnée.

The road from Llanberis to Pen-y-Pass follows a spectacular glaciated valley in bare Caradoc volcanics with a classic perched block on the N.E. side.

From the hotel at Pen-y-Pass the Pen-y-Gwryd (P.y.G.) track leads to the summit of Snowdon. For the first mile the track gradually ascends the E.-W. side of Tal-y-Llyn over a complex of steeply-inclined, acid intrusions cutting across cleaved vitric tuffs of the Lower Rhyolitic Series and the easily-weathered calcareous pumice-tuffs of the overlying Bedded Pyroclastic Series.

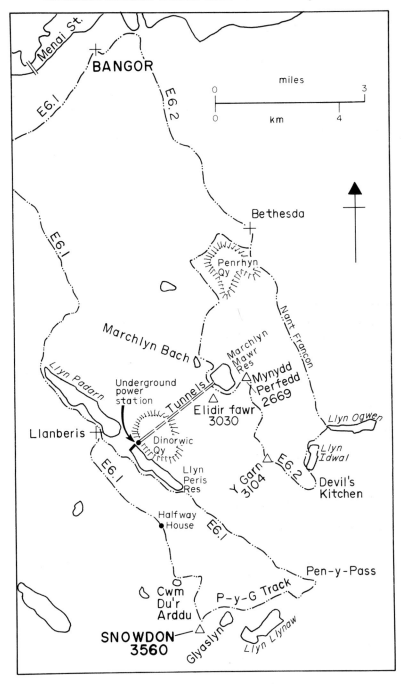

Fig. 6.2 Route map for Snowdon and Nant Francon areas.

From the crest of Tal-y-Llyn the moraine-dammed Llyn Llydaw can be seen,
backed by the cliffs of Lliwedd, composed of slabby vitric-tuffs of the
Lower Rhyolitic Series. In the next mile (1.6 km) the P.y.G. track runs
along the strike of the green, calcareous pumice-tuffs, close to the axis
of a tightly-compressed syncline. About 300 yds. (275 m) N. of the rock
tarn of Glaslyn the track follows the contact between the E. edge of a
steeply-inclined "felsite" intrusion and vitric tuffs and then passes onto
the intrusion.

As the track winds upwards across andesitic and basaltic pumice-tuffs there
is a fine view of the synclinal structure of Snowdon in the crags of
Clogwyn-y-garnedd. In the final ascent, parallel to the railway, thin
layers of rhyolite and rhyolitic tuffs, of the Bedded Pyroclastic Series
occur; ashy sediments within this series, just below the Summit Hotel,
yield <u>Dinorthis</u> and other brachiopods. From the peak of Snowdon, 3560 ft.
(1085 m) there is a magnificent view, extending in very clear weather to
the Leinster granite mountains (E6.31). Descend N. for about a mile (1.6 km)
parallel to the railway then swing W. across the large moraines which enclose
Cwm Du'r Arddu. The Cwm and the small lake are backed by cliffs showing an
impressive syncline in the Lower Rhyolitic Series. On the E. side this can
be seen to have been thrust over the Bedded Pyroclastic Series of the
Snowdon Syncline.

Continue N. across moraine to rejoin the main Snowdon path at the Halfway
House, 150 yds. (138 m) N. of which, in a railway cutting, blue-grey
Llandeilo slates contain large Orthograptids. Further N.N.W., in a cutting
between the Halfway House and a nearby reservoir, blue-black Llanvirn
slates yield <u>Didymograptus murchisoni</u>; the junction of the two series is
masked by a dolerite sill.

Quarter of a mile below the Halfway Station, where the railway crosses the
path, the Llanvirn slates are underlain by Arenig cleaved grits. About
quarter of a mile below the bridge a N.E. fault brings up rusty-weathering
Maentwrog Slates (Cambrian) which extend for about a mile (1.6 km) down the
path to an E.-W. fault on the upthrow side of which are the Bronllwyd Grits
forming a fine waterfall. The Grits are also well seen in the car park at
the terminus station.

On the opposite mountainside the Lower Cambrian slates with Ordovician
metadolerite dykes were worked on a vast scale. Inside the mountain one
of the world's largest cavern-complexes has been excavated for the Dinorwic
Pumped Storage Scheme. Facilities for a fascinating visit are offered by
the Central Electricity Generating Board.

E6.2   Nant Francon, Llyn Idwal and Bethesda (Fig. 6.2)

Cambrian sediments. Ordovician sediments, volcanics and intrusions.
Glacial features.

<u>Access</u>:- By bus or private transport to head of Nant Francon; return from
Bethesda. Walking on mountain tracks and ridges. Not to be undertaken in
bad weather.

<u>Maps</u>:- <u>Topo</u>:- 115 <u>Geol</u>:- EW 106, EW 119.

<u>Walking distance</u>:- 8 miles (13 km).

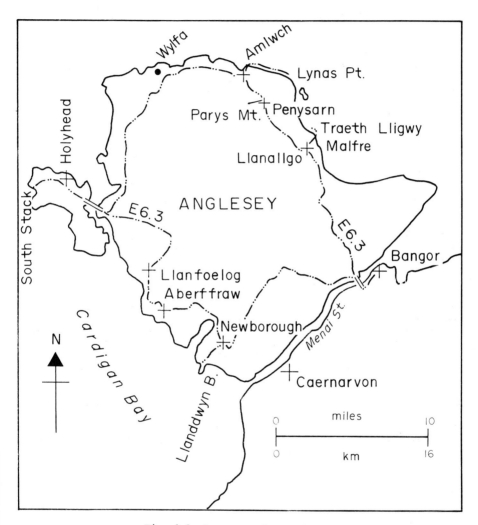

Fig. 6.3 Route map for Anglesey.

Itinerary:- From Bethesda the main road ascends the spectacular glaciated
valley of Nant Francon.

Leave the road at the W. end of Llyn Ogwen, to the S. of which Caradocian
volcanics form the rugged peaks of Tryfan. From the road a track rises
gently to the N. shore of Llyn Idwal where there is a magnificent view of
the synclinal form of the Snowdon Volcanic Series in the crags at the head
of the lake; the narrow chasm of the Devil's Kitchen (Twel Du) is eroded
along the axis of the fold. The volcanic succession is traversed on the
way to the foot of the Kitchen. At the entrance to the Kitchen rhyolitic
tuffs, locally showing columnar structure, is conformably overlain by green
calcareous pumice-tuffs of the Bedded Pyroclastic Series. The cairned track
ascends S.E. along a bedding-plane ledge, locally ripple-marked, near the
base of the Pyroclastic Series, to a plateau where it turns N. to the top
where andesitic basalts display a rude pillow structure.

From the Kitchen, walk N. across the Snowdon Volcanic Series, in descending
sequence, to the summit of Y Garn, 3104 ft. (946 m). Continue N.N.W. then
W. along the ridge, on the edge of which there are exposures of Gwastadnant
Grits and grey Llandeilo slates with rhyolite and keratophyre flows, and a
sill of dolerite.

About half a mile (0.8 km) N. of Y Garn the Llandeilo slates pass downwards
into blueish Llanviron slates, altered by the Bwlch-y-Cywion Caledonian
granite. Beyond the granite Arenig slates form the ridge, indurated to the
N. by the Moel Mynydd granite intruded along the faulted Ordovician-
Cambrian junction.

From Mynydd Perfedd a rough descent leads to the S. end of Marchlyn Mawr,
passing over the Cambrian Ffestiniog Grits and near the lake the underlying
Maentwrog Slates. These show minor folds with steep N.E. cleavage crossing
the bedding of thinly-alternating slates and siltstones.

Marchlyn Mawr is a corrie-lake, blocked by a moraine on which has been built
a rock-fill dam forming the Upper Reservoir of the Dinorwic Pumped Storage
Scheme (see also E6.1). Follow the road from the dam N.W. to Marchlyn Bach
at the head of which there is a spectacular section of graded Bronllwyd
Grits. Near the outlet these are underlain by Green Slates at the top of
the Lower Cambrian Slates succession. Continue N. to the S. end of the huge
Penrhyn Quarry where the slates, the continuation of those of Dinorwic
(E6.1), are still worked. They are cut by green Ordovician metadolerite
dykes and by a black Tertiary dolerite dyke. It may be possible to obtain
permission to view part of the workings.

Descend to Bethesda.

E6.3   Anglesey (Fig. 6.3)

Precambrian. Ordovician and Silurian sediments. Lower Old Red Sandstone
and Carboniferous.

Access:- To undertake the whole of this excursion private transport is
necessary. It is, however, possible to travel by public bus to Amlwch and
to study the exposures in this area by walking only; this bus service continues to Holyhead which can also be reached by train from Bangor. From
Holyhead it is possible to walk to the South Stack exposures.

Maps:- Topo:- 114 Geol:- 1:63,360 Special Sheet Anglesey.

Walking distances:- Using private transport 12 miles (19.2 km). Using
public transport - Amlwch area 8 miles (12.8 km); Holyhead area 7 miles
(11.2 km).

Itinerary:- Leave Bangor to the S.W. and cross the Menai Bridge. The
important north-easterly Dinorwic Fault, which is still slightly active
and causes minor earthquakes, partly coincides with the Menai Straits and
partly runs slightly to the S.E. of the Straits.

On Anglesey follow the main road to Llanallgo and turn down to Malfre on
the coast to examine exposures of Carboniferous Limestone. Continue to
Traeth Lligwy; on the N. side of this bay there are exposures of red sandstones of Old Red Sandstone age, probably Upper, although this is not
certain.

Continue N. along the main road and park transport just N.W. of Penysarn
before reaching Amlwch. The public bus from Bangor can also be left here.
Ascend Parys Mountain. This was once the site of one of the world's most
important copper workings which can still be seen. Mineralisation is in
sediments near the Ordovician/Silurian junction and is associated with a
felsite sill. Some pyrite and chalcopyrite may still be found.

Continue to Amlwch then examine the coast section to the E. in mica-schists
of the New Harbour Series (Precambrian Mona Complex or Monian). Half-a-mile
(0.8 km) S. by E. of Lynas Point these have been translated southwards by
the Carmel Head Thrust (traceable for 11 miles (17.6 km) to the W.) over
Arenig sediments.

Return to Amlwch and continue to Holyhead. On the way the Wylfa Nuclear
Power Station, founded on the mica-schists can be seen. Drive (if private
transport is available) or walk W. by S. to South Stack. Here the quartzose
South Stack Series, regarded as the oldest part of the Monian, shows magnificent folding.

After returning to Holyhead those with private transport should continue to
Llanfaelog, near which there are rather sparse exposures of the Precambrian
Coedana Granite. Follow main road to Newborough then walk along a track
S.W. to the coast at the N.W. corner of Llanddwyn Bay. Here spilite-lavas
of the Precambrian New Harbour Group show beautiful pillow-structure. This
is a Nature Reserve, and hammers should not be used. The "right-way up" of
the lavas can easily be recognised.

Return to the Menai Bridge and Bangor.

E6.4 Blaenau Ffestiniog, Tan-y-grisiau, Stwlan Dam and the Moelwyns (Fig. 6.4)

Cambrian sediments. Ordovician sediments, volcanics and intrusives.
Glacial features.

Access:- From Bangor by road, or by rail changing at Llandudno Junction.
Blaenau can also be reached by narrow-gauge, steam-operated railway from
Portmadoc.

Some geologists may prefer to stay in Blaenau, where there is a fair amount
of accommodation, although limited for groups.

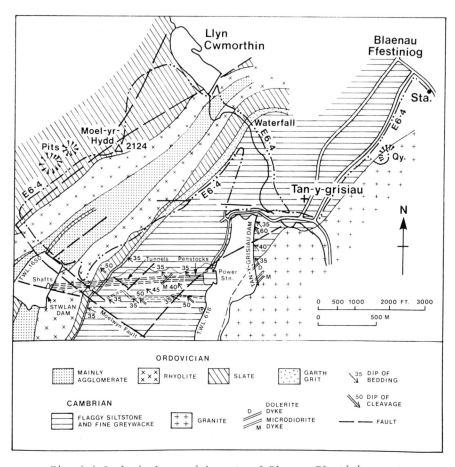

Fig. 6.4 Geological map with route of Blaenau-Ffestiniog area.

Maps:- Topo:- 115, 124 Geol:- 1:25,000 sheet Central Snowdonia.

Walking distance:- 6 miles (9.6 km).

Itinerary:- From Blaenau railway station follow the road S.W. Numerous exposures occur of flaggy alternations of fine greywacke, siltstone and slate probably of Tremadoc (Upper Cambrian age) with bedding dipping N.W. at 35° and cleavage N.W. at 60°. About half a mile (0.8 km) S.W. of the station a large quarry (now partly filled with rubbish) was formerly worked in the Tan-y-Grisiau Microgranite. This has the appearance of a laccolite although perhaps it has more the shape of an ellipsoidal, truncated cone. A jagged roof of hornfelsed siltstone is seen, and below it a distinctive roof-facies of fine-grained, vesicular granite.

From the quarry continue past Tan-y-Grisiau and the dam holding back the Lower Reservoir of the Ffestiniog Pumped Storage Scheme, founded partly on spotted hornfels and partly on granite. From N. of the Power Station ascend the construction road which winds up the mountainside to the Upper Reservoir. Near where the road doubles back to the S.W. the Cambrian flaggy beds are abruptly overlain by white, current-bedded gritty sandstone - the Garth Grit at the base of the Arenig. In upward succession Arenig flaggy sandstones are overlain by slates (?Llanvirn). A 10 ft. (3 m) dolerite dyke is marked by a slot in the cliffs.

The slates are sharply overlain by rhyolite (quartz-latite) which continues to the Upper or Stwlan Dam, of which it forms most of the foundations; the N.E. end is on baked shale and agglomerate. The rhyolite shows flow-banding, flow-folding and autobrecciation and is probably a sill rather than lava-flows as previously thought. The Upper Reservoir was formed by deepening a corrie-lake. From its N.E. side climb N.W. to the ridge N.E. of Moelwyn Mawr. The ascent provides a good cross-section of the Moelwyn Volcanic Group (Caradoc), highly varied but consisting largely of rhyolitic tuffs and agglomerates. Columnar-jointed felsite is probably another sill. A break in the ridge marks a N.W. fault downthrowing S.W. which was a factor in the construction of the dam.

Walk N.E. along the ridge to Moel-y-Hydd, 2124 ft. (647 m), avoiding large pits above disused underground workings in slates. From the summit there is a steep but easy descent on a northerly bearing to Llyn Cwmorthin occupying a glacial basin with striking roches moutonnées in the Moelwyn Volcanics. A track down to Tan-y-Grisiau crosses a descending Ordovician succession, and at a waterfall the white Garth Grit is seen.

Return to Blaenau. At the Llechwedd Quarries, on the N. outskirts, tourist visits to underground workings are offered.

Barmouth Centre (Fig. 6.5)

Barmouth is a pleasant coastal town with a number of hotels etc., although accommodation of groups may be difficult at holiday times. It is on the railway from Shrewsbury and Aberystwyth to Pwllheli.

E6.5   W. side of Harlech Dome

Cambrian sediments.

Access:- By train or bus N. from Barmouth or by private transport on the coast road, then by walking on roads and moorland.

Maps:- Topo:- 124 Geol:- EW 135, EW 149.

Walking distances:- 12 miles (19 km). With private transport 7 miles (11.2 km).

Itinerary:- Leave train or bus at Llanbedr. Walk E. up minor road over exposures of the Llanbedr Slates, the second oldest subdivision of the Cambrian succession. The oldest member, the Dolwen Grits, (the base of which is not seen) occurs only in the core of the Harlech Dome. Those who drive to Barmouth on the inland road S. of Ffestiniog can see the Dolwen Grits W. of the road S. of the Trawsfynydd Reservoir. The Llanbedr Slates at Llanbedr come up in the Coastal Anticline on the W. flank of the main dome.

About 650 yds. (595 m) upstream from Llanbedr the slates pass upwards into the Rhinog Grits, the coarser beds of which show graded bedding, indicating that the E. by N. dip is "right-way up", and the finer beds of which show cleavage at a higher angle than the bedding. About $1\frac{1}{4}$ miles (2 km) E. by N. of Llanbedr the Rhinog Grits are succeeded by the Manganese Shales with a disused working in the ore bed at their base. An E.N.E. fault displaces outcrops to the W.S.W. on its N.N.W. side.

Further exposures of the Rhinog Grits, on the S.S.E. side of the fault, can be seen further up the stream over which there is a bridge $2\frac{3}{4}$ miles (4.4 km) upstream from Llanbedr. By following a narrow road S.W. to Llanenddwn a further study can be made of the Rhinog Grits and underlying Llanbedr Slates; the latter formation is well seen along the main road to a point 1 mile (1.6 km) N.N.W. of Barmouth. The Rhinog Grits then outcrop to the town.

Along or near the road S.S.E. then E., N. of the beautiful Barmouth Estuary, the following upwards Cambrian succession occurs within 1 mile (1.6 km); Manganese Shales, Barmouth Grits, Gamlan Flags and Grits (metadolerite intrusion), Clogau Shales, Vigra Flags. The last form the lower part of the Maentwrog Slates (cf.E6.1 and E6.2).

Return to Barmouth.

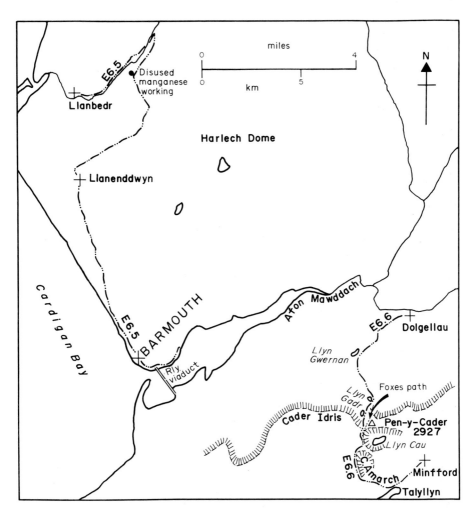

Fig. 6.5 Route map for Barmouth Centre.

E6.6   Cader Idris

Ordovician volcanics, sediments and intrusions. Glacial features.

Access:- Preferably by private transport to Dolgellau and back from
Talyllyn. Possible by public bus to Dolgellau and back from Minffordd.
Walking on rough and steep mountain paths. Not to be undertaken in bad
weather. If there are transport difficulties a return may be made from
the summit of Cader Idris to Dolgellau.

Maps:- Topo:- 124 Geol:- EW 149.

Walking distances:- From Dolgellau to Talyllyn, 7 miles (11.2 km). From
Dolgellau to Minffordd. 7½ miles (12 km). From Dolgellau to summit of
Cader Idris and back. 8 miles (12.8 km).

Itinerary:- From Dolgellau S.W. for 1¾ miles (2.8 km) to the N.E. end of
Llyn Gwernan. The road follows a depression marking the north-easterly
Dolgellau Fault. From the lake follow a path S. by E. to Llyn-y-Gadr.
Isolated outcrops of rhyolitic ashes and lavas (Lower Acid Group) of the
Arenig are seen but the stratigraphy is complicated by numerous small faults
and dolerite intrusions. To the N.E. lies the large dolerite mass of
Mynydd-y-Gader.

There is a steep rise S. by W. over cleaved pillow-lavas with interbedded
ashes of the Lower Basic Group (Llanvirn) to the corrie-lake of Llyn-y-Gadr
held back by a beautiful moraine. From its N.E. shore the steep, loose-
surfaced Foxes Path ascends S.E. over granophyre. At a sharp bend it
reaches the Llyn-y-Gadr Mudstones and Ashes (Llandeilo) overlain by a
dolerite sill. At Pen-y-Cader, 2927 ft. (892 m), the highest point of
Cader Idris, spilite pillow-lavas of the Upper Basic Group are well displayed.

From the summit the route continues S.W. then S. along the ridge of Craig
Cau with a fine view into a deep corrie occupied by Llyn Cau. Mudstones
are crossed, overlain, due to the southerly dip, by the highest volcanics
of the Cader Idris range, the Upper Acid Group. The path swings S.W. then
S. round the corrie of Cwm Amarch before descending a steep grass slope to
the glaciated basin of Talyllyn with its hanging tributaries. Those who
need to catch a bus should turn N.W. on approaching Talyllyn towards
Minffordd.

Church Stretton Centre (Fig. 6.6)

Church Stretton is picturesquely situated at the summit of the main road and
railway routes from Shrewsbury to Hereford.

Some accommodation is available but booking is necessary for parties. An
alternative is to stay in Shrewsbury, 12 miles (19 km) to the N. Both
places can be easily reached by public or private transport.

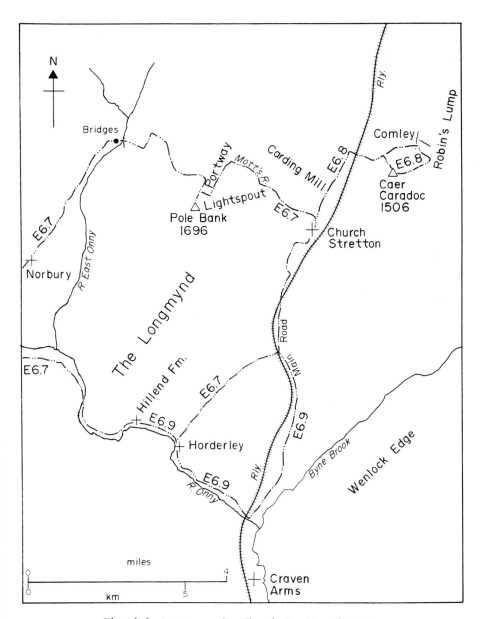

Fig. 6.6 Route map for Church Stretton Centre.

Three excursions within easy range of Church Stretton are described but
there are a number of others, well-documented by Dean (1968). The
succession of Precambrian rocks is as follows:-

Wentnor "Series" or Western Longmyndian	Bridges Group
	Bayston/Oakswood Group
	Portway Group
Stretton "Series" or Eastern Longmyndian	Lightspout Group
	Synalds Group
	Burway Group
	Stretton Shales
	Helmouth Grits
Uriconian	Volcanic

E6.7   The Longmynd

Precambrian. Silurian.

Access:- Walking on roads, tracks and hillsides. Return possible by
private transport from Bridges.

Maps:- Topo:- 137 Geol:- EW 166 1:25,000 sheet Church Stretton.

Walking distances:- 10 miles (16 km). If private transport is used to
return from Bridges, 5 miles (8 km).

Itinerary:- At the N. outskirts of Church Stretton turn N.W. up the
Carding Mill Valley road which cuts outcrops of the Stretton Shales.
About $\frac{1}{4}$-mile (0.4 km) above the main road a quarry shows the Buckstone
Rock, a grey tuff of cherty appearance at the base of the Burway Group,
the highest part of which consists of the Lower Cardingmill Grit, a green-
grey sandstone with current bedding which indicates that the succession
becomes younger to the N.W. the direction of the regional dip. Here the
stream section rather than the road should be followed.

The succeeding Synalds Group consists of the Upper Cardingmill Grit followed
by purplish shales and slates with two andesitic tuff horizons, the Batch
Volcanics, near the top. The Lightspout Group is named after the waterfall
near which the stream forks. At the base is the light-weathering Haddon
Hill Grit. Above the Spout a rough track, Mott's Road, beside the N.W.
fork, should be followed. Beside the lower path of the road grey-green
flags forming the upper part of the Lightspout Group are seen and higher
up purple sandstones of the Portway Group.

Beyond the head of the valley the ancient road, the Portway, little better
than a track, is reached. Follow this to the S.W. for about $\frac{1}{2}$-mile
(0.8 km) then turn off on a track to the N.W. A short diversion leads to
the highest point of the Longmynd, Pole Bank, 1696 ft. (517 m).

The track to the N.W. descends to Bridges. Purple and red sandstones are
seen with red conglomerates forming the Bayston-Oakswood Group. At Bridges
the valley of the R. East Onny is cut in the Bridges Group, mainly purple
shales and siltstones with some sandstones showing ripple-marks and graded-
bedding; quartz-dolerite dykes are present.

Those without private transport should return by the same route from
Bridges. With transport, Church Stretton can be reached by road, via
Norbury where Llandovery strata with Pentamerus and other fossils are seen.

E6.8   Caer Caradoc and Comley

Precambrian. Cambrian.

Access:- By road walking and hill walking. Private transport can be used
in part.

Maps:- Topo:- 137 Geol:- EW 166 1:25,000 sheet Church Stretton.

Walking distances:- 7 miles (11.2 km). With private transport to Comley
4 miles (6.4 km).

Itinerary:- Branch off the main road 1¾ miles (2.8 km) N.N.E. of Church
Stretton to Comley on the Cardington road. Follow the Comley road ¼-mile
N. from the hamlet to its confluence with the stream from Shootrough Wood
where the Orusia Shales (U. Cambrian) with calcareous nodules are poorly
exposed. Rare trilobites (Parabolina) occur. Return to the Cardington
road, just N. of which, 250 yds (228 m) E. of Comley, there is a very poor
field exposure of M. Cambrian sandstones, dipping E., with Paradoxides and
brachiopods. Further E. opposite Shootrough Farm, a small roadside quarry
shows the Hoar Edge Grits (Caradoc), vertical or slightly overturned, owing
to the proximity of the Cwms Fault, a major component of the Church Stretton
fault-system.

Return to Comley and examine Comley Quarry, famous as the locality from
which Lapworth demonstrated the first L. Cambrian trilobites known in
Britain. These were first attributed to Olenellus but later reassigned to
Callavia. Most of the fossiliferous blocks present are of weathered
L. Cambrian limestone but some are of Callavia sandstone.

Those with private transport will find it convenient to send it back to
Church Stretton from here.

Walk along a minor road S.E. from Comley for 500 yds. (457 m) then S.W. for
400 yds. (366 m) to the feature of Hill House ridge where a small quarry
shows fossiliferous blocks of L. Cambrian sandstone in a sandstone matrix
with M. Cambrian fossils. A little further S. the conical hill of Robin's
Lump shows M. Cambrian sandstone resting unconformably on Comley Limestone
(L. Cambrian). Fossils are not easily collected at this and the previous
locality.

An easy ascent follows to the summit of Caer Caradoc, 1506 ft. (460 m).
Uriconian volcanics include flow-banded rhyolites near the top and tuffs,
andesites and rhyolites along the ridge to the N.W. in the vicinity of the
Three Fingers Rock.

The summit ridge affords a good view of the through valley which follows
the Church Stretton Fault System. Movement along this complex system took
place at intervals until the Triassic. The system continues to the S.W. as
the Carreg Cennen Disturbance and marks the South-East Caledonian Front, a
view supported by Lower Palaeozoic facies changes across it. Rocks of the
Midland Kratogen facies continue, however, further W. to the Linley-Pontesford
Fault and the intervening strata may all form part of a Border Thrust zone.

Descend N.W. to the Cardington road; the Uriconian block is bounded to the N.W. by a component of the Church Stretton Fault System which separates it from poorly-exposed Silurian.

E6.9   The Onny Valley

Precambrian. Ordovician. Silurian.

Access:- By private transport. By public bus to junction of Hereford and Bishop's Castle roads. Walking mainly on roads.

Maps:- Topo:- 137. Geol:- EW 166 1:25,000 sheet Craven Arms.

Walking distances:- From road junction to Hillend and back 7 miles (11.2 km). Returning from Hillend by private transport 3½ miles (5.6 km).

Itinerary:- A road cutting on the Bishop's Castle road 650 yds. (592 m) from the main road shows Wenlock Shales with Monograptus and Dalmanites. About 600 yds (550 m) further on, at the Batch Gutter, the road should be left for the N.E. bank of the Onny. A short walk downstream leads to a small cliff, the lower half of which is composed of Caradoc mudstones with abundant fossils including trilobites, brachiopods and molluscs. These beds are overlain unconformably by the Purple (Hughley) Shales of the Upper Llandovery. The Silurian contains shelly fossils, mostly small. This part of the Onny River valley has been barred by local farmers.

Move upstream and cross the river N.W. of the confluence with the Batch Gutter. On the S.W. bank of the Onny grey, sometimes nodular, mudstones and shales (Caradoc) are well-exposed. They contain numerous shelly fossils, including trilobites.

Upstream from the Batch Gutter a good section was formerly exposed along the line of a disused railway but exposures are now poor. However, some compensation is afforded by a new road cutting S.E. of Horderley. This shows a long section in the Cheney Longville Flags (Caradoc). Large brachiopods (Dolerorthis, Bancroftina) are abundant.

In a large roadside quarry N.W. of New House the Middle Horderley Sandstone (Caradoc) is exposed. Just by this quarry the Cwms Fault is crossed, on the W. side of which the strata show much steeper dips. Thus the flaggy Lower Horderley Sandstone is almost vertical.

Leave the main road at Rock Cottage and follow a track crossing the river to an old quarry in the Hoar Edge Grits and Coston Beds (base of Caradoc). These conglomeratic sandstones, dipping steeply S.E., rest at the N.W. end of the quarry on Western Longmyndian arkoses.

Return to the main road and continue towards Horderley. Beside the road below Horderley, East Longmyndian rocks are seen bounded to the S.E. by the Lawley Fault and to the N.W. by the Church Stretton Fault, which runs approximately along the valley N.E. of Horderley. To the N.W. Silurian strata occur. Fossiliferous Wenlock fragments can be found near The Bank.

Further up the valley, to the N. of the road and S.W. of Hillend Farm, fossiliferous limestones and mudstones of the Pentamerus Beds (U. Llandovery)

have been quarried. Further N.W. slightly lower arenaceous and conglomeratic strata rest unconformably on Longmyndian.

Return to junction with main Hereford-Shrewsbury road.

Builth Wells Centre (Fig. 6.7)

Builth Wells, on the R. Wye, offers varied accommodation and can be reached
by bus, or by train to Builth Road station 1½ miles (2.4 km) to the N.W.
The classic early Ordovician shoreline described by Jones and Pugh (1949)
occurs nearby.

E6.10   Southern part of the Llandrindod Wells - Builth Ordovician Inlier

Ordovician sediments, volcanics and intrusions. Silurian.

Access:- By walking, or partly by private transport. Permission to enter
quarries should be sought.

Maps:- Topo:- 147   Geol:- 1:25,000 sheet Llandrindod Wells - Builth
Ordovician Inlier.

Walking distances:-   7 miles (11.2 km). With private transport 3½ miles
(5.6 km).

Itinerary:- Cross the bridge across the R. Wye to the Llanelwedd Quarries
in the Builth Volcanic Group. The E. quarry is bounded to the W. by a
strike-slip fault with slickensides and to the S. by an E.-W. fault. This
brings up black shales, probably below the Pebbly Feldspar Ash of the
quarry; these bedded agglomerates and tuffs have conspicuous shale clasts
near the base. In the major quarry to the W. spilitic lavas occur in the
face; these are divided into upper and lower divisions by the "Felsite
Agglomerate" which outcrops near the N.W. corner of the quarry. This is
probably a welded tuff and contains felsite blocks up to 0.5 m in diameter
in a matrix of predominantly flattened clasts and subordinate spilite
fragments. Weathered Upper Spilites with pillow-structure outcrop beside
the track beyond the W. end of the quarry.

Traverse W. across the hillside to examine crags of keratophyre and then
crags of Llanvirn sediments belonging to the base of the Newmead Group.
A short distance further W. a small quarry shows the Grey Feldspar Sands
which are products of the weathering of the Builth Volcanics. Above the
back of the quarry a conglomerate contains large boulders of spilite.

Walk N. for 765 yds (700 m) along a lane which follows approximately the
unconformable junction between the Newmead Group and Llandeilo Shales, to
a quarry behind Tan-y-Graig Cottage in Grey Feldspar Sands with Hesperorthis.

Travel north-westwards on A479 to Gwern-yfed-fach Cottage, just S.E. of
Builth Road station. The R. Wye section here shows Llandeilo shales and
dolerite intrusions. Road widening has revealed an exposure of shales with
Trinucleus fimbriatus and other fossils. The contact with an overlying
sill is seen. On the way N.E. up to Pen-Maenau Rocks alternating exposures
of the Wellfield Dolerite Complex and baked black shales are seen. From
the top of the rise, adjacent to a wall, a conspicuous line of crags outcrops down the hillside. The crags consist of gabbro to which several
dolerites are connected. It is considered that the Wellfield Complex is a

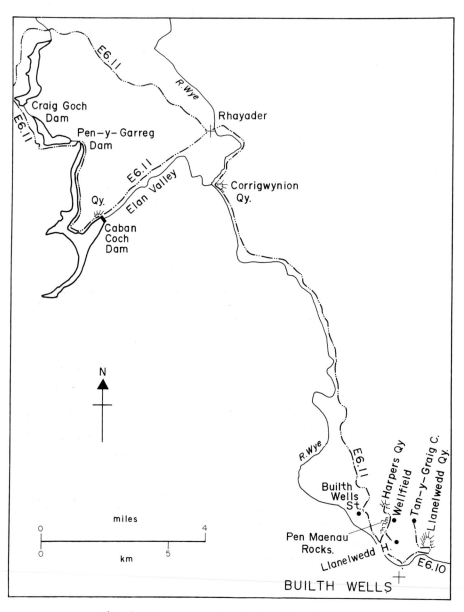

Fig. 6.7 Route map for Builth Wells Centre.

Cedar-tree laccolith.

A short detour down a lane leading to Llanelwedd Hall reveals exposures of
basal Wenlock with the trilobite Leonaspis sp. but the junction with the
Ordovician is difficult to locate. Return past Pen-Maenau Rocks and walk
towards Wellfield, crossing more exposures of the Complex, truncated near
Wellfield House by the Carneddau Fault Complex. Further N.W. Harper's
Quarry shows Llanvirn black shales, dipping N.E., underlain by a dolerite
intrusion.

Return to Builth Wells.

E6.11   Rhayader and the Elan Valley

Silurian.

Access:- Private transport is necessary to follow the whole of this
excursion, but it is possible to take a bus to Rhayader and to walk to
the Pen-y-Garreg Dam and back.

Maps:- Topo:- 147 Geol:- EW 179.

Walking distances:- Rhayader to Pen-y-Garreg Dam and back 13 miles
(20.8 km). With private transport 1 mile (1.6 km).

Itinerary:- By Wye Valley to Rhayader then up Elan Valley road. North of
the Caban Coch Dam a large, disused quarry exposes the Caban Conglomerates
(M. Llandovery). The individual conglomerate bodies have strongly scoured
bases and in some cases occupy steep-sided channels. They have been
interpreted as fluxo-turbidites which accumulated as fans at the foot of a
submarine canyon-complex. Between the Caban Coch Dam and the Pen-y-Garreg
Dam there are outcrops of M. Llandovery shales and flaggy, fine greywackes.

At the site of the uppermost or Craig Goch Dam there is a proposal to
raise a very much higher dam enclosing a very large reservoir. The rocks
in this area are shales, siltstones and fine greywackes. Close examination
shows that these rocks are microturbidites with graded bedding. Sharp N.E.
folds are present with cleavage dipping steeply N.W. Linear hollows on the
hillside mark faults. Sparse specimens of Monograptus sedgwickii
(Llandovery) occur.

Continue along the road to the N. end of the reservoir. Here the flat
floor of the glaciated Elan Valley shows classic meanders. Follow a road
S.E. back to Rhayader, stopping to walk S.W. to stream and hillside exposures of coarser greywackes of the Llandovery Series.

On the way down the Wye Valley back to Builth Wells a stop can be made at
Corrigwynion Quarry in coarse L. Llandovery grits.

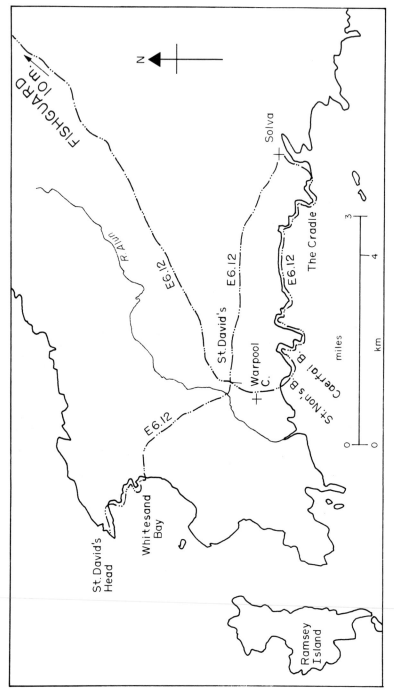

Fig. 6.8 Route map for Fishguard Centre.

Fishguard Centre (Fig. 6.8)

Fishguard is at the W. end of main rail and road routes through South Wales
and is a port for Rosslare in S.E. Ireland. Geologists travelling by this
route to Ireland might find it attractive to study the Precambrian and the
Lower Palaeozoic in the Fishguard district (see also E10.12). These show
considerable differences from the facies in North Wales (E6.1 to E6.6).
There is a fair amount of accommodation at Fishguard and at St. David's,
15 miles to the S.E.

In the town square a plaque commemorates the last invasion of Britain, a
landing nearby by Napoleon's troops in 1797.

E6.12   Fishguard and St. David's districts

Precambrian. Cambrian. Ordovician. Ordovician intrusions.

Access:- From Fishguard by private transport or public bus to St. David's
and on to Solva, returning from St. David's. Cliff-path and road walking.
More will be seen at low tide, but the latter is not essential.

Maps:- Topo:- 157 Geol:- 1:25,000 sheet St. David's.

Walking distances:- 10 miles (16 km). With private transport 6½ miles
(10.4 km).

Itinerary:- At Fishguard a good exposure in the Old Harbour shows shales
of the D. bifidus zone (Llanvirn) overlain by volcanics and cut by dolerite
intrusions. At Lampit Bach, on the W. side of the Old Harbour, volcanic
breccia and pillow lavas (Llanvirn) are intruded by a dolerite sill with
columnar jointing. Overlying rhyolites are followed by Llandeilo sandstones.

From Solva walk S. on path on W. side of harbour where there are exposures
of Upper Cambrian sandstones and shales (Lingula Flags) with hornblende-
porphyry intrusions. Follow cliff-path W. The Upper Cambrian sandstones
are folded on N.E. axes but there is a general downward succession to the
W. so that by Caerfai Bay (inner part) sandstones and shales (probably
M. Cambrian) are seen. In the next inlet to the W., St. Non's Bay, a
Cambrian basal conglomerate rests on Precambrian acid volcanics.

Follow track and minor road N. to St. David's. Around Warpool Court there
is Precambrian granophyre but it is poorly exposed.

From St. David's continue N. by private transport or by walking to Whitesand
Bay. The point at the N. end consists of Upper Cambrian sandstones and
shales (Lingula Flags), dipping steeply N.W. and overlain by Arenig shales
with Tetragraptus containing a bed of rhyolitic tuff. The next headland
marks a sill of coarse dolerite or gabbro; and in the bay beyond Llanvirn
shales with Didymograptus bifidus appear. St. David's Head is formed of
another coarse dolerite sill.

Return to St. David's.

6.2   The Lake District (Figs. 6.9 and 6.10)

The lakes lie in radiating, deep, glaciated valleys separated by high ridges.
It is a region of beauty and varied colour with four peaks over 3000 ft.
(915 m); one; Sca Fell Pikes, 3210 ft. (978 m), is the highest point in
England. Yet it is a compact district, easily reached from the large
cities. The main London-Carlisle trunk railway and road pass its E. margin
over Shap Pass and a coastal road runs round its S., W. and N. flanks.
Roads follow many valleys but excursion planning involving private transport
should take into account local prohibitions of coaches; the police should be
consulted.

Three centres are suggested, Keswick in the N., Windermere (or Kendal) in
the S. and Ravenglass (or Whitehaven) in the W. All are easily reached by
road. Windermere or Kendal, and Ravenglass (or Whitehaven) are on rail,
and Keswick is on a bus route, connecting with main-line trains at Penrith.
A bus service runs between Keswick and Windermere.

The region is very popular, and in summer enquiries for accommodation should
be made in advance, especially for groups.

The Lake District mountains can be dangerous, particularly in bad weather,
and are often snow-covered in winter. The succession is as follows:-

> Triassic
> Permian
> (unconformity)
> Carboniferous
> (unconformity)
> ? Lower Old Red Sandstone

	Ludlow Series & later	Kirby Moor Flags / Bannisdale Slates / Coniston Grits / M. & U. Coldwell Beds
SILURIAN	Wenlock Series	L. Coldwell Beds / Brathay Flags
	Llandovery Series	Browgill Beds / Skelgill Beds Stockdale Shales
	Ashgill Series	Coniston Limestone Group including Ashgill Shales at top BASAL UNCONFORMITY
	Caradoc (in part)	Drygill Shales (all boundaries faulted)
ORDOVICIAN	lower Caradoc Series & Llandeilo Series	Borrowdale Volcanic Group UNCONFORMITY
	Llanvirn Series (in part)	Eycott Group
	Arenig Series	Skiddaw Slate Group

A Northern Belt consists of about four anticlines involving the Skiddaw
Slates. To the N. they are succeeded by the Eycott Volcanics, probably
older than the Borrowdale Volcanics, almost unfossiliferous but accepted
as Llandeilo-Caradoc. The relationship of these to the Skiddaw Slates is
controversial; in places there appears to be a thrust, in others a transi-
tional junction, and in others an unconformity has been claimed. The
Borrowdale Volcanics are disposed in several major anticlines and synclines
and along the northern edge of a Southern Belt are overlain uncomformably
by a fairly thin Caradoc/Ashgill succession. This is followed to the S.
by thick Silurian also disposed in several major folds.

The intra-Lower Palaeozoic unconformities are evidence for early Caledonian
movements but the main structures are late Silurian, trending N.E. to E.N.E.
In the S.W. of the Southern Belt an anticline brings up the Skiddaw Slates
and the Borrowdale Volcanics again.

Following Carboniferous deposition there was relatively minor doming
followed by Permo/Triassic deposition. In the Palaeogene a major dome
was formed on which the radial drainage was initiated, later to become
superimposed on the Caledonian trend.

The Lower Palaeozoic rocks are cut by acid to ultrabasic plutonic and
hypabyssal intrusions, some Ordovician and some end-Silurian. Around the
complex Carrock Fell intrusion there is considerable mineralisation.

Comparison with other regions:- The extensive Llanvirn-Llandeilo-Caradoc
vulcanicity is comparable to that in much of Wales (6.1) but contrasts
with the much less extensive vulcanicity in these epochs in the Southern
Uplands. On the other hand the facies of the Silurian and the nature of
the Caledonian intrusive activity is comparable in the two regions. These
differences and similarities may be explained by the width of the diminishing
Iapetus Ocean in Ordovician times and its closure in the Silurian - the
suture probably lies under the Solway Firth.

The Carboniferous succession, although not complete, is generally similar
to that in most of England and Wales. The Permo/Triassic sequence, deposited under arid conditions, resembles that of most of England and
Scotland, though the Magnesian Limestone (7.4) sea transgressed W. of the
Pennines to a very limited extent.

Keswick Centre

E6.13   Skiddaw area

Access:- By hill walking. Can be extended if private transport is arranged
for return from Mosedale back to Keswick.

Maps:- Topo:- 90 Geol:- EW 23 EW $\frac{1}{4}$-inch to mile 3.

Walking distances:- 12 miles (19 km). 13 miles (21 km) to Mosedale.

Itinerary:- From Keswick N. by main road and minor road to point E. of
Applethwaite. By climbing up steep slope E. of Applethwaite Gill a section
of the Skiddaw Slates can be seen; these include siltstones and grits as
well as rather poorly cleaved slates. In the broad ridge of Jenkin Hill

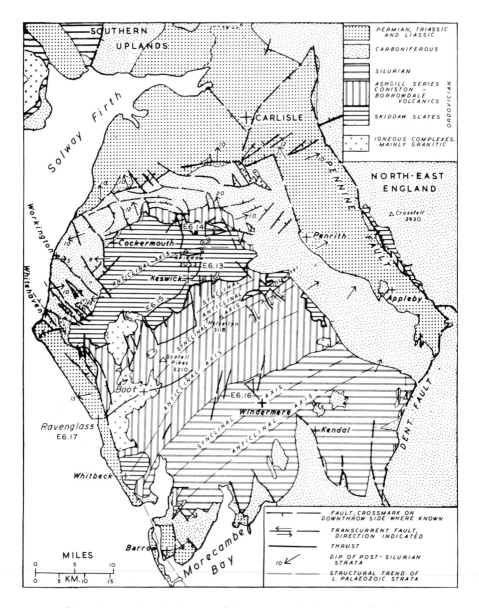

Fig. 6.9 Geological map, with routes, of the Lake District.

the wide Skiddaw tourist path is intersected and should be followed N.W.
Here the slates are clearly within the Skiddaw Granite aureole and carry
whitish chiastolite crystals. An anticline with N.E. axis is crossed near
the summit which is covered with frost-shattered hornfelsed slate. From
the top of Skiddaw, 3053 ft. (931 m) the rugged scenery of the Borrowdale
Volcanics to the S.W. contrasts with the smoother terrain of the Skiddaw
Slates. Far to the N., across the Solway Firth, rise the Southern Uplands.

Descend tourist path for $1\frac{1}{4}$ miles (2 km) then go down hillside E. by N. to
Sinen Gill. Here coarse, whitish granite outcrops, the most southerly of
the three surface appearances of the Skiddaw Granite. The slates are
altered to intensely hard, spotted hornfels which, about $\frac{1}{4}$-mile (0.4 km)
above the confluence of Sinen Gill with the main stream, can be seen to
overlie the granite.

Descend along right bank of Glenderaterra Beck to old lead mine $\frac{3}{8}$-mile
(0.6 km) downstream from Sinen Gill where minerals may be found on dumps.
Cross to path near left bank then recross near Derwentfolds and return to
Keswick by minor road on N. side of R. Greta.

If private transport from Mosedale can be arranged, and it is desirable to
save a day, it is possible to turn N. at the foot of Sinen Gill over a low
watershed and descend the Calder Valley. In this way at least some of the
exposures described in the next itinerary can be visited.

E6.14   Carrock Fell and Eycott Hill

Skiddaw Slates. Eycott Volcanics. Carrock Fell Complex. Skiddaw Granite.
Mineralisation. Carboniferous Limestone Series.

Access:- By private transport to Mosedale. By public bus on Keswick-
Penrith road to Liscaw then walking to Mosedale.

Maps:- Topo:- 90 Geol:- EW 23.

Walking distances:- From Mosedale $6\frac{1}{2}$ miles ($10\frac{1}{2}$ km) and $1\frac{1}{2}$ miles (2.4 km)
if Eycott Hill is included. From Liscaw on main road 13 miles (21 km).

Itinerary:- Ascend Carrock Fell by the path up its S.E. spur, leaving the
Swineside road about 0.4 mile (0.6 km) from Mosedale. Slightly metamorphosed arenaceous Skiddaw Slates, sharply folded about E.S.E. axes, are
first seen, then the contact with quartz-gabbro, crossed where the altered
slates contain garnets. At the head of the steep ascent bear N. by E. to
White Crags over outcrops of ilmenite-rich and biotite-bearing gabbro.
From there a N.W. route crosses further outcrops of ilmenite-rich gabbro,
some showing fluxion structure, then passes over hybrid rocks before reaching
granophyre around the summit, 2174 ft. (663 m).

From the summit walk W. to Round Knott just S. of which chilled granophyre
is in contact with "diabase". Cross the "diabase" N. to Rae Crags to
examine the northern outcrop of granophyre. Descend N.W. to Drygill,
where the Drygill Shales (Caradoc but structural relations unknown) contain
trinucleid trilobites of the genus Broeggerolithus and are intruded by
felsite dykes. The area is mineralised and the dumps of old mines contain
numerous minerals.

Walk S. to Brandy Gill recrossing diabase, granophyre and gabbro. At the confluence with Grainsgill Beck tips from old mines yield wolfram, mispickel, molybdenite, apatite and other minerals. The veins are intruded in greisen, a modified form of the Skiddaw Granite. Descend to the R. Calder. A short distance upstream of the confluence with the Grainsgill Beck exposures of the granite may be seen and ¾-mile (1.2 km) upstream chiastolite-slates. Descend the Calder Valley by the track on the N. side to Swineside noting further outcrops of metamorphosed Skiddaw Slates and continue back to Mosedale.

With private transport it is possible to make a diversion to examine the Eycott Volcanics and the Carboniferous Limestone. From Mungrisdale a road should be followed N.E. for one mile (1.6 km) then another road S.E. for 2 miles (3.2 km) to the N.W. side of Berrier where there are old quarries in the Carboniferous Limestone. A S.W. traverse can then be made to the top of Eycott Hill, 1131 ft. (345 m) crossing andesite lavas, some flow-banded.

From Berrier continue S.E. to main road and return to Keswick. At Threlkeld a short diversion may be made to the S. to a large quarry in microgranite.

E6.15   Borrowdale, Honister Pass and Buttermere

Skiddaw Slates. Borrowdale Volcanics. Ennerdale Granophyre. Glacial geology.

Access:- By private transport. By walking to Borrowdale or Honister Pass and back.

Maps:- Topo:- 90, 97 Geol:- ¼-inch to mile EW 3.

Walking distances:- To Borrowdale and back 12½ miles (20 km). To Honister Pass and back 17 miles (27 km). With private transport 1 mile (1.6 km).

Itinerary:- Leave Keswick on Borrowdale road. At Castle Head, an intrusive plug of dolerite marks the site of a vent for Borrowdale Volcanics. About halfway along the E. side of Derwent Water the lowest Borrowdale lavas and tuffs may be studied on the hillside and in a stream section. South of the lake a diversion to the W. side of the river over Grange Bridge, where excellent roches moutonnés may be seen, makes it possible to reach a crag about 200 yds S.E. of Hollows Farm. Here Borrowdale andesite rests with apparent conformity on cleaved Skiddaw Slates.

Return to road and continue S. Just N. of the Bowder Stone, a large fallen block, a disused quarry shows well cleaved Borrowdale ashes. At Resthwaite there are prominent moraines.

From Seatoller climb W. to the Honister Pass. Here cleaved tuff lava in the Borrowdale Volcanics is extensively worked for green slate; bedding traces are seen on many specimens.

With private transport the excursion may be continued by descending the fine U-shaped valley to Buttermere; note change of scenery where the Borrowdale Volcanics give place to Skiddaw Slates again. Across the N. end of the lake the Red Ennerdale Granophyre can be seen. At Buttermere village turn N.E. on road to Stair along which are exposures of Skiddaw Slates. At

Stair follow road to N. to Braithwaite then E. to Keswick. About ¾-mile
(1.2 km) E. of Braithwaite in Hodgson How Quarry, fossiliferous Skiddaw
Slates are exposed.

Windermere Centre

E6.16   Troutbeck and High Skelgill

Upper Ordovician. Lower Silurian. Glacial geology.

Access:- By public bus on Ambleside road to Low Wood Hotel, then walking.
By private transport to same locality then walking before rejoining
transport at Troutbeck Bridge.

Walking distances:- 6½ miles (10½ km). With private transport 5 miles
(8 km).

Itinerary:- From the back of the hotel a path leads to High Skelgill
(Skelghyll) Farm. In the main beck (stream) 350 yds E.N.E. of the farm
shales, dipping S.S.E., form the type locality of the Skelgill Beds (base
of Llandovery). Graphtolites, mainly monograptids, are well preserved.
In the highest part of the Llandovery red beds occur.

Descend on S.E. side of road for ½-mile (0.8 km) to minor road and follow
this E. then N. by E. to W. side of Troutbeck. On the hillside to the W.
the Coniston Limestone (Ashgill) is well exposed consisting of dark,
impure limestones and calcareous shales. The bedding dip is S. at 30°, and
the cleavage is inclined very steeply N. The overlying Ashgill Shales
with orthid brachiopods are sharply overlain by dark Skelgill Shales with
long monograptids (base of Llandovery). Looking E. across the valley it
can be seen that the line of the Coniston Limestone outcrop is shifted
½-mile S. by a N.-S. fault.

Descend S. by W. by road on W. side of the Troutbeck valley which shows
fine glacial features, noting outcrops of Brathay Flags (Wenlock). From
Troutbeck Bridge follow main road back to Windermere.

Ravenglass (or Whitehaven) Centre

Accommodation is limited in Ravenglass; it may be necessary, particularly
in the holiday season, to stay in Whitehaven, further N. Both may be
reached by road or rail round either the N. and W. margins of the Lake
District from Carlisle or round the S. and W. margins. Whitehaven is on
the Cumbrian Coalfield where workings extend under the sea, and to the S.
the overlying Permo-Triassic outcrops in the St. Abbs area. On the southern
route note should be taken of the extensive Carboniferous Limestone outcrops
N.W. of Bowder in which replacement iron ore deposits were mined on a large
scale.

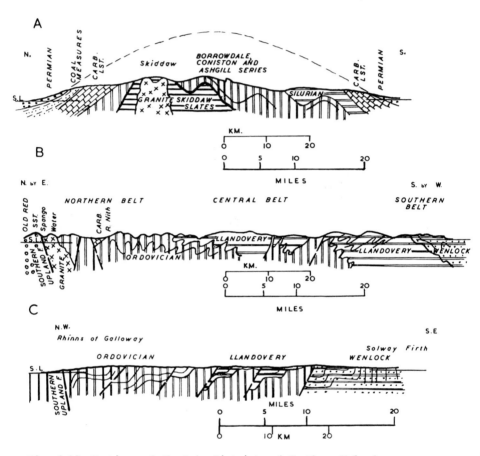

Fig. 6.10   Sections of the Lake District and Southern Uplands.
            A. Lake District.   B. Southern Uplands (after Peach and Horne).
            C. Southern Uplands (after Walton).

E6.17   Eskdale

Caledonian Eskdale Granite. Borrowdale Volcanics. Glacial geology.

Access:- By narrow gauge railway from Ravenglass to Eskdale station. By private transport to Boot.

Maps:- Topo:- 89, 96 Geol:- EW 37. $\frac{1}{4}$-inch to mile EW 3.

Walking distances:- 8 miles (13 km).

Itinerary:- Half-a-mile (0.8 km) N.W. of Ravenglass the contact of Triassic and L. Palaeozoic is crossed. The route to Dalegarth station or to nearby Boot follows the glaciated Esk valley, eroded in pink Eskdale Granite.

From Boot walk E. along the road to the Woolpack Inn, still in granite which on Great Barrow to the N. and on Hartley Crag and Gate Crag to the S. is capped by Borrowdale andesite. By turning down a narrow lane leading to Penny Hill, just W. of the Woolpack Inn, good exposures of granite can be seen in the R. Esk.

Continue along the road, which becomes very steep and rough, to the Hard Knott Pass, 1281 ft. (390 m). One mile (1.6 km) beyond the Woolpack Inn the contact of the granite and the Borrowdale Volcanics is crossed. On the ascent to the Hard Knott Pass there are many exposures of Borrowdale andesite-lavas and tuffs. To the N. there are good views of the Sca Fell range made up mostly of tuffs forming part of the Sca Fell Syncline.

Return by same route.

6.3   The Southern Uplands (Figs. 6.10 and 6.11)

The Southern Uplands, with a number of hills over 2500 ft. (762 m) high and rising to 2764 ft. (840 m) at the Merrick, form a considerable barrier between England and the Scottish heartland. There is good public transport by rail and road near the E. coast and along the two main routes from Carlisle to Glasgow, via Beattock and via Sanquhar. Elsewhere public transport is thin, and for some of the excursions described private transport is virtually a necessity.

Four centres are suggested, Dunbar, Dumfries, Ayr and Stranraer. With extra travel the Dunbar excursions can be done from Edinburgh. Ayr is also one of the centres for the Midland Valley (Ch. 7.5).

The succession is as follows:-

	Triassic
	Permian
	(Unconformity)
	Coal Measures
Carboniferous	Millstone Grit
	Carboniferous Limestone Series

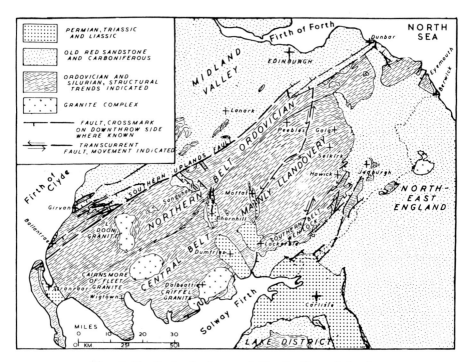

Fig. 6.11 Geological map of the Southern Uplands.

Calciferous Sandstone Series

Devonian	Upper Old Red Sandstone (Unconformity) Lower Old Red Sandstone (Unconformity)
Silurian	Ludlow Series Wenlock Series Llandovery Series
Ordovician	Ashgill Series Caradoc Series Llandeilo Series Llanvirn Series (Unconformity) Arenig Series

The structural units comprise a Northern Belt, a Central Belt and a Southern Belt. It has long been held that the Northern Belt and part of the Central Belt form an anticlinorium and the southern part of the Central Belt and the Southern Belt a synclinorium, each made up of numerous isoclines. Recent workers have shown that the structural pattern is essentially that of N.W. facing monoclines with long S.E. limbs, often isoclinally crumpled, and strike-faults bringing up older strata to the N.W.

Three large, post-tectonic calc-alkaline plutons of very late Silurian or early Devonian age occur in the S.W., and there are numerous smaller intrusions of the same age.

Comparison with other areas:- Unlike most of the other Lower Palaeozoic Blocks in the British Isles, the Southern Uplands show no rocks older than the Arenig, beneath which there may be a decollement. The contrast between the thick, coarse turbidites of the Northern Belt and the thin shale sequence of the Central Belt (E6.22) was made classic by Lapworth; the contrast becomes less marked in the Silurian. In general the Lower Palaeozoic sediments resemble those of the other Lower Palaeozoic regions, apart from the carbonates of the North-West Highlands. Vulcanicity was intense in the Arenig, but only sporadic in later Ordovician times, unlike other British areas. The Ordovician faunas of the Southern Uplands differ considerably from those of the Lake District and Wales, which can be explained by the presence of a diminishing Iapetus Ocean. Nevertheless the graptolite fauna of the Central Belt has some relationship with that of the Anglo-Welsh area and of Europe. The shelly faunas of the Northern Belt, and of Girvan (7.5), are of Appalachian aspect. Post-tectonic Caledonian granites are similar to those of the Lake District and of the Highlands.

The unconformity between Lower and Upper Old Red Sandstone shows that the late Caledonian movements of the Midland Valley extended into the Southern Uplands. The Carboniferous succession, though thinner, is similar to that of the Midland Valley (Ch. 7.5), and in the presence of coals in the Lower Carboniferous (E6.19) also resembles the Scremerston Coal Group (Ch. 7.4) of North-East England. The Permian succession is like that of Northern England (apart from the absence of the Magnesian Limestone) and much thicker than in the Midland Valley which, however, it resembles in the occurrence of basalt lavas.

Dunbar Centre (Fig. 6.12)

Dunbar, a good centre for accommodation, can be easily reached by rail or road from Edinburgh or from Newcastle and further south.

E6.18   Siccar Point and Cove

Silurian. Hutton's unconformity. Upper Old Red Sandstone. Lower Carboniferous. Glacial overflow channel.

Access:- By bus or private transport from Dunbar then by walking.

Maps:- Topo:- 67 Geol:- S 33 and S 34.

Walking distance:- 7 miles (11 km). Reduced if private transport is used.

Itinerary:- Alight from public bus on the A1 road 1¼ miles (2 km) S.E. of Cockburnspath. Walk along the Eyemouth road, crossing the bridge over the post-glacial gorge of the Pease Burn. Branch off to S. (private buses should be left here) then follow a narrow road E. (S. of ruins of St. Helen's Church) through well-defined marginal glacial channel to Old Cambus Quarry (disused) in Llandovery greywackes and shales. Excellent ripple-marks can be seen; monograptids may be found with difficulty from shales behind the old weigh-bridge.

Up path to N.E. and across field to top of cliff above Siccar Point where the unconformity of Upper Old Red Sandstone on Llandovery greywacke and siltstone was used by Hutton in his Theory of the Earth. The shore can be reached by descending a steep grass slope.

Return to Old Cambus Quarry and walk W.N.W. along road S. of St. Helen's Church to Pease Bay which is rimmed with cliffs of Upper Old Red Sandstone. Continue along road to Cove, then walk down narrow road to Cove Harbour past thick sandstone belonging to the Calciferous Sandstone Series. At the Harbour two thin limestones contain abundant but fragmentary marine fossils. These are underlain by the Heathery Heugh Sandstone dipping N.W. at 55°. A hollow in the shore marks the outcrop of shales and thin coals; the latter can be seen in the cliff. To the E. the succession continues downwards with another thick sandstone, a thin calcareous bed with the bivalve mollusc, Sanguinolites and cementstones and shales which have a conformable junction with the Upper Old Red Sandstone.

Return through Cove to the main road where an Edinburgh bus can be caught back to Dunbar.

E6.19   St. Abbs, Eyemouth and Berwick

Silurian. Lower and Upper Old Red Sandstone. Lower Carboniferous.

Access:- By private transport. Bus services are available to Eyemouth and Berwick, and train services to Berwick.

Walking distances:- Reston - St. Abbs - Eyemouth 8 miles (13 km). Shore N. of Berwick 4 miles (6.5 km). Using private transport 8 miles (13 km).

Maps:- Topo:- Geol:- S 26, S 34.

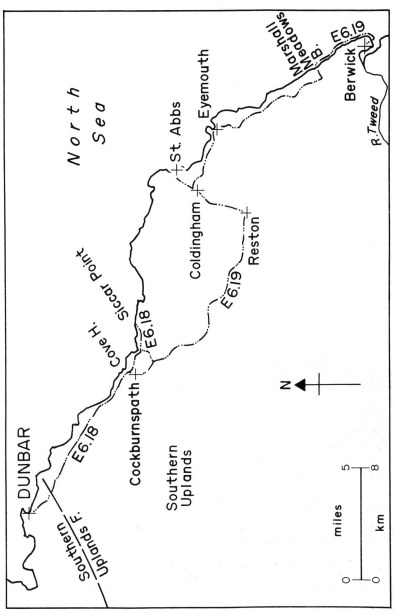

Fig. 6.12 Route map for Dunbar Centre.

Itinerary:- Turn off the main road (A1) at Reston or leave public bus at
same place. Follow road to Coldingham where it is probably best to leave
private transport. Continue to St. Abbs near which the coastal section
shows Lower Old Red Sandstone resting unconformably on Llandovery and
Upper Old Red Sandstone discordant on the Lower.

Along cliff path to Coldingham Bay above cliffs of Llandovery greywackes
and shales at the N.E. end of the Southern Belt of the Southern Uplands.
Walk inland on minor road towards Coldingham where private transport can be
rejoined to go on to Eyemouth. Those walking should, however, turn off at
the N.E. outskirts of Coldingham and follow a minor road near the coast to
Eyemouth.

The coast section to the N.W. of Eyemouth again shows Lower Old Red Sandstone
unconformable on Llandovery.

From Eyemouth private transport should be used to a point on the main road
2½ miles (4 km) N. of Berwick; transport should then be sent into the town.
Walkers may find a suitably timed bus to take them to the same spot. Walk
down a minor road to Marshall Meadows Bay. The tide should be low. The
shore section shows sandstones of the Scremerston Coal Group overlain by
a thin coal and the Dun Limestone at the base of the Lower Limestone Group.
South-east of the Needle's Eye the higher group forms the whole of the shore
section then around Sharper's Head three faults, downthrowing S.E. or S.
bring down the Middle Limestone Group. Between here and the mouth of the
Tweed these rocks, which include the Eelswell Limestone, are strongly folded
on N.E. axes. Sedimentary structures in the arenaceous rocks show that the
palaeocurrents were towards the S.W. or S. (Smith 1967).

Walk into Berwick and return to Dunbar by private transport, train or bus.

Dumfries Centre (Fig. 6.13)

Dumfries is on one of the main routes from Carlisle to Glasgow and is served
by good rail and public bus services. Accommodation can easily be found.

E6.20   Thornhill and Sanquhar Basins and Nith Valley

Ordovician and Lower Silurian. Coal Measures. Permian sediments and lavas.
Glacial features.

Access:- By bus to Carronbridge or by private transport.

Maps:- Topo:- 78 Geol:- S 9 and S 5.

Walking distances:- 11 miles (17.5 km) Carronbridge to Sanquhar, 6 miles
(9.5 km) using private transport.

Itinerary:- For 6 miles (9.6 km) out of Dumfries the Sanquhar road crosses
Permian sediments and for the next 4 miles (6.4 km) Silurian. Exposures
are very sparse owing to the cover of fluvioglacial sands and gravels,
often moundy and from their content derived from the melting of the Nith
glacier. The Thornhill Carboniferous/Permian Basin is entered 10 miles
(16 km) from Dumfries but the journey should be continued to Carronbridge.

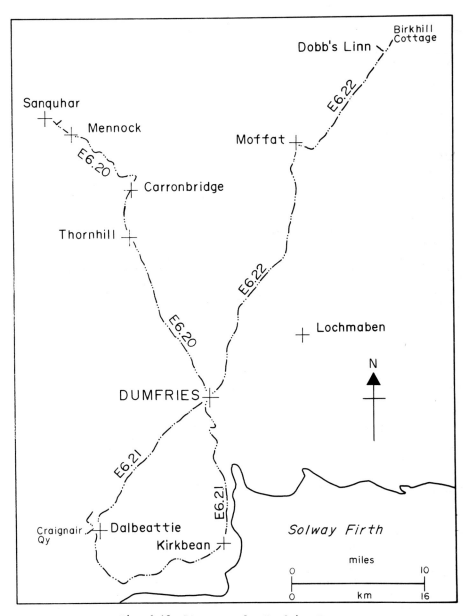

Fig. 6.13 Route map for Dumfries Centre.

Walk along the E. fork of the road for about a mile (16 km) and examine
exposures of Permian sandstones in the Carron Water upstream from the
bridge. Walk W. along minor road towards Sanquhar road. Immediately N.
of a loop northwards before the main road, there are exposures of Permian
basalt lava.

Continue on main road, on the E. side of which, ½-mile (0.8 km) N. of its
junction with the minor road, arenaceous sediments of Millstone Grit or
Coal Measures age outcrop, overlying Llandovery sediments. To the N. there
are two marginal glacial channels and W. of the road a lateral moraine.

From here to Mennock the road runs close to the R. Nith in which there are
exposures at intervals of greywackes and shales of Caradoc age, striking
N.E. Opposite Burnmouth these are cut by several N.E. basic dykes of
probable Lower Devonian age. Between the two furthest upstream of these
dykes an anticline brings up dark shales of Llandeilo age; graptolites are
present but difficult to find.

One mile (1.6 km) S.E. of the centre of Sanquhar a diversion should be made
up a small, S.S.W.-flowing stream to examine the Coal Measures, including
two thin coal seams separated by a fault. Two small outliers of Permian
basalt lavas flank the stream, 0.3 mile (0.5 km) above the road. At
Sanquhar a small agglomerate vent of Permian age cuts the Coal Measures E.
of the railway at the disused station.

Return to Dumfries by public bus or private transport.

E6.21   Dalbeattie (Criffel) Granite

Caledonian Granite and its metamorphic aureole in Silurian.

Access:- By public bus to Dalbeattie or by private transport.

Maps:- Topo:- 84 Geol:- S 5.

Walking distance:- 5 miles (8 km).

Itinerary:- From Dalbeattie walk on the Castle Douglas road to Craignair
Hill. Above the right bank of the Urr Water upstream from the road bridge
granite has been worked since 1824. The rock can be seen in outcrops and
in some of the abandoned workings but permission is needed to visit working
quarries. Petrographically the granite is a coarse quartz-diorite with
hornblende. Basic zenoliths occur, and aplites and pegmatites are present.

Return to crossroads 0.3 mile (0.5 km) N.E. of bridge then walk one mile
(1.6 km) along road running N. by W. From this point the irregular granite
margin can be followed E.N.E. up the hillside for ¼-mile (0.4 km), then the
contact aureole in Llandovery greywackes and shales studied by walking
N.N.W. for 0.6 mile (1 km) over a low hill; dykes of porphyrite occur.

Descend to road and walk back to Dalbeattie.

Those with private transport can return to Dumfries by Kirkbean. Along the
coast of the Solway Firth, N.E. of Craigmouth Point, there are exposures
of Wenlock strata in contact with the granite. To the N.E. Upper Old Red
Sandstone and Lower Carboniferous occur S.E. of the Silurian but exposures
are few owing to the cover of fluvioglacial sand and gravel.

E6.22   Moffat and Dobb's Linn

Ordovician and Silurian of Central Belt.   Glacial features.

<u>Access:-</u> By road via Lockerbie and Moffat; private transport virtually essential. The excursion can also be undertaken from Edinburgh, suggested routes being outwards via Broughton and returning via Selkirk.

<u>Maps:-</u> <u>Topo:-</u> 73, 78 <u>Geol:-</u> S 16.

<u>Walking distance:-</u> 3 miles (4.8 km).

<u>Itinerary:-</u> The route N.E. from Dumfries crosses a wide expanse of moundy fluvioglacial deposits masking Permian sediments. Beyond Parksgate the road traverses the N.W. end of the Lochmaben Permian Basin, and Moffat lies at the N. end of the Annan Basin.

At Moffat follow the Selkirk road which soon reaches Moffatdale, a straight, glaciated valley along a N.E. fault. The Grey Mare's Tail is a waterfall on a hanging tributary.

It is preferable to park at this stream, particularly with a coach, and to walk up the road to Dobb's Linn, really the name of a waterfall (hidden from the road) on the most north-westerly of the small streams. The strata form an anticline with a N. 5° E. axis which oscillates about the vertical so that the strata are inverted in opposite directions in different places.

The sequence is as follows:

SILURIAN

Gala Grits	10 ft. (3 m)
Upper Birkhill Shales - grey and black shales Monograptus convolutus, M. sedgwicki and M. maximus	77 ft. (23 m)
Lower Birkhill Shales - black shales M. triangulatus	60 ft. (18 m)

ORDOVICIAN

Upper Hartfell Shales - green, yellow and black mudstones D. complanatus and D. anceps	50 ft. (15 m)
Lower Hartfell Shales - black shales C. Wilsoni and D. clingani	45 ft. (14 m)
Glenkiln Shales - cherty black shales	15 ft. (4.6 m)

Owing to strike faulting near the axis of the anticline the Glenkiln Shales are seen only above the right bank of the southerly-flowing main stream 146 yds. (128 m) below where it is joined by the easterly-flowing Linn branch. The Hartfell Shales are well-exposed in the main cliff above the right bank of the main stream and the Birkhill Shales above both banks of the Linn branch downstream from the waterfall (Linn) over the Gala Grits. Owing to scree and cliff instability it is not possible to collect from a single section.

Radiometric determination from bentonite from the Lower Birkhill Shales
gave an age of 437 My (Ross and others 1978, 365).

Before returning to the transport it is worth walking ¼-mile up the road
to the watershed to view a plaque at Birkhill Cottage commemorating
Lapworth's work.

Ayr Centre (Fig. 7.21)

E6.23   Loch Doon

Ordovician of the Northern Belt; the Loch Doon Caledonian Pluton. Lower
Old Red Sandstone; glacial geology.

Access:- This excursion is best undertaken by private transport, preferably
cars. There is a bus service from Ayr to Dalmellington but this approach
involves a long walk, even if the southern end of the excursion is omitted
(see below).

Maps:- Topo:- 77 Geol:- S 8 and S 14.

Walking distances:- From Dalmellington to Craigmulloch (within N. end of
complex) and back 16 miles (25.5 km). From junction of Loch Doon and main
roads and back 19½ miles (31.2 km). From junction to Craigmulloch and back
13 miles (21 km). From Landoughty and back 9½ miles (15.2 km).

Itinerary:- Leave Dalmellington on the Carsphairn road above which, on the
S.E. outskirts of the town, a faulted dolerite sill of probable Permian
age penetrates Millstone Grit outcrops. The fault brings up Lower Old Red
Sandstone ½-mile (0.8 km) from Dalmellington. Sandstones, conglomerates,
basic lavas and a felsite sill can be seen N.E. of the road dipping steeply
N.W. on the S.E. margin of the Midland valley and disturbed by faulting.
The Southern Uplands Fault, hidden under boulder clay, occurs near the
junction with the minor road leading to Loch Doon. Coaches should be left
here but it should be possible to drive by car part of the way up the W.
side of the loch.

The Loch Doon complex (Gardiner and Reynolds 1932) consists of a core of
granite, a transition zone and an outer zone of tonalite (quartz-diorite).
Norite occurs on the N.W. margin. On its N.E. and S.W. sides (not seen on
the excursion) intrusion has strongly deflected the strike of the Ordovician
sediments.

The glaciated scenery is the most spectacular in the Southern Uplands. The
highest mountains are formed of the metasediments of the aureole.

On the way to the loch and along its W. side, Upper Ordovician sediments,
striking N.E., consist of the mainly greywacke facies characteristic of the
N.W. part of the Northern Belt.

Cars should be left at Landoughty and about 0.7 mile (1.1 km) further on a
traverse made a short distance above the track and the morainic drift to
study the 0.6 mile (1 km) wide aureole outside the margin of the quartz-
diorite.

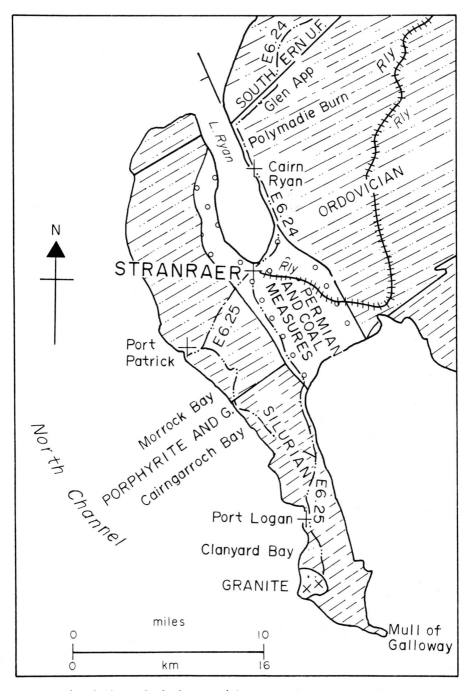

Fig. 6.14 Geological map, with routes, for Stranraer Centre.

The Carrick Lane (stream) and its tributary the Eglin Lane should be
followed S.W. Just above the confluence the Transition Zone between
quartz–diorite and granite is entered. Where the Black Garpel flows in
from the W. a traverse (compass needed in mist) should be made to the E.
to the Gala Lane to cross the N. end of the granite core. Descend the
Gala Lane to the head of Loch Doon, over moraine, and return to the track
on the W. side of the Loch.

Stranraer Centre (Fig. 6.14)

Stranraer (accommodation available) can be reached by rail or road from
Glasgow, Ayr and Girvan and is the port for the main ferry route to Northern
Ireland.

E6.24   East side of Loch Ryan and Glen App

Ordovician sediments; Southern Uplands Fault; raised beaches.

Access:- By walking to Glen App and returning by public bus or vice–versa.
By private transport. Low tide necessary. The excursion can also be undertaken when on transit from Ayr or Girvan to Stranraer.

Maps:- Topo:- 76, 82 Geol:- S 3, S 7.

Walking distance:- 9 miles (14½ km). With private transport 2 miles (3 km).

Itinerary:- By main road N. along E. side of Loch Ryan; raised beaches
occur at approximately the 25 ft. (7.6 m), 50 ft. (15 m) and 100 ft. (30.5 m)
levels. The highest beach is floored mainly with clays, containing Arctic
shells but exposures are not likely to be available unless excavation is
going on.

From ¾-mile (1.2 km) N. of the port of Cairn Ryan to Glen App there is a
long shore section, which should be followed on foot, in Caradoc shales and
greywackes of the Northern Belt dipping to S.E. at high angles. About 1¼
miles (2 km) N. of Cairn Ryan, 20 yds. (18.3 m) S. of the mouth of the
Polymadie Burn and 30 yds. (27.4 m) N. of the mouth, dark streaks in grey
shales intercalated with grits contain graptolites including Didymograptus,
Diplograptus, Climacgraptus, Dicranograptus and Cryptograptus indicating
the Glenkiln horizon (E6.22). A N.E. porphyritic dyke, probably L. Devonian,
occurs at the N. end of the section.

A large disused quarry on the E. side of the road about ½-mile (0.8 km) S.
of the sharp turn into Glen App exposes greywackes. Good examples of
sedimentary structures representing almost complete boumer cycles can be
studied as can typically Southern Upland fold structures.

Long, straight Glen App follows the N.E. striking Southern Uplands Fault.
The sediments to the N.W. are of the same age as most of those to the S.E.
but are coarser with large conglomerate lenses suggesting that this part of
the fault was an Ordovician submarine feature. Exposures can be studied in
small disused quarries along the roadside where the main road to Ballantrae
climbs northwards out of Glen App. These exposures are, however, further
from Stranraer than those already mentioned and require private transport.

Return to Stranraer.

E6.25   Portpatrick and W. coast of Mull of Galloway

Ordovician and Silurian, Lower Devonian intrusions; raised beaches.

Access:- By public bus to Portpatrick (and back) then walking. By private transport. Low tide necessary.

Maps:- Topo:- 82 Geol:- 51 and 53.

Walking distances:- 14 miles (22.5 km). With private transport 5 miles (8 km), including southwards extension (see below).

Itinerary:- Immediately S.E. of Portpatrick Harbour, and S.W. of the disused railway line, there are coastal exposures of Caradoc/Ashgill greywackes and shales, striking N.E. These are cut by Lower Devonian Lamprophyre dykes. Return to town then follow road, branching S.E. off Stranraer road at disused railway station, to near Morrock Bay. Walk down to Bay where, between tide-marks, there are good exposures of Arenig mudstones, tuffs and cherts followed by shales with well-preserved graptolites indicative of the Glenkiln/Hartfell horizon (E6.22).

The beds are repeated by isoclinal folds, with axes inclined S.S.E. at high angles and complicated by strike faults and N.E. porphyrite dykes.

Return to road and continue to Cairngarroch and walk down to bay of same name. On the S.E. side of the bay Llandovery fine greywackes and shales are exposed. To the N. the cliffs show an intrusion of porphyrite followed further N. by a small irregular mass of granite; a contact aureole occurs in the sediments: there are also numerous Lower Devonian porphyrite and lamprophyre dykes.

Walk back to Portpatrick to catch bus for Stranraer. However, those with private transport can drive further S. towards the Mull of Galloway. Raised beaches are well-developed N.E. of Port Logan. Further S. a short walk down a track leads to Clanyard Bay where two anticlines bring up black shale of the Upper Birkhill horizon with graptolites including Monograptus cometa and M. spinigerus. Further S. still, centring on Knockencille, a granite intrusion has produced a considerable aureole in the Silurian sediments.

6.4   The "Southern Uplands" in Ireland (Fig. 6.15)

Lower Palaeozoic strata continue from the Southern Uplands to North-East Ireland where they form a large, triangular outcrop S.W. of Belfast; they occur both in Ulster and in Eire. The region has lower relief than the Southern Uplands, and parts are heavily covered with glacial drift. Excursions, for those mainly interested in the solid geology, are rather unrewarding. However, one excursion, mainly to coast exposures, is suggested starting from Belfast, which is also a centre for visits to Tertiary Igneous Complexes (Ch. 9.2).

The Lower Palaeozoic rocks of the region range probably only from Caradoc to Wenlock. Outliers of younger formations up to Tertiary lavas occur. One large Caledonian granite (Newry) and three important Tertiary Complexes (Ch. 9.2) are present. Further mention of Silurian strata is made in the description of the Mourne Mountains (E9.14).

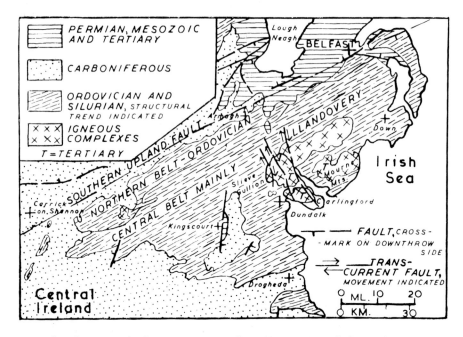

Fig. 6.15 Geological map of the "Southern Uplands" in Ireland.

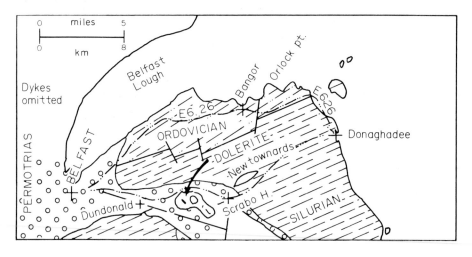

Fig. 6.16 Geological map, with route, for Belfast Centre.

Comparison with other areas:- The facies of the Lower Palaeozoic strata is very similar to that of the S.E. part of the Northern Belt of the Southern Uplands.

Belfast Centre (Fig. 6.16)

E6.26   East of Belfast

Ordovician. Silurian. Triassic. Tertiary intrusions. Glacial geology.

Access:- By private transport. By public bus to Donaghadee, thus missing out Scrabo Hill exposures (see below) and returning to Belfast by bus or train.

Maps:- Topo:- $\frac{1}{2}$-inch Ireland 5 Geol:- Northern Ireland 29 and 37.

Walking distances:- 9 miles (14.5 km). Using private transport 3 miles (5 km).

Itinerary:- From Belfast E. to Dundonald where moundy fluvioglacial sands and gravels are widespread, and where the second spillway of glacial Lake Belfast is seen, through which the water discharged into Strangfjord Lough.

Continue to Newtonards then S.W. for about $\frac{3}{4}$-mile (1.0 km), pass a dismantled railway bridge and park just to the S. Climb to a series of quarries on the E. face of Scrabo Hill. Follow these quarries S., examining Bunter sandstones (Triassic) with well-displayed sedimentary structures. Thin Tertiary basalt sills occur, and a thick dolerite sill, transgressing S., caps the hill. Three N.W. Tertiary dykes cut the sills.

Return to transport and follow the road to Donaghadee and Coalpit Bay. The shore section to the S. shows Lower Llandovery massive grits, dipping S. by E. at 75°, underlain by 10 ft. (3 m) of black shales, with Rastrites and several species of Monograptus, resting on 35 ft. (10.7 m) of unfossiliferous mudstones beneath which are more fossiliferous black shales.

Return to Donaghadee where the shore section S. of the Lifeboat Station shows massive grits, with graded bedding, folded on E.-W. axes. Two and a half miles (4 km) to the N.E., where the coast road turns inland, the Ordovician is brought up by an E. 25° N. fault occupied by a lamprophyre dyke.

A little further on leave the road and walk to the beach at the S. side of Orlock Point, and just S. of another lamprophyre dyke. Sandstones and shales contain flinty bands yielding species of Climacograptus, Dicellograptus and Dicranograptus. Return to road and continue to Bangor and Belfast.

6.5   South-East Ireland (Fig. 6.17)

South-East Ireland is a hilly to mountainous region consisting very
dominantly of Lower Palaeozoic rocks, invaded by the Caledonian Leinster
granite batholith, the largest in the British Isles, a mica-schist roof-
pendant of which forms the summit of the highest mountain, Lugnaquillia,
3039 ft. (926 m). The block is bounded to the E. and S. by the sea and to
the N. and W. mainly by Lower Carboniferous (Ch. 7). The block lies S. of
Dublin but the Howth and Portrane inliers N. of the capital are included in
the account. Good coastal and hill sections are available but the lower
ground is frequently covered with glacial deposits. A fair number of roads
traverse the region with public buses serving the towns. Rail services
roughly follow the coast via Bray and Arklow from Dublin to Rosslare and
Waterford; from Waterford there is also an inland route to Dublin. From
Britain there are boat services from Holyhead to Dun Laoghaire (for Dublin)
and from Fishguard to Rosslare and Waterford. There are air services from
several British cities to Dublin.

Four centres are suggested: Dublin, Bray, Arklow and Waterford.

The geological succession is as follows:-

> Carboniferous Limestone Series
>   (unconformity)
> Old Red Sandstone
>   (unconformity)
> Silurian
>   (unconformity)
> Ordovician
>   (unconformity)
> Cambrian
>   (unconformity)
> Precambrian

Precambrian structures, ranging back over 2000 My, occur near Rosslare.
Breaks in the Lower Palaeozoic succession are evidence of early Caledonian
disturbances but the main Caledonian movements were late Silurian. In a
broad sense, N.N.E. to N.E. anticlinoria were produced in the N. and S. with
a synclinorium in the Arklow district in the centre. In the N. there is
evidence that the Devil's Glen Series and the overlying Bray Series (shown
by microfossils to be L./M. Cambrian) of the Bray district (E6.29 and E6.30)
form a block thrust over younger Lower Palaeozoic.

Comparison with other areas:- The Lower Palaeozoic facies and succession,
which includes widespread volcanics, are broadly similar to those of Wales,
(Ch. 6.1). However the production of mica-schist in the aureole of the
Leinster Granite is not matched on any scale around other British Caledonian
granites. The Rosslare-Precambrian rocks are older than any of the Precambrian rocks of Wales and in fact have been compared with those of the
Laxfordian, (Ch. 4).

Dublin Centre (Fig. 7.23)

E6.27   Donabate and Portrane

Ordovician sediments and igneous rocks. Silurian and Old Red Sandstone
sediments.

Access:- By public bus to Donabate or by private transport.

Maps:- Topo:- ½-inch Ireland 13 Geol:- 1-inch Ireland 102.

Walking distance:- 5 miles (8 km). With private transport 1 mile (1.6 km).

Itinerary:- The route N. from Dublin crosses Carboniferous Limestone,
mostly thickly mantled in boulder clay. A large quarry (permission
required) on the W. side of the railway at Donabate is worked in porphyritic
andesite lava forming blocks enclosed in non-porphyritic andesite. This may
be the site of an Ordovician volcanic neck.

Continue to Portrane. At the N. end of the shore section W. of a point
conglomerates of Old Red Sandstone type unconformably overlie Ordovician
basic lavas to E. At the point these are overlain by agglomerates with
black shales in which graptolites have been found. To the S. these are
overlain by interbedded coral limestones and black shales of Ashgill age.
The section also includes conglomerates considered to be of slump origin.
Grits, followed further S. by limestones, are considered to be Silurian;
it is doubtful whether the junction with the Ordovician is an unconformity
or a thrust. Further S., across a sandy bay, Ordovician basic lavas again
appear. The Lower Palaeozoic rocks lie on the Lambay-Portrane Caledonoid
Axis.

Return to Dublin.

E6.28   Howth Peninsula

Cambrian turbidites with large scale slumping.

Access:- By rail or bus to Howth.

Maps:- Topo:- ½-inch Ireland 16 Geol:- 1-inch Ireland 112.

Walking distance:- 5 miles (8 km).

Itinerary:- From Howth E. by coast road then cliff path to Nose of Howth
examining exposures of quartzites, quartzitic siltstones and cleaved shales
of the Nose of Howth Group of the Bray Series. These are turbidites in
which graded bedding indicates that the S.S.E. dip is normal. Chaotic
sedimentary breccias with very large blocks are due to slumping generated
from the S.W. (van Lunsen and Max 1975). The same formation continues along
coast S. from Nose of Howth for about a mile (1.6 km). Near the S. end of
this section the steep S.S.E. dip can be shown to be inverted. Further S.
the Censure Group occurs consisting of quartzites, siltstones and mudstones.
Large blocks of quartzite have become detached; the bases of the blocks have
slump features, while their tops generally do not - a contrast with the
breccias of the Nose of Howth Group.

Climb W. from coast to road which leads N. to Howth. A short diversion
should be made to W. to the top of Ben of Howth, 563 ft. (171 m), a good
viewpoint along the coast. Quartzite of the Nose of Howth Group is again
seen.

The broad structure of Howth is a syncline with sinuous E.-W. axis.

Bray Centre

The coastal town of Bray provides plentiful accommodation. However, the
two excursions described can also be undertaken from Dublin, connected with
Bray by road and by frequent trains.

E6.29   Killiney and Bray Head

Northern end of Leinster Granite and its aureole. Bray Group (Lower/Middle
Cambrian)

Access:- By train from Bray to Killiney, then walking.

Maps:- Topo:- ½-inch Ireland 16 Geol:- 1-inch Ireland 112, 121.

Walking distance:- 8 miles (13 km).

Itinerary:- Walk a short distance N. from Killiney station, then to top of
Killiney Hill, 480 ft. (146 m) which is in a public park. The granite here
is slightly gneissose and contains numerous, generally sharply-defined,
aplites and pegmatites with spodumene and beryl. Walk down to Killiney
Bay at the extreme north end of which the granite is seen in contact with
mica-schist which outcrops for some distance S. along the coast and in
places contains conspicuous andalusite; garnet and staurolite also occur.
The schists have been formed from Lower Palaeozoic sediments by the effects
of thermal metamorphism and movement during the cooling of the granite.

Southwards by coast road through Bray and along cliff path to Bray Head.
Along the path and on the shore are continuous exposures of the Bray Group,
consisting of turbidites-graded greywackes, siltstones, shales and
quartzites. Many beds show slump features. The bedding dip is steeply N.W.
The graded bedding and the relationship of the steep cleavage show that the
strata become younger in this direction, although there is some minor
folding.

Return to Bray.

E6.30   Great Sugar Loaf

Bray Group (Lower/Middle Cambrian). Glacial overflow channel.

Access:- By walking from Bray.

Maps:- Topo:- ½-inch Ireland 16 Geol:- 1-inch Ireland 121.

Walking distances:- 10½ miles (16.8 km). This can be reduced by using
private transport for part of the excursion.

Itinerary:- Walk W.S.W. from Bray to S. outskirts of Enniskenny then
continue to Killough; the solid rocks are almost completely covered by
boulder clay. From Killough a path rises S.W. over moorland to a broad
ridge which leads S. to the summit of the Great Sugar Loaf, 1654 ft.
(504 m). There are abundant exposures of quartzites of the Bray Group
dipping E. by N. on the W. limb of the syncline which is the main structure
of the area (Bruck and Reeve 1976). The summit is an excellent viewpoint
towards the Leinster Granite mountains to the S. and the coast to the E.

Descend E. to the road S. of Kilmacanooge, noting that the strike of the
quartzites swings round the syncline to the E.N.E. To the S.E. the Glen
of the Downs is conspicuous; this is a glacial overflow channel.

Return to Bray.

Arklow Centre

The coastal town of Arklow is a good centre for accommodation and can be
easily reached by road or rail either from Dublin or from Rosslare, the
port for the crossing from Fishguard.

E6.31   Lugnaquillia

Ordovician volcanics and intrusives. Leinster Granite and its aureole.
Glacial features.

Access:- By private transport via Rathdrum to Drumgoff Bridge or Glenmalure.
By train to Rathdrum, thence by hired car to Drumgoff Bridge. From latter
by road, path and hill walking.

Maps:- Topo:- ½-inch Ireland 16   Geol:- 1-inch Ireland 129, 130.

Walking distance:- 11 miles (17½ km).

Itinerary:- About half-way from Arklow to Rathdrum the route passes close
to Avoca where sulphide ores, including pyrites, chalcopyrites, galena and
zincblende in the Ordovician slates, are worked on a large scale. The ores
are considered to be of Caledonian date. On the approach to Rathdrum outcrops occur of acid pyroclastics and intrusive metadolerites, of probable
Caradoc age, striking N.E.

Transport should be left at Drumgoff Bridge. A well-marked terminal
moraine is seen here; Glenmalure is strongly glaciated. Walk up the rough
road in this glen for about 1½ miles (2.4 km) to a conspicuous tributary
on the S.W. side. Follow a zig-zag track then a faint path to head of
tributary. Exposures of muscovite-biotite-granite occur on the S. side
of the stream. From the head of the stream there is a short climb to a
grassy ridge which leads W. and then for a short distance S. to the top of
Lugnaquillia, 3039 ft. (926 m). Mica-schist occurs around the top, forming
a roof-pendant.

A deep corrie cuts into the E. side of the mountain. In clear weather it
is possible to look across the Irish Sea to the mountains of North Wales.

Return by the same route.

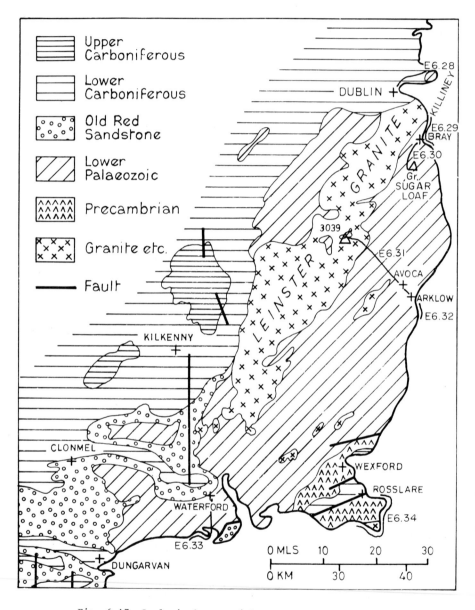

Fig. 6.17 Geological map, with routes, of S.E. Ireland.

E6.32   Coast S. of Arklow

Ordovician volcanics, intrusives and sediments.

Access:- By walking along coast.

Maps:- Topo:- ½-inch Ireland 19   Geol:- 1-inch Ireland 139.

Walking distance:- 5 miles (8 km).

Itinerary:- Walk S. along the coast-line which for the first mile (1.6 km)
is low and backed by extensive blown sand. The ground then rises to Arklow
Head, and there are exposures of acid volcanics (mainly pyroclastics) and
metadolerite intrusives of probable Caradoc age.

The next headland to the S. is formed of cleaved siltstones; the steep
bedding and the cleavage strike N.E. Fossils are rare.

Return by the same route.

Waterford Centre

Waterford, with plentiful accommodation, can be reached by road or rail
from Dublin or from Rosslare Harbour, and is itself the terminal of a boat
service from Fishguard.

E6.33   Waterford and Tramore

Ordovician sediments and volcanics. Unconformity of Upper Old Red Sandstone
on Ordovician.

Access:- By private transport or by public bus to Tramore, thence by
coastal walking. For the shore-sections low tide is needed.

Maps:- Topo:- ½-inch Ireland 23   Geol:- 1-inch Ireland 168, 179.

Walking distance:- 8 miles (13 km).

Itinerary:- Before leaving Waterford a spectacular unconformity of Upper
Old Red conglomerate on steeply-dipping Ordovician acid lavas and cleaved
ashes should be seen. The unconformity is displayed in a quarry on the
S.W. side of the R. Suir on the W. outskirts of the town, on the opposite
bank and on rising ground W. of the railway station.

From Waterford travel to Tramore, and then follow the coast section S.S.W.
Immediately beneath the town the dark Tramore Shales (Arenig) outcrop on
the shore. At Doneraile Cove, which can be reached by a footpath on the
outskirts of the town, the 50 ft. (15 m) thick Tramore Limestone (Caradoc)
is seen, overlain by acid volcanic with graptolitic shales of Nemograptus
gracilig Zone and Climacograptus pettifer Zone forming part of the Lower
Tramore Volcanic Formation. Dolerite sills also occur.

The coastal cliffs are diversified by stacks and arches. At Newton Cove
and in a small roadside section immediately above, dark, calcareous shales
and impure nodular limestones are exposed. The purer beds of limestone
abound in the stromatoporoid Monticulipora petropolitana, and the shale
beds are full of rather fragmentary brachiopods, trilobites and bryozoans.

Further W. at the feature known as the Metal Man and at Garrarus Strand, the Upper Tramore Volcanic Formation is seen including acid pyroclastics (in places extremely coarse and agglomeratic), flow-banded felsites and black shales.

A return should be made by a road somewhat set back from the coast which crosses a platform at about 250 ft. (76 m) O.D.

E6.34   Rosslare, Carnsore Point and Kilmore Quay

Polycyclic Rosslare Complex (Precambrian - at least 1600-1700 m.y. old; possibly Archean). Ordovician.

Access:- For the whole excursion private transport is needed but it is possible to travel by train to Rosslare Harbour and to examine the shore section S.E. to St. Helens by walking. Only small parts of the shore section can be seen at high tide.

Maps:- Topo:- ½-inch Ireland 23 Geol:- 1-inch Ireland 170, 180, 181.

Walking distances:- From Rosslare Harbour to St. Helens and back 5 miles (8 km). At Carnsore Point and Kilmore Quay an additional 3 miles (4.8 km).

Itinerary:- At Rosslare Harbour (where transport should be left) walk S.E. along the shore. The section is discontinuous, and there is a thick boulder clay cover. Exposures are mainly of amphibolitic gneisses which are generally more deformed than at Kilmore Quay. Alternations of deformed gneiss and derived chloritic mylonites occur on all scales. Of particular interest are two downfaulted segments of unmetamorphosed sandstones and conglomerates similar to those associated with fossiliferous Lower Ordovician near Tagoat. They occur approximately 700 and 1100 m. S.E. of the Harbour. The unmetamorphosed state of these rocks indicates the pre-Ordovician age of the Rosslare Complex metamorphism (Baker 1955, 65). (It is possible to continue along the shore beyond Greenore Point to St. Helen's where the Complex is again exposed - generally much deformed - in a 1 mile (1.6 km) long section.)

Walk back to Rosslare Harbour.

Those with vehicles should then travel W. then S.E. to Netherton. The Carnsore Granite is well-exposed for about ½-mile (0.8 km) southwards to Carnsore Point. This has large felspar megacrysts and abundant xenoliths. The latter vary from almost unmodified insular relics of the Rosslare Complex and younger dykes to felspathised and recrystallised discoidal types with strong parallel orientation. Narrow aplite veins occur irregularly.

Return to vehicle at Netherton and travel W. to Kilmore Quay. Walk W. from harbour.

The westernmost exposures on the west-facing shore N. of Forlorn (Cross-farnoge) Point consist principally of foliated quartz-oligoclase-biotite gneisses (with microscopic garnet) of probable meta-sedimentary origin. The rather fine-scale foliation results from an early deformation, and occasionally shows small scale folds. Subordinate bands of darker amphibolite and of pegmatite are intercalated, together with crossing-

cutting basic dykes. The latter are developed best near Forlorn Point where
they intrude a complex of granitic gneisses. Metamorphism and deformation
of these dykes is indicated by their amphibolitic composition and their
schistosity (marginal only in some), but they are clearly younger than the
main foliation of the gneisses.

Return to the harbour, E. of which amphibolitic gneisses with less well-
defined foliation occur on the lower part of the shore. Closer to the low
water-mark, a foliated granite with some xenoliths has been interpreted as
the northern margin of the Saltees granite.

Return to Waterford.

References

Baker, J. W. 1955. Pre-Cambrian Rocks in Co. Wexford. G.M., 92, 63-68.
Baker, J. W. and Hughes, C. P. 1979. Summer (1973) field meeting in
    South Central Wales. P.G.A., 90, 65-79.
Brindley, J. C. and Gill, W. D. 1958. Summer Field Meeting in Southern
    Ireland. 1957. P.G.A., 69, 244-261.
Brüch, P. M. and Reeves, T. J. 1976. Stratigraphy, Sedimentology and
    Structure of the Bray Group in County Wicklow and South County Dublin.
    P.R.I.A., 76B, 53-77.
Charlesworth, J. K. and Preston, J. 1960. Geology around the University
    Towns: North-East Ireland - The Belfast Area. Exc. Guide No. 18.
    Geol. Assoc.
Dean, W. T. 1968. Geological Itineraries in South Shropshire. Guide No.
    27 (rev. ed.). Geol. Assoc.
Eastwood, T. 1953. British Regional Geology: Northern England (3rd ed.)
    M.G.S.
Gardiner, C. J. and Reynolds, S. H. 1932. The Loch Doon "Granite" Area,
    Galloway. Q.J.G.S., 88, 1-34.
George,T. N. 1961. British Regional Geology: North Wales. (3rd ed.)
    M.G.S.
George,T. N. 1970. British Regional Geology: South Wales. (3rd ed.)
    M.G.S.
Jones, O. T. and Pugh, W. J. 1949. An early Ordovician shore-line in
    Radnorshire near Builth Wells. Q.J.G.S., No.5, 65-99.
Lunsen, H. A. van and Max, M. V. 1975. The geology of Howth and Ireland's
    Eye, Co. Dublin. Geol. J., 10, 35-58.
Mitchell, G. H. 1970. The Lake District. Guide No. 2. Geol. Assoc.
Natley, C. E. and Wilson, T. S. 1946. The Harlech Dome, North of the
    Barmouth Estuary. Q.J.G.S., 102, 1-40.
Pocock, R. W. and Whitehead,T. H. 1971. British Regional Geology: The
    Welsh Borderland. 3rd. ed. revised by Earp, J. R. and Harris, B. A.
    M.G.S.
Pringle, J. 1971. British Regional Geology: The South of Ireland.
    3rd ed. revised by Greig, D. C. and others. M.G.S.
Roberts, B. 1979. The Geology of Snowdonia and Llyn. Bristol.
Ross, R. J. and others. 1978. Fission-track dating of Lower Palaeozoic
    Volcanic Ashes of British Stratotypes. U.S. Geol. Surv. Open File
    Rep. 78-701, ed. R. E. Zartman.
Smith, T. E. 1967. A preliminary study of sandstone sedimentation in the
    Lower Carboniferous of the Tweed Basin. Scot. Journ. Geol., 3, 282-305.
Ramsay, D. and Ramsay, J. G. 1959. Geology of some Classic British Areas:
    Snowdonia. Guide No. 28. Geol. Assoc.

CHAPTER 7

Hercynian Terrains

7.1   South-West England (Fig. 7.1)

This is the whole of England W. of a line from Gloucester to Lyme Bay,
roughly coinciding with the base of the main outcrop of the Jurassic.
There are, however, considerable outcrops of Mesozoic and Permian rocks
further W., partly covering the Upper Palaeozoic strata which dominate the
region.

The succession is:

> ? Pliocene
>
> Oligocene
>
> Cretaceous
>
> Jurassic
>
> Triassic
>
> Permian
>
> Upper Carboniferous
>
> Lower Carboniferous
>
> Upper Devonian
>
> Lower Devonian
>
> Silurian
>
> Ordovician
>
> ? Cambrian
>
> Precambrian

The region contains six large granite outcrops, upshoots from a huge W.S.W.
orientated Armorican batholith. There are associated acid dykes and sheets.
Basic Devonian intrusions (greenstones) are widely distributed.

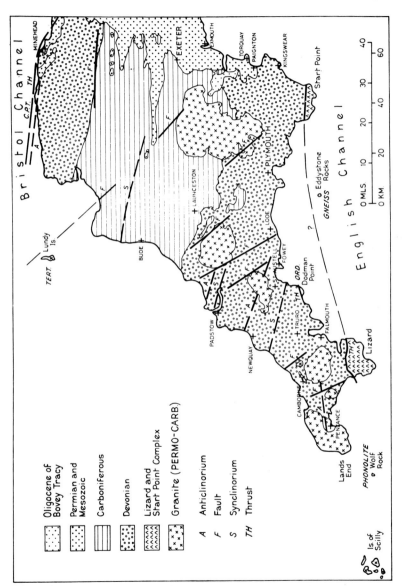

Fig. 7.1 Geological map of South-West England.
C.P.T. Cannington Park Thrust.

South-West England is on the Armorican Arc which continues into Southern
Ireland (Ch. 7.7) and most of the region is S. of the Hercynian Front.
Consequently all pre-Permian rocks show strong folding, thrusting and often
cleavage. Broadly, there is a Central Synclinorium of Carboniferous flanked
by Anticlinoria of Devonian. Even in the part of the region N. of the Front,
which includes the Bristol Coalfield, there is complex folding and some
thrusting.

Metal mining, particularly for tin, goes back to Prehistoric times. There
are two mines still in production, and others may be re-opened. There are
numerous dumps at abandoned mines on which minerals can still be found, but
after generations of collecting the chances of obtaining good specimens are
slight.

Fine cliffs diversify the coast, and the granite with its tor weathering
forms pictuesque moorlands.

Centres suggested are: Penzance, St. Agnes, St. Austell, Bude, Paignton
and Bristol. All have plentiful accommodation but all, except Bristol, are
in the major holiday counties of Devon and Cornwall, and bookings should be
made well ahead for parties.

Comparison with other areas:- The region is the type-locality for the
Devonian, mostly of marine facies which contrasts strongly with that of the
Devonian (Old Red Sandstone) further N. (Chs. 5.3, 5.4, 7.2, 7.3, 7.5).
The shale/sandstone Culm facies of the Carboniferous is also very different
from that of the Carboniferous further N. (see same Chs.). South-West
England and Southern Ireland (Ch. 7.7) are the only parts of the British
Isles S. of the Hercynian Front, with structural effects mentioned above.

The large Hercynian granites are unique in the British Isles as are the large
tin deposits.

Permian and Mesozoic rocks are similar to those of much of the rest of the
country.

Penzance Centre (Fig. 7.2)

Penzance is the terminus of a main railway line from London, and there are
local train and bus services.

E7.1   Land's End

Hercynian granite and its margin.

Access:- By public bus to Land's End and back from St. Just or by private
transport. Low tide required for Cape Cornwall.

Maps:- Topo:- 203. Geol:- E W 358.

Walking distances:- 11.5 miles (18.4 km). With private transport 3.5 miles
(5.6 km).

Itinerary:- At Land's End the influence of jointing on the cliff scenery is
strikingly displayed. The granite here is coarse, with large orthoclase
phenocrysts, but a finer variety is also present e.g. in the cliffs next to

the 'First and Last House in England'.

A walk along the path to Sennen Cove gives further fine views of the granite
cliff scenery; numerous quartz-tourmaline veins are seen.

From Sennen Cove to main road then N. to road junction on E. side of St.
Just. About a mile (1.6 km) along the road to the E. a chimney marks the
disused Balleswidden tin mine (tin is still worked at the Geevor Mine on
the coast to the N.E.). There are also china-clay pits (New Consolidated
Mines of Cornwall) along a N.W. belt in which the granite is thoroughly
kaolinised.

Through St. Just and along road W. to a point overlooking Priest's Cove on
the S. side of Cape Cornwall. In the cove there are spotted slates and
hornfelses. Walk N. to the beach at Porth Sedden, on the N. side of Cape
Cornwall. At low tide the contact between granite and pelitic hornfelses
can be seen. Near the N. and S. ends of the beach the granite at the contact is altered to a dark bluish quartz rock.

Return to St. Just and Penzance.

E7.2   Newlyn and Mousehole

Hercynian granite and its margin. Metadolerite. Low tide required.

Access:- Road and shore walking.

Maps:- Topo:- 203. Geol:- E.W. 358.

Walking distance:- 7.5 miles (12 km).

Itinerary:- South through Newlyn to Mousehole. Along rough shore to the
S. where an irregular contact is exposed between the granite and veined
hornfelses.

Return to Mousehole and walk up road to Paul. Beyond this village branch
off on road to W. which passes top of the very large Penlee Quarry worked
in metadolerite.

This high point gives a fine view of the Cornish coast to the E. including
the platform of the Lizard Peninsula (E7.3) at about 245 ft. (75 m).

Descend to Newlyn and return to Penzance.

E7.3   The Lizard

Igneous and metamorphic rocks of Lizard Complex. Devonian.

Access:- By private transport. Low tide required.

Maps:- Topo:- 203, 204. Geol:- E.W. 359.

Walking distance:- 7 miles (11.2 km)..

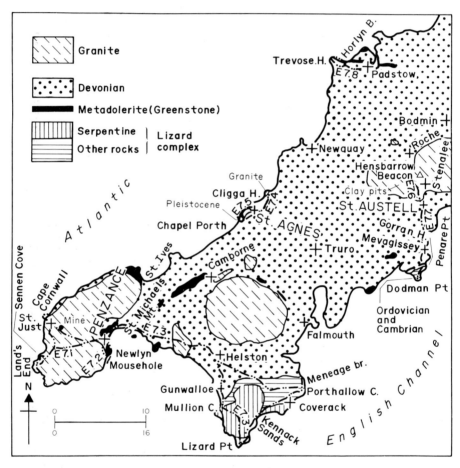

Fig. 7.2 Geological map, with routes, of Penzance, St. Agnes and
St. Austell centres.

Itinerary:- A coastal tour from W. to E. is outlined.

Branch off from the road S. of Helston to Gunwalloe Church and walk N. along
path to Jangye Ryen. This rocky cove shows alternating slate and greywacke
belonging to the Gramscatho Beds.

On to Mullion Cove then walk N. along path to Polurian Cove. On the N.
side the Gramscatho Beds occur. These are separated by fault-breccia from
felspathic hornblende-schist of the Lizard Complex. This reversed fault
contrasts with the Meneage Breccia on the N. side of the Complex further E.
(see below).

Return along path to Mullion Cove, noting bodies of fine-grained serpentinite
in the hornblende-schist. S. of Mullion Cove the latter is faulted against
the main mass of the Lizard serpentinite.

Continue S. to Kynance Cove where the two varieties of the serpentinite-
tremolite-serpentinite and bastite can be seen, separated by a fault.
Both varieties cut by several bodies of granite-gneiss and epidiorite.

Through Lizard village past the Lizard and Caerthilian Hotels to the end of
the road then down a path to the shore at Pistil Ogo.

Further E. the shore can be reached by taking a path which leads from the
"most southerly point of England" (signposted) to the old lifeboat station
at Polpeor Cove. At both localities muscovite, chlorite, epidote and
hornblende schists, together with impure quartzites, belonging to the Old
Lizard Head Series, are seen. At the extremity of Lizard Point the mica-
schists are cut by the Man of War gneiss.

E. by road to Landswednednack Church then walk down to Church Cove in which
are exposed hornblende-schists which differ from those at Mullion Cove in
the presence of greenish-yellow epidote.

N.E. to Kennack Sands where banded gneiss shows alternations of a dark basic
component and a light acid component. An exposure on the W. side of Kennack
should be studied. Some observers consider this shows an intrusive sequence
of serpentinite/gabbro dyke/basalt dyke/gneiss but there are difficulties
about this interpretation. Talc and asbestos occur at this locality.

Continue N.E. by road to Coverack. Bastite-serpentine is cut by a large
number of dykes and other minor intrusions, all basic.

Still further N.E. massive gabbro is worked at Dean Quarry (Cawoods Road
Materials Ltd.). Veins of calcite, analcite, natrolite and green botryoidal
prehnite occur.

North to Porthallow Cove, on the N. side of which there is a $1\frac{1}{4}$ mile (2 km)
section northwards through the Meneage Breccia, here forming the north
margin of the Lizard Complex. This puzzling formation consists predominantly
of a dark, well-cleaved breccia containing fragments of slates and many
other rock types, some of very large size. The breccia has been variously
interpreted as tectonic, sedimentary or pyroclastic.

Return to Penzance.

St. Agnes Centre

St. Agnes can be reached by public bus service from Truro. There is a fair
amount of accommodation.

E7.4   Cligga Head

Access:- Road and cliff top walking. Road distance can be reduced by using
private transport.

Maps:- Topo:- 200, 203, 204. Geol:- E.W. 346.

Walking distance:- 6 miles (9.6 km).

Itinerary:- Walk E. then N. from St. Agnes to about 1 mile (1.6 km) S.W.
of Perranporth then along a track to E. (N. of aerodrome) to buildings near
the "Contact Shaft" of the Cligga Head Mine. The dumps contain both
granite and hornfelsed slate but the only minerals readily found are
copper pyrites and tourmaline.

Walk N.W. to a small quarry in the Cligga Head Granite, which has spectacular
greisen (quartz-mica veins) dipping steeply N.

The granite is also seen at Cligga Head itself and along the cliffs for 0.3
mile (0.5 km) to the S. At Hanover Cove, however, where the cliff line
bends S.W., hornfelsed Lower Devonian slates are seen, with quartz porphyry
intrusions.

Return by the same route.

E7.5   St. Agnes Beacon (Fig. 7.2)

Lower Devonian. Hercynian granite. Ore bodies. Marine Pleistocene.

Access:- Road, path and coast walking. The walking distance may be
shortened by using private transport but it should be noted that coaches
are not allowed more than $1\frac{1}{4}$ miles (2 km) along the Chapel Porth road.
Low tide required.

Maps:- Topo:- 203, 204. Geol:- EW 346.

Walking distance:- 6 miles (9.6 km).

Itinerary:- From the S. outskirts of St. Agnes follow the road to Chapel
Porth. Walk N. along the beach; the cliffs consist of hornfelsed Lower
Devonian slates with psammitic beds. About 650 yds. (600 m) N. of Chapel
Porth a disused engine house at the top of the cliff marks the Towanreath
Lode, which produced tin with a small amount of copper. A steeply-dipping
quartz-porphyry dyke about 7 ft. (2 m) thick is associated with the lode
which is marked by haematite reddening.

Return to Chapel Porth and back along the St. Agnes road then turn to the
N. on Beacon Drive. Near where the drive turns sharply N.N.E., exposures
of granite may be seen. The total outcrop is small, but the granite, to
judge from the width of the aureole, must be much more extensive at a
shallow depth.

About 880 yds. (800 m) along the drive from the St. Agnes road a path to
the W. leads to the restored engine houses of Wheal Coates. From the
remains of a mine 275 yds. (250 m) E.N.E. of the upper engine house an
opencast working extends along a lode which cuts tourmalinised hornfels.

Continue along Beacon Drive for 820 yds. (750 m) to N.W. then follow
metalled road for 340 yds. (300 m) before branching off on a path to N.E.
This leads to shallow pits in the St. Agnes Beds, unfossiliferous sands
and clays at about the 400 ft. (122 m) contour, believed to be marine
Pleistocene.

Return to St. Agnes by the N. section of Beacon Drive, passing the extensive
dumps of the Polperro tin and copper mine.

St. Austell Centre (Fig. 7.2)

St. Austell is on the main railway line from London to Penzance, and there
are a number of bus services. There is a fair amount of accommodation.

E7.6   North of St. Austell

Hercynian granite and its aureole. China-clay workings.

Access:- By public bus or private transport to Roche. Walk back to St.
Austell on roads and tracks.

Maps:- Topo:- 200. Geol:- EW 347

Walking distance:- 7.5 miles (12.0 km).

Itinerary:- At the S. side of Roche the rock rises as a high tor of a
satellite of the St. Austell Granite. Scattered exposures of hornfelsed
Lower Devonian occur in the vicinity.

Walk S. from Roche for one mile (1.6 km) then follow W. fork of road. Near
the summit of the road branch off along track to Hensbarrow Beacon, 1026 ft.
(313 m) on the St. Austell Granite. Return to road and walk S., downhill,
for 600 yds. (549 m) before turning S.E. on a rough track which further on
becomes a road joining another road about half-a-mile (0.8 km) S.W. of
Stenalee, passing on the way, the Gunheath China-clay Quarry. Permission
must be sought to enter this and other working china-clay pits but a good
view can be obtained from the road. At the road junction S.W. of Stenalee
a disused pit can be seen.

About 650 yds. (594 m) S. from the junction turn E. along a track which
merges into a minor road passing a disused pit flooded with bright-green
water. The granite on the dumps contains abundant schorl. Further E.
this road joins another road ¼-mile (0.4 km) N. of Carclaze. China-clay
workings with very large dumps can be seen.

Southwards along the road to St. Austell there are exposures of hornfelsed
Lower Devonian on the S. margin of the St. Austell Granite.

E7.7   Mevagissey and Dodman Point Area

? Cambrian. Ordovician. Lower Devonian. Metadolerite.

Access:- By public bus to Mevagissey then road and coastal walking. By
private transport to Mevagissey and on (car only) to Gorran Haven then
mainly coastal walking.

Maps:- Topo:- 204. Geol:- EW 353.

Walking distances:- 10 miles (16 km). With private car 3.5 miles (5.6 km).

Itinerary:- From Mevagissey Harbour walk N. along the cliff path to Penare
Point. The cliffs are of Lower Devonian slates and grits folded on E.N.E.
axes. Return to Mevagissey then follow minor road through St. Gorran Haven.
Here quartzite which has yielded Orthis, Calymene, Phacops and other fossils
regarded as Llanvirn/Llandeilo, outcrops on the shore. The quartzite is
followed to the N. by slates and radiolarian chert (Veryan "Series") with
a thick sheet of metadolerite.

Follow a track S. then W. to the cliffs on the S. side of Maenease Point.
These have been eroded in the Dodman "Series", consisting of silty phylites
tentatively assigned to the Cambrian.

Return to St. Austell.

E7.8   Padstow, Harlyn Bay and Trevose

Upper Devonian. Metadolerite.

Access:- By private transport to Harlyn Bay, W. of Padstow. By train to
Bodmin Road, thence by connecting bus to Padstow. Those not wishing to
travel from St. Austell and back in a day can readily find accommodation
in Padstow. Fairly low tide required for Harlyn Bay section.

Maps:- Topo:- 200. Geol:- EW 335.

Walking distances:- From Padstow and back 9.5 miles (15.2 km). From
Harlyn Bay and back 4.5 miles (7.2 km).

Itinerary:- If starting from Padstow, walk W. to E. end of Harlyn Bay.
Private transport can be parked nearby.

At the E. end of the Bay, Upper Devonian slates outcrop. These lie near the
N. margin of the Southern Anticlinorium. Cleavage is the dominant discon-
tinuity but bedding traces can be seen. There has been at least one fold-
phase following that which produced the cleavage, for the latter is folded
into a syncline pitching roughly W.

Along minor road to Trevose Head, crossing large deposit of blown sand.
Beside the lighthouse a metadolerite sill in sandstone dips S.; quartz
veins are frequent. The bedding, and cleavage in nearby slates, are
parallel, and both are folded into an E.W. syncline. In a disused quarry
on the W. side of the headland, a metadolerite sill, displaced by a fault,
is intruded into quartzose beds. Some of the slates in the Trevose Head are
calcareous.

Walk along cliff path E. from Trevose Head, noting metadolerite sills.
Soon after the path turns S.E. fine stacks are seen in sandstone. Near
Cataclews Point there is a disused quarry in metadolerite; a roof of altered
sandstone is seen, dipping N.W. at 30°.

Continue along coast to Harlyn Bay.

Bude Centre (Fig. 7.3)

Bude has plentiful accommodation, and there is a public bus service which
connects with trains at Exeter St. Davids station. Some excursions are
described by Dearman and others (1970).

E7.9   North of Bude

Upper Carboniferous.

Access:- Shore and cliff top walking. Low tide essential for shore
walking.

Maps:- Topo:- 190. Geol:- EW 307, 308.

Walking distance:- 4 miles (6.4 km).

Itinerary:- From the Summerleaze Beach Car Park walk along the shore to
the swimming-pool, then follow the platform in front of the retaining wall
and go round a rocky ridge to examine the "Bude Fish Bed", 14 ft. (4 m) of
black shales; about 4 ft. (1 m) above its base clay/siderite nodules
contain palaeoniscid fish and an acanthodian. Between this locality and
the N. end of Maer Low Cliff, the succession is repeated many times by
folds plunging W. The folds on the foreshore are cut by a dextral wrench
fault marked by a gully. Sandstones in this area show well-displayed
sedimentary structures.

Given a low and still ebbing tide round the small headland into Maer High
Cliff Beach, where gently westward-plunging folds in thick sandstones are
well seen. If the tide is still low, pass two sandstone buttresses and
continue to a rock-slide known as The Earthquake. On the N. side of the
buttress facing the slide (best exposed at the cliff-face) sedimentary
structures are well seen. A 3 ft. (1 m) layer of sandstone between beds
of shale of about the same thickness is of particular interest. From the
top of the sandstone, numerous sinuous to wedge-shaped "dykes" and "sills"
of sandstone intrude the shale. During folding the dykes were reorientated
sub-parallel to the cleavage. At the rock-slide large blocks of sandstone
have slipped down the bedding of the syncline.

An easy path leads to the cliff top. From the path back to Bude the folds
are spectacularly displayed in plan on the foreshore.

E7.10   Okehampton and N. edge of Dartmoor

Carboniferous (Dinantian and Namurian) Dartmoor Granite and its aureole.
Tor topography.

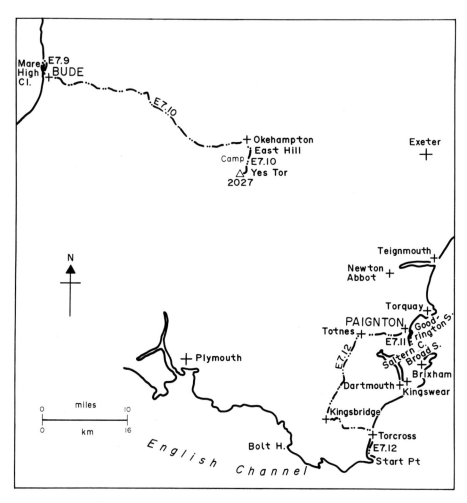

Fig. 7.3   Route map for Bude and Paignton Centres.

Access:- By private transport or public bus to Okehampton. Road, track and moorland walking.

Maps:- Topo:- 191. Geol:- EW 324.

Walking distance:- 6.5 miles (1.4 km).

Itinerary:- Walk S. from Okehampton then E. by S. to sharp bend on road on W. side of East Hill where there are outcrops of Dinantian shales dipping N.N.W. at 20° intruded by dolerite. Continue S. up road and along a track on E. side of army camp. South of the camp there are a few exposures of hornfelsed Namurian shale. Walk S. across moor to the picturesque granite scenery of West Mill Tor and Yes Tor, 2027 ft. (619 m).

Return by same route.

Paignton Centre (Fig. 7.3)

Paignton is the terminus of a main rail route and there is plentiful local transport by road and rail. There is abundant accommodation both in Paignton and adjacent Torquay.

E7.11 Paignton to Brixham

Devonian (mainly carbonate facies).

Permian.

Access:- Topo:- 202. Geol:- EW 350.

Walking distance:- 5 miles (8 km).

Itinerary:- From the sea-front in the centre of Paignton walk S. then follow a cliff-path on Roundham Head to study almost horizontal red Permian sandstones and breccias. Walk along Goodrington Sands between the S. end of which and the next headland Lower Devonian sandstones and shales occur, with two small Permian outliers. At the headland, Upper Devonian cleaved shales are faulted up and can be seen to be unconformably overlain by Permian. A short distance further S., at the S. end of Saltern Cove, Middle Devonian limestones appear from under the Permian and then at the N. end of Broad Sands there is a faulted contact between these two formations. In the centre of Broad Sands Bay the same fault brings down the Upper Devonian.

At the S.E. end of Broad Sands fairly open folds, trending E. by N., are well seen in the Upper Devonian limestones which continue to Elberry Cove. Corals can be found, although recrystallisation has obscured their structure. Elberry Cove has been eroded in Upper Devonian shales with steep cleavage striking E. by N. From the end of the Cove follow a path above cliffs of Middle Devonian limestone to Churston Cove. Walk into the centre of Brixham.

Return by bus to Paignton.

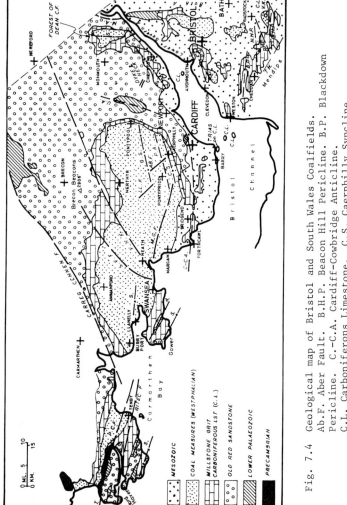

Fig. 7.4   Geological map of Bristol and South Wales Coalfields.
Ab.F. Aber Fault. B.H.P. Beacon Hill Pericline. B.P. Blackdown
Pericline. C.-C.A. Cardiff-Cowbridge Anticline.
C.L. Carboniferous Limestone. C.S. Caerphilly Syncline.
G.S. Gelligaer Syncline. J.T. Johnston Thrust. LL.F. Llanwonno
Fault. M.A. Malvern Axis. M.G.F. Moel GilauFault. N.H.P. North
Hill Pericline. P.A. Pontypridd Anticline. P.C.F. Pembrokeshire
Coalfield. P.H.P. Pen Hill Pericline. S.V. Swansea Valley Fault.
TH. Thrust V.K. Vobster Klippe, V.N.F. Vale of Neath Fault.

E7.12   Start Point

Lower Devonian. Start Point metamorphics of doubtful age.

Access:- By private transport. Coastal walking. Low to fairly low tide
necessary.

Maps:- Topo:- 202. Geol:- EW 356.

Walking distance:- 3.5 miles (5.6 km).

Itinerary:- Travel W. to Totnes then S. to Kingsbridge before turning E.
to Stokenham and Torcross where transport should be left. To the N. a long
spit encloses a lagoon.

Walk along shore. At the point immediately S. of Torcross, Lower Devonian
slates and sandstones are cut by metadolerite dykes, and a few more exposures
of Lower Devonian strata, dipping N., are seen in the next 1½ miles (2.4 km)
to the S. Then, across an unexposed contact, there is a narrow outcrop of
hornblende-and chlorite-schists followed by mica-schists which continue to
Start Point. The obvious structure is an E.-W. schistosity but this is
superimposed on earlier structures which are evidence of a complex history.
Ages ranging from Precambrian to (metamorphosed) Devonian have been suggested.

Bristol Centre (Figs. 7.4, 7.5)

Bristol is on main rail and road routes from London and elsewhere, and there
are numerous public transport services. There is plentiful accommodation
of all types, some of it on the city outskirts.

It may seem surprising that an excursion to the type-section of the Avonian
is not included, but the Avon Gorge is not now suitable for geological
parties owing to heavy traffic on the dual carriageway.

Excursions (E8.17 and E8.18) in the Mesozoic from Bristol are described in
the next chapter. Some additional excursions in the Bristol district are
given by Savage (1977).

E7.13   Cheddar Gorge and the Mendip Hills

Carboniferous Limestone. Triassic. Karst phenomena.

Access:- By public bus or by private transport to Cheddar. It should be
noted that during the summer months coaches may only descend the Cheddar
Gorge.

Maps:- Topo:- 182. Geol:- 1:63,600 Special Sheet Bristol.

Walking distance:- 8.5 miles (13.6 km). With private transport 7 miles
(11.2 km).

Itinerary:- A start should be made at Cheddar, on the S. limb of the
Blackdown Pericline. Above Cheddar village the grey, crinoidal Hotwells
Limestone (D), dipping S. at 20°, is exposed high up, but access is difficult.
Walk up gorge, noting a number of springs, to Gough's show cave which can be
visited at most times.

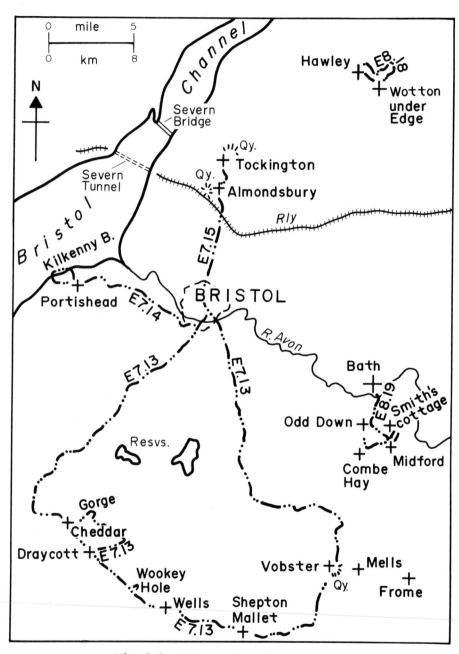

Fig. 7.5 Route map for Bristol Centre.

The section between here and the Horseshoe Bend is in the upper part of the
Clifton Down Limestone (S2), banded calcite mudstones with fine-grained
limestone and chert stringers. Algal beds occur, and Lithostration is
common. Several small caves are seen. From the Horseshoe Bend to a
reservoir the current-bedded Cheddar Oolite outcrops. The fauna includes
Composita, Lithostration and Chonetes. At the reservoir the oolite is
underlain by the Cheddar Limestone (S2) which continues to the Black Rock
Gate. Many of the top beds are rich in fossils, including, in addition to
those mentioned above, Palaeosmilia and Productus. At Black Rock Gate the
Brinington Oolite (C2) outcrops from a point 160ft. (50 km) N. of the gate.
It comprises fossiliferous crinoidal limestones and oolites.

Continue along the road S. then S.E. across an upward succession. Exposures
are scattered, but some Karst features can be seen. South of a road fork
the Hotwells Limestone (not readily reached in the gorge) dips N.W. at 18°
with the pitch of the Pericline. 1.2 miles (1.9 km) S.E. of the fork, turn
S.W. on a narrow road. 0.9 mile (1.4 km) S.E. of the road junction the
South-Western Overthrust (not exposed but seen as a feature) is crossed,
bringing the Black Rock Limestone over the Hotwells Limestone. To the S.W.
the road follows the unconformity with overlying Dolomitic Conglomerate
(Triassic).

Descend to Draycott and return to Cheddar passing on the way a small inlier
of the Carboniferous Limestone.

However, it is suggested that those with private transport should arrange
to meet it at Draycott and follow the Wells road to the famous Wookey Hole
(open to public) in the Carboniferous Limestone immediately below the
Triassic.

Continue through Wells, Shepton Mallet and Doulting to a road junction
1.2 miles (1.9 km) W. of Mells. Carboniferous Limestone, steeply folded
on an E.-W. axis is exposed in a large disused quarry. This is the largest
of the Vobster klippes and is surrounded by Coal Measures. These are not
exposed here but their presence is evidence from old tips. The mines
showed the Coal Measures to be contorted.

Return to Bristol.

E7.14   Coast S.W. of Portishead

Old Red Sandstone. Carboniferous. Triassic.

Access:- By public bus or private transport to Portishead. Road, path and
shore walking. Low tide is necessary for shore exposures near Portishead
Pier but others can be seen at almost any state of the tide.

Maps:- Topo:- 172. Geol:- EW 264. 1:63,600 Special Sheet Bristol.

Walking distance:- 3.5 miles (5.6 km).

Itinerary:- From Portishead town centre N. to the Royal Hotel, W. of the
pier. Descend steps to beach where there are exposures of Pennant Sandstone
(Upper Carboniferous, cf. E7.20) brought down by an exposed fault against
Black Rock Dolomite Carboniferous Limestone to S. The Pennant mostly dips
steeply N.N.E. but at the E. end of the section is overturned. At the foot

of the steps unconformably overlying Dolomitic Conglomerate is seen.

Along road to W. then down to shore again at Portishead Point where the
Black Rock Dolomite appears from under the Triassic. Southwards the
Dolomite and limestones and siltstones of the Lower Limestone Shale are
involved in sharp E.-plunging, S.-verging folds which are well worth
detailed study.

Fossils are abundant in both the Limestone Shale and in non-dolomitised
parts of the Black Rock Dolomite. To the S. a fault brings up the Portishead
Beds of Upper Old Red Sandstone age.

Walk along coast to the S.W. end of Kilkenny Bay. Here the Woodhill
Conglomerate at the base of the Portishead Beds overlies the Black Nore
Sandstone (Lower Old Red). There is no angular disconformity but the base
of the Conglomerate is irregular and pebbles fill scour hollows in the Black
Nore Sandstone. Along the coastal section there are patches of unconformably
overlying Triassic conglomerate.

Leave shore and follow path across the downs back to Portishead.

E7.15   Cattybrook, Almondsbury and Tockington

Carboniferous. Triassic.

Access:- By private transport. By public bus to Almondsbury. Road walking.
Permission to visit the Cattybrook brick pit must be obtained from the
Manager, Cattybrook Brick Co. Ltd., Almondsbury, Bristol.

Maps:- Topo:- 172. Geol:- 1:63,360 Special Sheet Bristol.

Walking distances:- 7 miles (11.2 km). With private transport 2 miles
(3.2 km).

Itinerary:- With private transport travel on the M5 then branch off on the
B4055 which should be followed to the bridge over the South Wales-London line
S.W. of Almondsbury. Those travelling by public bus should alight on the S.
outskirts of Almondsbury and walk S.W. to the railway. By looking into the
railway cutting (do not enter) N.W. of the bridge it can be seen that the
Upper Cromhall Sandstone (local top of the Carboniferous Limestone) has been
thrust N.W. over Coal Measures sandstones. The strata in the brick-pit are
mudstones with subordinate sandstones belonging to the Modiolarius zone.
The complex tectonic structures (space prevents full description) are worth
detailed study. The main structure is an anticline plunging S.W. and verging
N.W. Near the entrance a cover of Dolomitic Conglomerate (Triassic) is seen.
Follow the roadway to the "Blue Face" near the N.W. corner where an inverted
sequence of mudstones, with sandstones in the middle, is exposed. Clay ironstone conglomerates, which are known to occur at the base of Coal Measures
sandstone units, indicate that the sequence is inverted. Thin lenticular
coal seams occur within the mudstones which yield well-preserved plant-fossils
from beds immediately and stratigraphically above the sandstones. These
include Calamites, Neuropterus, Mariopteris and Alethopterus.

To the S.W. the "Common" face shows mudstones, some with plant fossils in the
core and on the flanks of the main anticline.

On leaving the quarry follow the B4055 N.E., passing on the S.E. side of
Almondsbury close to the contact of the Carboniferous Limestone and unconformably overlying Dolomitic Conglomerate to the N.W. After crossing the
M4 turn off to the N.W. and descend to Tockington noting exposures of red
Keuper marl. Walk up a narrow road to about $\frac{1}{4}$-mile (0.4 km) N. of Tockington
to a large disused quarry in fossiliferous Clifton Down Limestone (S2),
dipping S. by E. at 15°.

Return to Almondsbury and Bristol.

7.2   South Wales and Pembrokeshire Coalfields and their Borders (Fig. 7.4)

The South Wales coalfield is topographically rugged for the hard Pennant
sandstone in the Coal Measures forms hills rising to nearly 2000 ft.
(610 m), deeply dissected by the glaciated South Wales valleys. The
surrounding Carboniferous Limestone outcrops in picturesque scarps, and
to the N. the Old Red Sandstone reaches 2906 ft. (884 m) in the Brecon
Beacons. The Pembrokeshire coalfield, not now worked and separated by
Carmarthen Bay, is structurally the continuation of the South Wales field.

The South Wales valleys are often deeply floored with drift, as are the
coastal platforms, occurring at several levels. Spectacular shore sections
occur along the Bristol Channel forming the S. margin of the region.

There are numerous public transport routes in South Wales and a network of
roads.

Four centres are suggested: Cardiff, Swansea, Tenby and Milford Haven.

The succession is as follows:

> Liassic
> Triassic
> Coal Measures
> Millstone Grit
> Carboniferous Limestone
> Old Red Sandstone
> Silurian
> Ordovician
> Precambrian

Intrusive igneous rocks are virtually confined to the south-western part of
the region.

The dominant structure of South Wales is the E.-W. downfold of the coalfield
produced by post-Westphalian compression. This was preceded, however, by
movements throughout the Carboniferous, the most important being Sudetic uplift which is marked by a major break at the base of the Namurian. The uplift of the Usk Axis (E7.16) played a major role. The South Wales Coalfield
is a major downfold, which includes a number of minor folds, extending for
60 miles from the Burry Estuary in the W. to Pontypool in the E.

There are numerous faults. Two major zones i.e.the Neath and Swansea Valley
disturbances trend N.E. across the central portion of the Coalfield and then
cut across the Old Red Sandstone belt into Herefordshire, the Neath Disturbance

extending as far as the Malvern-Abberley line.

Further W., as the Hercynian Front is approached, the folding becomes more severe, and there is important thrusting. In fact, the Ritec Fault, along Milford Haven, is probably one of the major structures along the front itself. Here the Hercynian structures cut across the Caledonian (Ch. 6.1).

Comparison with other areas:- The Lower Carboniferous carbonate facies is closely comparable to that of the Bristol district (E7.13, E7.14, E7.15) and to that of Central England (E7.33, E7.34). The Upper Carboniferous also closely resembles that of other British Coalfields (Ch. 7.3, 7.4 and 7.5) but is very different from the Culm facies (E7.9) across the Bristol Channel.

The limited outcrops of late Precambrian (E7.27) and of Lower Palaeozoic in Pembrokeshire are like those a little further N. within the Caledonian Fold Belt (E6.1). On the other hand the Silurian E. of Cardiff (E7.16) is a shelf-facies like that of Central England (Ch. 7.3).

The Triassic to Liassic succession is mainly a continuation of that of Southern England (Ch. 8.2); of particular interest, however, are the shore-line Mesozoic deposits near Bridgend (E7.21).

Cardiff Centre (Fig. 7.6)

There are good local rail and bus services around Cardiff. Accommodation at all levels is plentiful, some on the city outskirts. Additional excursions are described by Anderson (1977).

E7.16   The Usk Silurian Inlier

Silurian. Old Red Sandstone.

Access:- By train from Cardiff Central to Pontypool Road station or by public bus (change at Newport) to New Inn. By private transport; the narrow roads across the inlier are not suitable for a coach which should be parked near the beginning of the excursion or sent round to Usk.

Maps:- Topo:- 171. Geol:- EW 232, 233, 249, 250; Special 1:25,000 sheet Usk-Cwmbran (southern part of inlier only).

Walking distance:- Pontypool Road station to Usk 7.5 miles (12.0 km). Pontypool Road station and back via Greenmeadow 15 miles (18.4 km).

Itinerary:- On the S.E. side of the approach road to Pontypool Road station red marls with sandstones, dipping fairly steeply W., are exposed. These are the Raglan Marls (Downtonian) which make up a considerable part of the Lower Old Red Sandstone of the district.

Walk one mile (1.6 km) S.E. along main road then ¼-mile (0.4 km) N. on a minor road before turning E. on road past Sluvad Farm. Descend to dam enclosing Llandegfedd Reservoir. At the W. end of the dam a road cutting exposes Ludlow impure limestones with corals and brachiopods dipping W.S.W. at $10°$ and at the E. end another cutting is in mudstones and siltstones at a slightly lower horizon with a similar fauna.

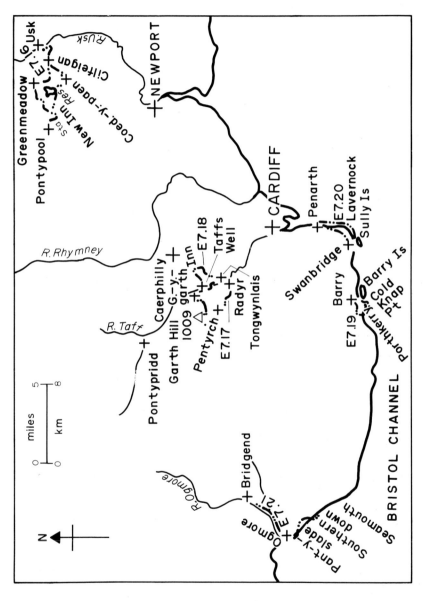

Fig. 7.6 Route map for Cardiff Centre.

Continue E. through Coed-y-paen, crossing one of several faults which cut
the inlier, then E.S.E. for 0.6 mile (1 km) to an old quarry in a wood on
the N. side of the road in the Wenlock Limestone. This dips gently S. and
contains rather sparse corals (<u>Favosites</u>, <u>Halysites</u>) and brachiopods.

N.E. to road junction near Cilfeigan. If transport has been sent to Usk the
excursion can be continued by walking E. to the R. Usk then N. along the
road on the left bank to the bridge. A cutting along this road shows Ludlow
impure limestones with abundant brachiopods. Alternatively, walk W. from
near Cilfeigan then N. to Greenmeadow. Roadside exposures occur of the
Wenlock Shales, in the centre of the Anticline; fossils are sparse. On the
angle of the road at Greenmeadow the Wenlock Limestone, dipping W., is again
seen.

Walk W. from Greenmeadow then S.W. back to Pontypool Road station.

E7.17   Radyr, Pentyrch and Garth Hill

Old Red Sandstone. Carboniferous. Glacial deposits.

<u>Access</u>:- Train from Cardiff (Central or Queen St.) to Radyr, or private
transport to Radyr. Return from Taffs Well. Walking on roads, rough tracks,
paths and hillside.

<u>Maps</u>:- <u>Topo</u>:- 171. <u>Geol</u>:- EW 263.

<u>Walking distance</u>:- 6.5 miles (10.4 km).

<u>Itinerary</u>:- At Radyr the station and adjacent sidings are situated on the
lowest terrace of the R. Taff. Walk up station road to the main road (Heol
Isaf) through the village which is on the second terrace. Cross the main
road and continue up road which passes golf-course, crossing remnants of a
third terrace. The terraces have all been eroded in glacial deposits of the
Taff valley. These deposits can be seen in typical moundy form at and around
the golf-house. They represent a zone of diffused melting at the Southern
limit of the Devensian ice. Turn right past golf-course along a track
leading into a road which continues across the M4 to a bridge over a dismantled railway. From the bridge a lane leads N.W. to a road on a ridge.
Hard, dull-red sandstone (Upper Old Red Sandstone) dipping N. is seen along
the ridge, which is formed of Quartz Conglomerate at the top of the Upper Old
Red Sandstone. In order to see the conglomerate, however, it is necessary
to follow the road down to another road S. of Pentyrch. Cross this road and
continue along a lane to yet another road immediately W. of which there is a
disused quarry in the conglomerate.

Return to the Pentyrch road (which follows a minor fault) and walk N. A
hollow corresponds to the Lower Limestone Shales (not seen). Beyond a
disused quarry on the E. side of the road in the main outcrop of the
Carboniferous Limestone, here dolomitosed and unfossiliferous. Thin veins
of barytes occur near the entrance.

Through Pentyrch on the W. edge of which beds near the top of the Carboniferous yield corals and brachiopods. Turn N. on narrow road. A hollow marks
the outcrop of shales (not seen) in the lower part of the Millstone Grit
and at a rise, sandstones, dipping steeply N., are exposed at the roadside.
The next hollow corresponds to the Lower Coal Series. Turn E. at a road
junction then N.E. on a rough track which angles up a scarp of Pennant

Sandstone dipping N. at 40°,coarsely quartzo-felspathic and with plant
markings, to the top of Garth Hill, 1009 ft. (308 m). The dip-slope N.
and the slope in the opposite direction seen from the summit are the topographical expression of the Caerphilly Syncline. Beyond are the hills and
valleys of the main Coalfield Basin.

Descend S. to road which should be followed E.N.E. to Gwaelod-y-garth.
Turn S.S.E. here. Between the road and the R. Taff small tips mark old
workings in the Lower Coal Series.

A footbridge crosses the river to Taffs Well station where train can be
caught back to Cardiff. Those who have left private transport at Radyr
catch a train to that station.

E7.18   Taffs Well and Tongwynlais

Old Red Sandstone. Carboniferous. Iron ore deposit.

Access:- Train from Cardiff (Central or Queen St.) to Taffs Well or
private transport. Walking on roads, rough tracks and paths.

Maps:- Topo:- 171. Geol:- EW 263.

Walking distance:- 4.5 miles (7.6 km).

Itinerary:- From Taffs Well walk S.E. on the road from the station through
the Taff Gorge which the river has cut along a fault through the Carboniferous Limestone. Dolomitised limestone, dipping steeply N. by W. is
exposed in a cutting on the trunk road and also in the high face of a
disused quarry where there is considerable brecciation due to the fault.
Near Tongwynlais an anticline brings up the fossiliferous Lower Limestone
shales (with ironstone band) at the base of the Series. These can be seen
to pass conformably downwards into sandstones and shaly marls with plants
at the top of the Upper Old Red Sandstone. The fold is one of several at
the S. margin of the limestone outcrop.

At Tongwynlais turn sharply N. on a secondary road. The large quarry on
the E. side shows sharp folds in the limestone but permission must be
obtained to enter (Tarmac Ltd.). Turn off to the W. on the road marked
Castell Coch then almost immediately branch off on a path which leads
steeply upwards and northwards to a forest track leading N.E. on the
limestone. About ½-mile (0.8 km) along the track, among the trees on the
N.W. side there are old, partly opencast, partly underground workings for
replacement iron ore. The workings, which are small, can be entered with
care, and goethite and calcite specimens obtained.

Continue along forest track to rejoin the road which should be followed N.
to just short of the Black Cock Inn, where the road should be left for a
rough path running S.W. and slanting down to a wooded valley. The path is on
a dip-slope of quartzitic sandstone (blocks of which may be seen) in the
upper part of the Millstone Grit. The presence of the Lower Coal Series is
shown by a disused colliery and tips across the valley below a scarp of
Pennant Sandstone. The quartzitic sandstone dipping steeply N. is seen in
a small disused quarry just after the path emerges from the wood. Follow
a track beside the stream in which Millstone Grit shales below the
quartzitic sandstone are exposed.

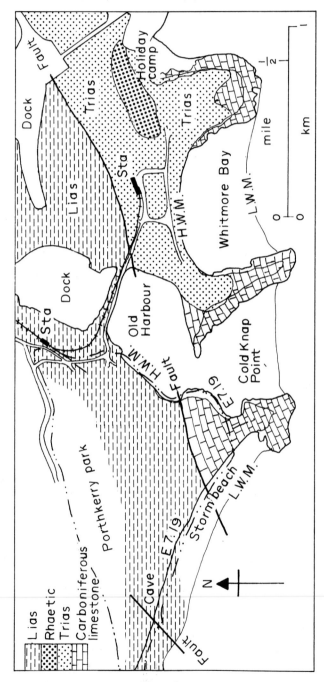

Fig. 7.7 Geological map of Barry District with route.

Continue along the track under a railway bridge then walk through a housing
estate back to Taffs Well.

E7.19   Barry   (Fig. 7.7)

Carboniferous Limestone. Triassic. Liassic.

Access:- Train from Cardiff Central or Queen St. to Barry. Coastal and
road walking. Low tide necessary.

Maps:- Topo:- 170, 171. Geol:- EW 263.

Walking distance:- 3.5 miles (5.6 km).

Itinerary:- From Barry station walk S. to the W. end of the causeway to
Barry Island then along the shore on the W. side of the Old Harbour,
examining a section in Liassic limestones and shales with Ostrea. The
strata are mostly flat but at one locality form a small dome and are cut
by small faults. The Liassic, considerably contorted, ends at a major N.E.
fault which brings up limestone and shale (Lower Limestone Shales, K zone)
near the base of the Carboniferous Limestone Series. These dip steeply
seawards and contain corals, brachiopods and other fossils. Across the
strait it can be seen that the Limestone is unconformably overlain by
Triassic which overlaps towards the N. At very low tide it is possible to
cross the sands to see the unconformity in detail.

Continue S. to Cold Knap Point where more massive limestones of the C
zone contain numerous fossils. Walk N.W. along the promenade and along a
striking storm beach. This covers the continuation of the N.E. fault and
the cliffs beyond are of Liassic limestone and shale near the base of the
formation. Round a headland, passing a cave along a small fault, to another
storm beach, damming an alluvium-filled inlet. Walk through Porthkerry Park,
mostly sited on this alluvium, back to Barry.

E7.20   Sully Island to Penarth

Carboniferous Limestone. Triassic. Rhaetic. Liassic.

Access:- Train from Cardiff Central or Queen St. to Penarth then bus to
Swanbridge, returning from Penarth. Low tide essential The crossing should
not be made to Sully Island unless the tide is still falling. Coastal
walking.

Maps:- Topo:- 171. Geol:- EW 263.

Walking distance:- 5.5 miles (8.8 km).

Itinerary:- Leave the bus at the road leading to Swanbridge and walk to the
coast. Cross to Sully Island over Keuper Marls; a fault trending E. by N.
and downthrowing about 12 ft. (3.5 m) can be followed across the isthmus.
Walk across Sully Island to the end where fairly fine Triassic red breccia
rests unconformably on Carboniferous Limestone. The limestone, which is
dolomitised, is traversed by haematite veins; some galena and haematite
also occurs.

Return to mainland and walk along the coast to the E. Triassic limestones,

sandy limestones and marls are cut by small normal faults. At the W. end of
St. Mary's Well Bay the Lavernock Fault (reversed) brings Triassic boulder
beds to the W. against Triassic tea-green marls, overlain Rhaetic marls, and
black shales with Pteria contorta. These are overlain by about 12 ft. (3.5 m)
of pale marls and limestones (White Lias) followed by Liassic limestones and
shales with Ostrea liassica, above which are limestones and shales of
Planorbis age succeeded, in the centre of a gentle syncline, by the Lavernock
Shales of Angulata date. On the E. limb of the syncline, near Lavernock
Point, the Rhaetic bone-bed, with teeth and scales of fishes and bones of
reptiles, outcrops near extreme low tide mark.

Round Lavernock Point and walk N. along the shore to Penarth Pier. Gentle
folding and some minor faulting affects a succession of Triassic red marls,
Triassic tea-green marls, Rhaetic and Liassic. Below Lower Penarth a syncline brings the Rhaetic bone-bed down to beach level. Here it is sandy
and has yielded ichthysaur vertebrae. Nearer Penarth the Triassic marls
contain gypsum.

From the pier walk up through a park to Penarth station.

E7.21   Ogmore-on-Sea and Southerndown, near Bridgend

Carboniferous Limestone. Triassic. Rhaetic. Liassic.

Access:- By private transport to Ogmore-on-Sea. By train from Cardiff
Central to Bridgend, then by public bus to Ogmore-on-Sea. Coastal walking.

Maps:- Topo:- 170. Geol:- EW 262.

Walking distance:- 5 miles (8 km).

Itinerary:- Walk E. along the shore examining exposures of Carboniferous
Limestone (C zone) with depressions infilled with coarse Triassic conglomerate.

West of the dry ravine of Pant-y-slade a shallow syncline contains thick-
bedded limestones of the C_2S zone with giant specimens of Caninia cylindrica,
and also Chonetes, Michelima and Syringothris.

Nearer Pant-y-slade littoral Liassic deposits rest with only slight discordance
on the Carboniferous Limestone. The lower part of these deposits - the
Sutton Stone - is about 20 ft. (6 m) thick with a conglomeratic base passing
upwards into pale-cream, massive limestone with lamellibranchs, gastropods
and corals. A short distance E. of Pant-y-slade the terrace of Carboniferous
Limestone ends abruptly, and the Sutton Stone is seen to be banked against a
Carboniferous Limestone cliff. The upper part of the littoral Liassic consists of the Southerndown Beds consisting of fine conglomerate with chert
and limestone fragments in a limestone matrix. They are sparsely fossiliferous; the few ammonites indicate the Angulata and Bucklandi zones. To the
E. these beds merge into normal Liassic limestones and shales.

Leave the shore here and walk along the cliff top to Southerndown. The
Liassic limestones and shales in the bay at Seamouth (Bucklandi zone) have
been overthrust by Carboniferous Limestone overlain by Sutton Stone. The
Liassic limestones and shales are sharply folded.

Return to Ogmore-on-Sea.

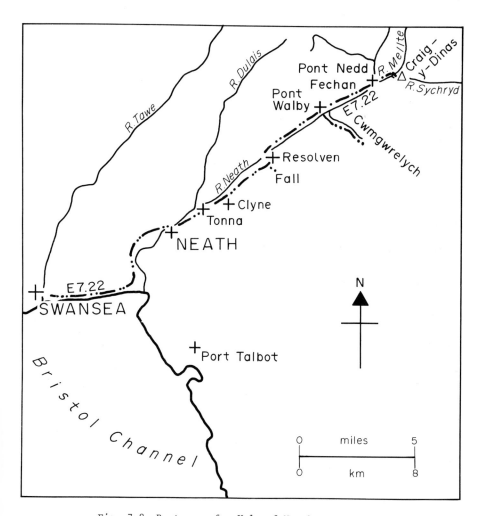

Fig. 7.8   Route map for Vale of Neath excursion.

Swansea Centre

The same remarks about accommodation and transport apply as for Cardiff.
Additional excursions are described by Owen and Rhodes (1960). E7.21 can be
undertaken from either Cardiff or Swansea.

E7.22   Vale of Neath (Fig. 7.8)

Carboniferous. Glacial features.

Access:- By private transport or public bus to Pontwalby. Road and stream
section walking.

Maps:- Topo:- 160, 170. Geol:- EW 231, 247, 248.

Walking distance:- 5.5 miles (8.8 km).

Itinerary:- The straight course of the Vale of Neath follows the Neath
Disturbance - a complex fault-zone (Owen, 1954). The Vale has been strongly
glaciated and retreat moraines occur at Tonna and Clyne. Stops should be
made to examine these if travelling by private transport; the better views
are from the road on the S.E. side. A stop should also be made at Melin
Court, ½-mile (0.8 km) S.W. of Resolven, and a path followed a short distance
up the left bank of the hanging tributary. Near the road Coal Measures sandstones show brecciation due to the Vale of Neath Disturbance. Further on
there is a spectacular waterfall over massive sandstone, on top of thin-
bedded sandstones and shales.

Cross the valley at Resolven and continue to Pont Walby to examine the section
in Cwmgrelych stream. S.W. of a railway viaduct the quartzitic sandstones
and siltstones of the local "Farewell Rock" are exposed (Westphalian A).
These are underlain by the Gastrioceras subcrenatum Marine Band, poorly
exposed 45 yds. (13 m) from the railway viaduct on the right bank. In a
small tributary the uppermost siltstones of the "Farewell Rock" yield plants
and are overlain by a marine band with Lingula and Productus. The overlying
shale/sandstone sequence contains clay-ironstones known as the "Rosser Veins".
On the right bank 600 yds. (180 m) above the viaduct, the Cnapiog Coal with
underlying seat earth is exposed and 7 ft. (2 m) beneath the coal dark blue
shales contain Carbonicala. An ascending sequence in the Coal Measures
(Westphalian B),which have been extensively worked, can be followed up the
main stream and three tributaries which come in from the S. as far as a
minor road leading to Rhigos.

Return to Pont Walby and walk up the road to Pont Nedd Fechan. At the confluence of the Mellte and the Sychryd the Upper Limestone Shales (D3) are
exposed on the left bank of the Mellte.

Continue up valley of R. Sychryd to Craig-y-Dinas. Here there is a spectacular anticline, Bwa Meen (Bow Rock), in Carboniferous Limestone (S_2, D1,
D2) split by the Dinas Fault, one of the main fractures of the Neath Disturbance. On the S.E. flank the Basal Grits of the Millstone Grit rest on
D2 and contain highly siliceous beds quarried for silica rock. They are
overlain by shales and then the "Farewell Rock".

Return to Swansea.

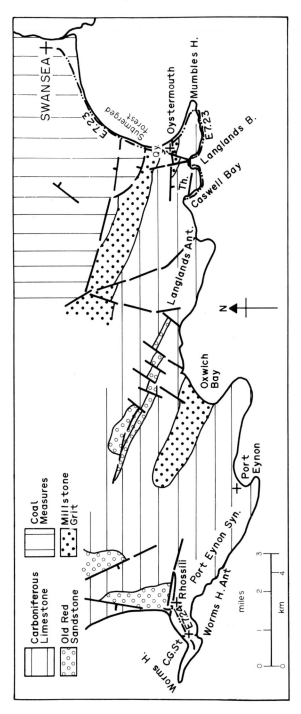

Fig. 7.9 Geological map of Gower with routes.

E7.23   Eastern Gower - Oystermouth to Caswell Bay (Fig. 7.9)

Carboniferous. Patella Beach. Submerged Forest.

Access:- By public bus to Oystermouth, returning from Caswell Bay. Coast
walking, mainly on roads or paths. Low tide helpful for parts of excursion
but not essential.

Maps:- Topo:- 159. Geol:- EW 247.

Walking distance:- 4.5 miles (7.2 km).

Itinerary:- At Oystermouth dark calcareous shales (Upper Limestone Shales =
D3) are seen in a large disused quarry. Fossils are plentiful, including
Spirifer and the coral Triplophyllites; Trilobite fragments may be found
(Griffithides).

Walk along the road to Mumbles Head, passing a syncline of Namurian marked
by a hollow. At very low tide it is possible to walk across the sand to
examine the peat and tree stumps of a mesolithic Submerged Forest. At
Mumbles the limestones are D1/D2; the half-tide islands forming the head
are separated by hollows following minor faults. On the road cutting the
limestone is partly dolomitised resulting in the formation of breccia.

Walk along coast path to Langlands Bay above cliffs of D1/D2 limestones
dipping S. off the Langlands Anticline which, in the inner part of the bay,
brings up the Caninia Oolite (C_1).

Walk round Langlands Bay and along cliff path to W. again above D1 cliffs.
Where the coast turns into Caswell Bay the Patella Beach is seen, of
Pleistocene age but preceding the earliest local glaciation.

An excellent section occurs on the E. side of Caswell Bay. In the seaward
half of the section a descending succession, partly repeated by a southward
dipping Thrust, from S_2 to C, is seen. The Laminosa Dolomites (C_1) form the
core of the Langlands Anticline which is followed to the N. by a syncline.
This contains the Caninia Oolite with an irregular surface on which rests
a Modiola-Phase - 15 ft. (4 m) of shales and calcite-mudstones - overlain by
crinoidal CS limestones with abundant corals. The syncline is overridden
towards the S. along the Caswell Thrust by the Laminosa Dolomites followed
by an upward succession.

Return from Caswell Bay to Swansea.

E7.24   Western Gower - Rhossili (Fig. 7.9)

Old Red Sandstone. Carboniferous.

Access:- By public bus or private transport to Rhossili, where there is a
car park. Coastal walking.

Maps:- Topo:- 159. Geol:- EW 246.

Walking distance:- 3 miles (4.8 km).

Itinerary:- Walk S.W. along track to Coast Guard Station, making a diversion
to the top of the cliffs to the N.W. to look at the Carboniferous Limestone
(C1) beds in the Port Eynon Syncline. If the tide is low, descend a path
to the intertidal platform between the station and Worms Head to examine in
the core of the Worms Head Anticline limestones of the Z zone.

Return to Rhossili and walk N. along a track above the cliff. At a spring
the Lower Limestone Shales (K) are exposed, with Camarotechnia and Productus.
To the N., across a fault, red conglomerates (Old Red Sandstone, probably
Upper) are seen in an old quarry. The K shales are distorted due to in-
competent reaction between the conglomerates and hard Z limestones to the S.
from which they are separated by the Port Eynon Thrust.

Return to Swansea.

Tenby Centre (Fig. 7.10)

Tenby can be reached by rail or road and has plenty of accommodation, although
it should be noted it is a popular resort.

E7.25   Tenby, Saundersfoot and Amroth

Carboniferous.

Access:- By train or private transport to Saundersfoot. Walk from Saundersfoot to Amroth then back to Tenby. Walking mainly on shore and cliff paths.
Low to fairly low tide necessary.

Maps:- Topo:- 158. Geol:- EW 228, 245.

Walking distance:- 11 miles (17.6 km).

Itinerary:- From Saundersfoot Harbour walk S. along the shore across flaggy
sandstones and shales of the communis zone of the Coal Measures showing
spectacular folding along E. by S. axes and thrusting. The structures include an overturned syncline with a thrust towards the N. above it and, to
the S., the much-photographed Lady's Cave Anticline. Current bedding rather
doubtfully suggests that the strata are inverted and the fold therefore an
antiform.

Turn back to harbour underneath which there is probably a thrust bringing
the communis zone beds over the Lower similis-pulchra zone beds to the N.
Walk N.E. along the shore. Folding and faulting are again well seen, and
an anticline brings up the Farewell Rock ½-mile (0.8 km) beyond Saundersfoot.
Further on, at the N.E. end of the bay at Wiseman's Bridge, a fault is well
seen in the Coal Measures forming the cliff.

Continue to Amroth; several coal seams can be seen and, 500 yds. (457 m)
beyond the fault mentioned above, the Amman Marine Band.

Return to Saundersfoot, walk up Tenby road for ¼-mile (0.4 km) then branch
off S. onto a path which leads to a cliff path. Follow this path to Waterwynch 1 mile (1.6 km) N. of Tenby, noting structures in cliffs. Descend to
shore at Waterwynch where sandstones, siltstones and mudstones of the basal
Coal Measures are exposed. The sedimentary structures in the N. cliff are
worth detailed study.

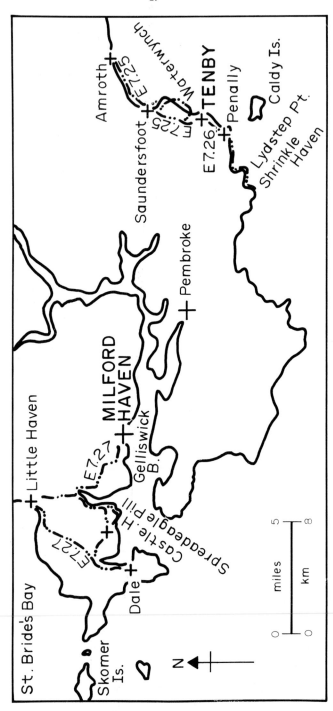

Fig. 7.10 Route map for Tenby and Milford Haven Centres.

Walk up minor road to main road and back to Tenby.

If the tide is low the shore below Tenby provides a good opportunity to
study both sedimentary structures and E.-W. orientated tectonic structures
in the Namurian. To the S., at the harbour, the W. by N. Ritec Fault
(E7.27), which has caused much shearing, brings up the Carboniferous
Limestone.

E7.26   Lydstep and Shrinkle Haven

Old Red Sandstone. Carboniferous

Access:- Road and coastal walking. Return by private transport from N. of
Shrinkle Haven or by public bus from Penally. Fairly low tide necessary for
some sections.

Maps:- Topo:- 158. Geol:- EW 245.

Walking distances:- 10 miles (16 km). With private transport 6 miles
(9.6 km).

Itinerary:- Walk along path beside railway to Penally then along road W.
before following path, which avoids War Department area, to coast E. of
Proud Giltar. Along path to W. above spectacular cliffs of S. to D zones
of Carboniferous Limestone dipping steeply S.

Round Lydstep Haven where Millstone Grit, poorly exposed, occurs in a syncline. At Lydstep Point, on the S. limb of the syncline, the D zone limestone is vertical or slightly overturned. Gash-breccia, probably Triassic,
occurs in pockets.

A descending sequence to the K zone occurs along the coast to the W.S.W.,
then at Shrinkle Haven a thin zone of sandstone is referred to the Upper
Old Red Sandstone. To the S.W. conglomerates, sandstones and marls, dipping
steeply N. or slightly overturned, belong to the Lower Division (cf. E7.27).

If transport has been arranged, walk N. to road, if not, return by same route
to Penally where bus can be caught to Tenby.

Milford Haven Centre (Fig. 7.10)

Milford Haven can be reached by road or rail. It lies in a typical ria
and was a base for Nelson's navy and for his meetings with Lady Hamilton.
There is a fair amount of accommodation but bookings for parties should be
made. For further details about the Silurian and Old Red Sandstone see
Allen and others (1981).

E7.27   North side of Milford Haven

Precambrian. Silurian. Old Red Sandstone. Carboniferous.

Access:- Private transport is necessary to carry out the itinerary. Road
and coastal walking. Low tide required for some sections.

Maps:- Topo:- 157. Geol:- EW 226 and 227.

<u>Walking distance</u>:- 6.5 miles (10.4 km).

<u>Itinerary</u>:- Travel W. then S. to Gelliswich Bay, then park at the W. end
of the road crossing the head of the inlet. Walk S. then W. to the steps
leading down to the beach beneath the oil jetty at Little Wich. Skirt
small rocky promontory to gain access to the cliffs and broad foreshore to
the W. Upward fining sequences of pebbly sandstone, sandstone and siltstone are seen belonging to the Gelliswich Bay Formation of late Silurian-
Lower Devonian age. Structurally they form parasitic folds on the S. limb
of the Burton Anticline.

Return to transport and continue W.N.W. then N. to a minor road across the
head of Spreadeagle Pill. The central part of a cutting is occupied by
slightly cleaved siltstones which dip N.E. off the fossiliferous limestones
of Wenlock age. At the partly overgrown N.E. end of the cutting the siltstones are abruptly succeeded by the Lindsay Bay Formation which consists
of conglomerates followed by siltstones, thin calcaretes and an air-fall
dust tuff. Towards the S.W. end of the cutting the Wenlock is in faulted
contact with the Sandy Haven Formation which lies above the Lindsay Bay
Formation. Both are of late Silurian-Lower Devonian age.

Follow road on W. side of inlet to Sandy Haven then walk S. to Castle Head.
Here there are cliff and shore-sections of sharply folded beds of the Sandy
Haven Formation which include the Townsend Tuff. This overlies bright red
sandstones and consists of about 17 ft. (5 m) of graded air-fall tuffs.

W. to St. Ishmael's where transport should be parked. Follow path S.E. to
Lindsay Bay. Here the Lindsay Bay Formation reaches its maximum thickness
of some 200 ft. (80 m) of probable alluvial origin. The conglomerate of the
formation can be seen to interfinger with the finer beds of the Sandy Haven
Formation.

Return to St. Ishmael's and continue W. then S. towards Dale. At a point
W. of the road, 0.6 mile (1 km) N. of Dale, keratophyre lavas of Silurian
age outcrop. The hollow at Dale marks the Ritec Fault (E7.25), one of the
main structures along the Hercynian Front. E.S.E. along road to Dale Point
Lower Old Red Sandstone is seen, dipping steeply S.

N. to Little Haven and along shore to S.W. where there are exposures of
Coal Measures at the W. end of the Pembrokeshire Coalfield. These are
overthrust from the S. by various older rocks, although the thrust over the
Coal Measures is not exposed. However, by walking a path from the sharp
bend W.N.W. for ½-mile (0.8 km) outcrops of Precambrian diorite can be seen,
and by descending a path to the shore the diorite can be observed over
crushed Carboniferous Limestone.

Return to Milford Haven.

<u>7.3   Central England</u> (Fig. 7.11)

Although Hercynian events largely shaped Central England, Alpine movements
played a greater role than was the case further N.; moreoever, older, Precambrian rather than Caledonian, structures exercised a significant influence
on the development of both the Hercynian and Alpine structures. The region,
as defined for the present purpose, lies S. of the Craven Faults and W. of the
base of the Liassic as far S. as Gloucester. From there an arbitrary W.N.W.

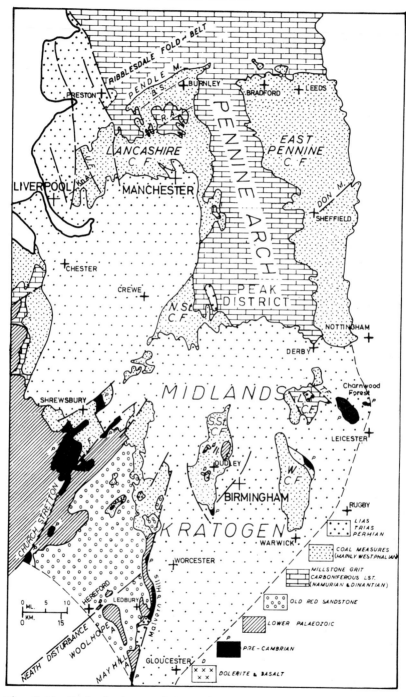

Fig. 7.11 Geological map of Central England. B.S. Burnley Syncline.
C.F. Coalfield. Kn.A. Knowsley Anticline. R.A. Rossendale
Anticline. U.F. Upholland Fault.

line to beyond Hereford runs to the edge of the Welsh Block. Northwards, to the sea at the Dee Estuary, Central England borders the same Block, partly along the Church Stretton Zone.

Along the Pennine Upfold the ground rises to over 2000 ft. (610 m) and hard outcrops further S., e.g. the Precambrian of the Malvern Hills, rise to over 1000 ft. (305 m), but much of the region, which includes the Midland Plain, lies below 400 ft. (122 m). Exposures on this lower ground and along the coast are sparse, owing to the widespread glacial drift, but there are numerous quarries.

The geological succession is as follows:-

> Liassic
> Triassic
> Permian
> Coal Measures
> Millstone Grit
> Carboniferous Limestone
> Old Red Sandstone
> Silurian
> Cambrian
> Precambrian

The Precambrian appears in a number of small inliers and probably underlies the S. of the region as part of Precambrian kratogen separated by the Church Stretton Zone from the Caledonian Fold-Belt.

Widespread public transport is available.

Four centres are suggested: Birmingham, Leicester, Buxton and Sheffield.

Comparison with other areas:- Central England differs from other regions described in this and the next chapter in being underlain at shallow depth by the Midland kratogen or platform of Precambrian rocks which show through in isolated exposures (E7.28, E7.29, E7.31, E7.32); Caledonian granite also appears (E7.32).

The Precambrian volcanic rocks can be correlated with the Uriconian (E6.8); part of the Charnian (E7.32) may be the equivalent of the Stretton Series in the Longmynd (E6.7). The Lower Palaeozoic is of shelf-facies as in eastern South Wales (E7.16).

Most of the Dinantian consists of thick carbonates, although in the N. of the region strata appear which are more characteristic of the Lower Carboniferous in North-East England (Ch. 7.4). The Upper Carboniferous outcrops include the type-locality of the Millstone Grit and Coal Measures broadly similar to those elsewhere in England and Wales.

The Permian and Triassic are generally of the continental "New Red Sandstone" facies, although the Magnesian Limestone just spreads into the N. (E7.36) from North-East England.

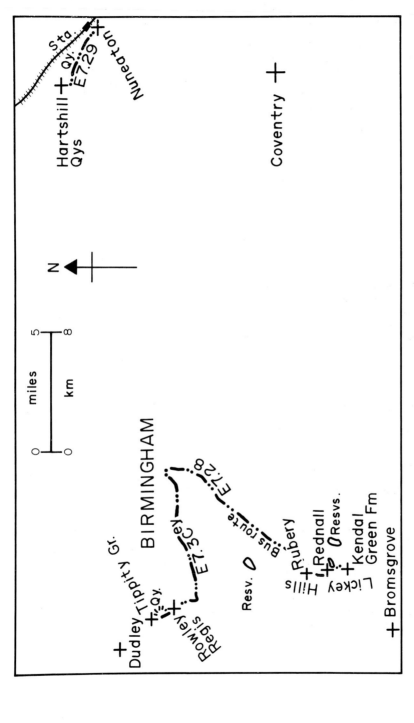

Fig. 7.12 Route map for Birmingham area.

Birmingham Centre (Fig. 7.12)

Accommodation of all kind is available both near the centre of Birmingham and the outskirts. All the excursions suggested can be undertaken either by public or private transport.

E7.28   Lickey Hills

Precambrian. Cambrian. Silurian. Carboniferous. Permian. Triassic.

Access:- By public bus (Rubery or Rednall) to Rubery or by private transport.

Maps:- Topo:- 139. Geol:- EW 183.

Walking distance:- 5 miles (8 km).

Itinerary:- At Rubery the Rubery Sandstone (Llandovery) is exposed at the S. side of the A38 road. The basal beds contain some pebbles of quartzite and rest unconformably on a slightly uneven surface of Lickey Quartzite (Cambrian). The quartzite can also be seen in an adjacent old quarry which is difficult of access owing to shops and houses.

Walk S. along road then ascend main ridge of Lickey Quartzite. Follow ridge then descend by footpath to E. side of the Rednall gorge. Along Barnt Green road S. for about 600 yds. (549 m) stopping at a disused quarry in the woods where purple-brown Lickey Quartzite shows a large overfold.

To Kendal Green Farm on the S. side of which Precambrian felsitic tuffs and grits are exposed. These occur in a fault-bounded block. A few yards N. of a fence marking the N.W. limits of the farm a section in a small S.W. draining stream shows red sandstones and clays of the Keele Beds (Upper Carboniferous). Walk N. to examine a weathered contact between these beds and the Bunter Pebble Beds (Triassic).

Follow the B4096 down the escarpment of the Bunter and after passing the grounds of a school, walk along the first path leading N. to two small streams draining artificial pools. Upstream traverses show Keele Beds overlain by Clent Breccia (Permian).

Return to Rubery.

E7.29   Nuneaton Ridge

Precambrian. Cambrian. Triassic.

Access:- By train or private transport to Nuneaton.

Maps:- Topo:- 140. Geol:- EW155, EW169.

Walking distance:- 6 miles (9.6 km).

Itinerary:- To Midland Quarry (permission from Midland Quarry Co. Ltd.). This is worked in the Hartshill Quartzite (Cambrian) with pink microdiorite sills overlain in the S.E. end of the quarry by Keuper sandstone.

Cross main road (A47) to Windmill Hill Quarry. Here the Hartshill Quartzite

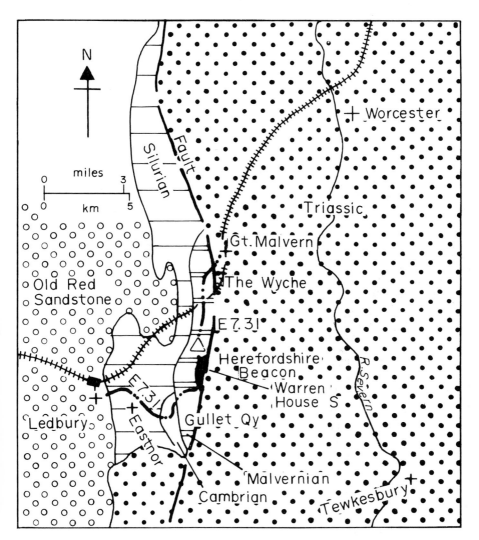

Fig. 7.13 Geological map of Malvern Hills, with route.

rests unconformably on Precambrian volcanics of the Caldecote Series;
Keuper sandstone and conglomerate can also be seen.

Walk along road N.W. to Hartshill village. Quarries to the W. of the road
are in Hartshill Quartzite with sills; those to the E. are in quartzite and
Caldecote volcanics.

At Hartshill village a quarry (Site of Special Scientific Interest) in the
highest beds of the Hartshill Quartzite reveals the <u>Hyolithus</u> Limestone with
abundant <u>Coleolides</u>. Continue along path to Mancetter Quarry (permission
from Man–Abel Quarries Ltd., Mancetter, Atherstone). Diorite sills occur
in Stockingford Shales (Cambrian).

Return to Nuneaton.

E7.30   <u>Rowley Regis</u>

Permian dolerite intrusion.

<u>Access</u>:- By public bus going to Dudley (via Blackheath) and alighting at
the Tippity Green road end.

<u>Maps</u>:- <u>Topo</u>:- 139. <u>Geol</u>:- EW 168.

<u>Walking distance</u>:- 1 mile (1.6 km).

<u>Itinerary</u>:-   Hailstone Quarry (permission from Tarmac) is worked in
fine-grained dolerite containing analcite. Rough columnar jointing is
seen.

E7.31   <u>Malvern Hills</u> (Fig. 7.13)

Precambrian. Cambrian. Silurian. Lower Old Red Sandstone.

<u>Access</u>:- By train to Ledbury, returning from Great Malvern.

<u>Maps</u>:- <u>Topo</u>:- 149, 150. <u>Geol</u>:- EW 199, Ew 216.

<u>Walking distance</u>:- 10 miles (16 km).

<u>Itinerary</u>:- From Ledbury station S. then turn on to first road to E. An
exposure of Lower Old Red Sandstone (Downtonian) marl is seen, then in an
old quarry fossiliferous Ludlow limestones and mudstones. At a fork, the
lowest path of the Ludlow is again seen in a road cutting.

Follow S. fork then turn down road to S.W. to examine a large disused quarry
in the Wenlock Limestone overlain by lowermost Ludlow. Wenlock fossils
include <u>Atrypa</u>, <u>Favosites</u> and <u>Calymene</u>. All these are Silurian beds sharply
folded on N.-S. axes.

Return to main road and continue through Eastnor to Hollybush on the W. side
of which the dark green glauconitic Hollybush Sandstone can be seen (Cambrian).

N., along E. side of Malvern scarp, to the Gullet Quarry (permission from
Hollybush Quarries Ltd., Ledbury, Herefordshire). All the rocks (Precambrian)
are intensely sheared and brecciated. Diorite is cut by granite, acid

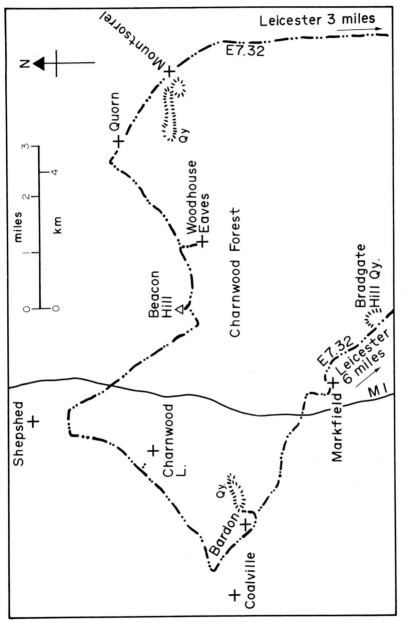

Fig. 7.14 Route map for Charnwood Forest.

pegmatites and metadolerite. At the uppermost level of the quarry Silurian (probably Llandovery) conglomerate rests on the Precambrian.

From the quarry follow the crest of the Malvern ridge N. On Swineyard Hill mica-schists are seen. Turn N.N.E. towards E. side of hills and Broad Down to examine exposures of the Warren House Series. This consists of spilitic basalts and pyroclastics. Their only slightly altered state shows that they are younger than the Malvernian. They may be of the same age as the Uriconian.

N.W., and onto the Malvernian again, to Herefordshire Beacon, 1114 ft. (340 m) with its hill-fort. From here there is a view to the W. to the Lower Palaeozoic hills of the Welsh Block and to the E. over the Triassic plain.

Descend to the road at Wynds Point, then rejoin the ridge and walk N. to the Wyche, examining on the way various components of the Malvernian, mainly granitic.

The road cutting at the Wyche follows a fault. To the S. are mica-schists, to the N. granite; there is also a highly sheared dolerite dyke.

Descend road N. to Great Malvern.

Leicester Centre (Fig. 7.14)

Leicester is easy of access and is a good starting point for the study of some of the rocks of the Midland kratogen which show through the Triassic.

E7.32   Charnwood Forest and Mountsorrel

Precambrian. Caledonian granite. Triassic.

Access:- Private transport is necessary to cover the excursion in one day. However, by taking a public bus to near Markfield and returning from Coalville, the first part can be carried out by walking. On a second day a visit to Mountsorrel may be made by using the Loughborough bus service.

Maps:- Topo:- 129, 140. Geol:- EW 155, EW 156.

Walking distances:- From near Markfield to Coalville on route described 11 miles (17.6 km). At Mountsorrel 1 mile (1.6 km). With private transport 5 miles (8 km).

Itinerary:- N.W. from Leicester to about 1½ miles (2.4 km) S.E. of Markfield road-end. Diorite (markfieldite) is worked (permission required) on Bradgate Hill, N.E. of the road, and the same rock can be seen round Markfield itself. In the New Plantation to the N.W. the Swithland Slate can be seen to overlie the Brand Grit and Quartzite. These are the youngest beds of the Precambrian succession. To the N. of Markfield, at the Altar Stone, the Slate Agglomerate further down in the sequence is exposed.

Continue N.W. along main road to Bardon Quarry (N.E. of road). Here the "good rock" is a welded orthobrecciated lava of Lacitic composition; the Slate Agglomerate occurs near the S. end of the quarry.

Continue N.W. along main road to E. outskirts of Coalville then turn N.E. on

road which leads towards Loughborough. About 2 miles (3.2 km) from the
road junction a short diversion along a drive leading to Charnwood Lodge
shows spectacular exposures of coarse agglomerates (Beacon Beds) of the
Maplewell Group.

At this point those without private transport should return to Coalville to
catch the bus back to Leicester.

With private transport continue N.E. to S. outskirts of Sheepshed then turn
S.S.E. on B5330 for 3 miles (4.8 km) before turning E.N.E. to Beacon Hill
where the Beacon Hill Hornstones are seen. Continue along same road to
Woodhouse Eaves. Here the Church Quarry shows the Brand Conglomerate
and Grit. It was in the underlying Woodhouse Beds (top of Maplewell Group)
in the North Quarry of Charnwood Golf Course at Woodhouse Eaves that the
first discovery was made of the problematical Precambrian fossil Charnia.

Follow the road N.E. to Quorndon (Quorn) then turn S.E. to Mountsorrel.
On the W. side of the road the Mountsorrel Granite (Caledonian) showing
a fossil landscape unconformably overlain by Keuper (Triassic) marl is
quarried on a large scale (permission required from quarry offices).

Return to Leicester.

Buxton Centre (Fig. 7.15)

Buxton, where there is plentiful accommodation, can be reached by rail or
road, and there is good local public transport. Additional excursions in
the area are described by Cope and others (1972).

E7.33   Buxton to Miller's Dale

Carboniferous sedimentary and volcanic rocks.

Access:- Walking from Buxton and returning from Miller's Dale by public
bus.

Maps:- Topo:- 119. Geol:- EW 111.

Walking distance:- 8 miles (12.8 km).

Itinerary:- Leave Buxton by Spring Gardens and go on to Ashwood Dale.
Continue E. to N.E. side of Buxton Gas Works where the cutting on the N.
side of the railway shows D_1 limestones resting on brown-weathering basalt
- the Lower Lava Flow of the area.

Beneath the lava and visible along the roadside as well as in the cutting
are the limestones of the Clee Tor Beds. A short distance E. of the Gas
Works a cart-track leads under the railway and into a small dry valley at
the entrance to which the Davidsonia septosa Band is exposed. This and
other fossils can be collected.

Continue E. along the main road to the Devonshire Arms Hotel opposite which
dolomitised limestones appear from under the Clee Tor Beds.

Walk past Pig Tor to Tapley Pike where a roadside section shows the dark
limestones of the Woo Dale Beds (S_1). After continuing for about 1 km
along the main road the spoil heaps associated with an abandoned quarry

in the Calton Hill vent will be seen. The exposures are now almost entirely confined to an olivine-dolerite though the Upper Lava flow can also be seen.

Continue along the main road to the R.A.C. box at the top of Blackwell (Sandy) Dale. Descend the dale by the road for about 550 yds. (500 m) to an outcrop of the Lower Lava Flow forming an inlier on the floor of the dale. The rest of the valley is in the Miller's Dale Beds (D_1). Further on dark limestones with Lithostration junceum of the Station Quarry Beds (D_2) are seen. These lie immediately below the Upper Lava Flow seen a little further down the road in a small disused quarry.

Descend the hill and cross the Wye by the road to the left before going on to the sidings yard at Miller's Dale station. Here a large washout in the Miller's Dale Beds is filled by basal Station Quarry Beds.

Return to Buxton by public bus.

E7.34   Peak Forest district

Carboniferous sedimentary volcanic and intrusive rocks.

Access:- The excursion is best carried out by private transport. A public bus can be taken to Peak Forest, returning from Castleton, but this involves a long walk.

Maps:- Topo:- 119. Geol:- EW 99.

Walking distance:- With private transport 5 miles (8 km). From Peak Forest to Castleton $15\frac{1}{2}$ miles (25 km).

Itinerary:- At Peak Forest village, to the E. and to the N. of Snelslow Plantation, a dolerite sill is exposed, intruded into beds of S_2 age. Along main road E., up a marked escarpment of D_1 limestones above the sill, before turning right to Tideswell. Through village and continue S. just beyond the point where the road from Litton enters from the left. Limestones and the Upper Lava Flow are exposed cut by a fault.

Continue S. to a small avenue of beeches on the left which should be followed to a Picnic area. The lava is seen, and just beyond, marmorised limestone in scattered exposures. The metamorphism is due to a dolerite sill which can be seen farther down the dale.

Walk S. along a grass-covered track through Tideswell Dale to a disused quarry in igneous rock up the slope to the W. Crumbly-vesicular basalt of the Upper Lava Flow is seen.

Retrace the route through the beech avenue and continue through the village to Litton. On reaching the main road turn to the left and go on for about 1 mile (0.6 km). At Lane Head, to the N. of Tideswell, follow a road running N.E. over a surface of gently dipping limestones of D_2 age. At the hamlet of Windmill the Hucklow Edge Vein is crossed, marked by old opencast workings. The scarp on the right is capped by the lowest grit of the Namurian.

Continue N. into Bradwell Dale where there is a good section in D_2 limestones overlain by the Nunlow Limestones of P age.

Through Hope to Castleton where a public bus can be caught to Buxton.

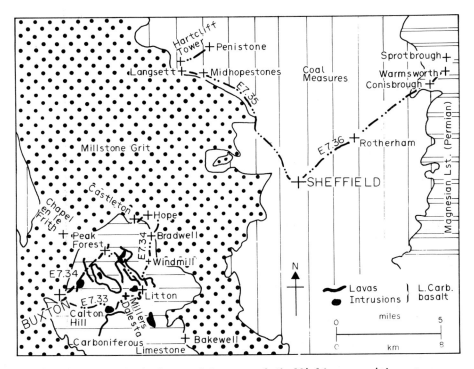

Fig. 7.15 Geological map of Buxton and Sheffield areas with routes.

Sheffield Centre (Fig. 7.15)

Sheffield can be quickly reached by road and rail and there is good local
public transport. For other excursions see Downie (1960).

E7.35   Langsett Area

Lower Coal Measures.

Access:- To Midhopestones road-end by public bus from Sheffield Midland
Station; return from Penistone by train. By private transport.Road walking.

Maps:- Topo:- 110. Geol:- EW 86, EW 87.

Walking distance:- 6 miles (9.6 km).

Itinerary:- From Midhopestones road-end walk along the main road W.N.W.
Features caused by the hard sandstone beds in the Lower Coal Measures are
well seen. About ¼-mile (0.4 km) from the road-end a small bridge to the
right of the road carries an old railway over a track and stream. In the
gully beyond, shales contain two "mussel-bands", 20 ft. (6 m) apart with
Carbonicola.

Return to road and continue towards Langsett. Turn left immediately before
the bridge over the old railway line. River exposures show coarse grits at
the top of the Rough Rock overlain downstream by shales with a thin coal.
The splintery black shale above the coal contains non-marine "mussels", but
the shale above the splintery shale yields marine fossils such as Dunbarella
Posidonia and Gastrioceras subcrenatum.

Just beyond the village of Langsett turn right up a steep hill formed by the
Middle Bed Rock, seen in a quarry. From the tip of Bullhouse Colliery good
uncrushed specimens of G. listeri and G. circumnodeum can be collected.
At the cross roads S. of the colliery turn right up a hill to Hartcliff
Tower, crossing features made by successively higher sandstone in the Lower
Coal Measures until large quarries in the Greenmoor Rock are reached. From
the Tower there is a fine view to the W. over the Pennines.

Down long dip-slope of the Greenoside Sandstone to Penistone.

E7.36   Conisburgh

Upper Coal Measures. Permian.

Access:- By public bus to Conisburgh. Walking on roads.

Maps:- Topo:- 111. Geol:- EW 87, EW 88.

Walking distance:- 6 miles (9.6 km).

Itinerary:- At Conisburgh the relationship is seen of the Magnesian Limestone.
(Permian) to the Upper Coal Measures in a large brick-pit S.E. of the main
road (A630) just before B6094 branches off. (Permission required) The
Upper Coal Measures contain plants and also a central bed with fish remains
and ostracods. The Permian starts with a pebbly bed overlain by marls
above which lie 25 ft. (7 m) of dolomitised oolitic limestone with

Bakevillia antiqua and Schizodus obscutus.

Continue N.E. along main road then turn N.W. at Warmsworth. A railway
cutting is crossed about ¼-mile (0.4 km) from the main road which shows
the Lower and Upper Magnesian Limestones separated by the Middle Permian
Marls. The same relationships can be seen at Sprotborough, ¾-mile further
on.

Turn S.W. to Cadeby where the "reef"-bearing Lower Magnesian Limestone is
worked in large quarries.

Return to Conisburgh and Sheffield.

7.4   North-East England   (Fig. 7.16)

North-East England is defined for present purposes as the region bounded to
the W. by the Pennine and Dent faults, to the S. by the North Craven Fault
and to the E. by the curving fault running from Hartlepool (Fig. 7.18).
It contains the northern part of the Pennines rising to 2930 ft. (893 m)
at Cross Fell and the high hills on the Scottish Border, including the
Cheviot, 2676 ft. (815 m). Near the North Sea, however, there is mainly
low-lying ground, heavily covered with glacial drift, which includes the
important North-East Coalfield.

The succession is as follows:-

                    Permian
                       (unconformity)
                    Coal Measures
                    Millstone Grit

            North   South

    Scremerston Coal Gr.   Limestones   } Carb.
    Fell Sandstone Gr.   Yoredale shale/   } 1st
    Cementstone Gr.   1st/sst. sequences   } Series

                    Upper Old Red Sandstone
                       (unconformity)
                    Lower Old Red Sandstone
                       (unconformity)
                    Silurian
                    Ordovician
                       (unconformity)
                    Ingleton = Precambrian (?)

Apart from the Cheviot/North Tyne area, North-East England is crossed by
numerous roads and most towns can be reached by public bus; there are good
train services along the East Coast Main Line, in the Durham/Newcastle
district and inland up the South Tyne valley.

For the N. part, excursions are described from Wooler and from Alnmouth,
and for the S. the Cathedral and University city of Durham is a good centre.
With some extra travel the Durham excursions can be undertaken from
Newcastle.

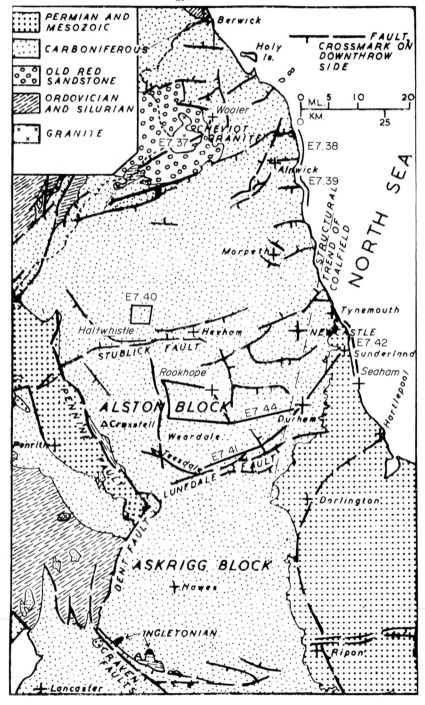

Fig. 7.16 Geological map of North-East England with routes.

Comparison with other areas:- The Lower Palaeozoic rocks are similar to those of the Lake District (Ch. 6.2). The thick Lower Old Red Sandstone lavas around the Cheviot (E7.37) are comparable to those of the Midland Valley (Ch. 7.5).

The Lower Carboniferous succession in the S. has considerable affinities with that of Central England (E7.32, E7.33) but the Yoredale shale sequence strongly contrasts with the strata of the same age further S. In the N. of the region the sequence shows considerable affinities to that of the Midland Valley (Ch. 7.5).

The Magnesian Limestone (E7.42) is the only marine Permian in the British Isles and spreads only to a very limited extent outside the region. In the S. part of the region the control of sedimentation by concealed blocks is effectively demonstrated. Here the Alston and Askrig Blocks (E7.41) have been shown by boring to contain Caledonian granites. Significant mineralisation is concentrated in the relatively thin sediments above these blocks. The mineralisation is epigenetic and of the lead/zinc/fluorine/barium type. The region was at one time an important lead/zinc mining field; some metal ores are still extracted but the main production is of fluorite.

The extensively quarried Great Whin Sill (E7.38, E7.40, E7.41) and a number of dykes are petrographically similar to the quartz-dolerites of the Midland Valley.

Wooler Centre

Wooler can be reached by public bus or private transport.

E7.37   North-Eastern Cheviot Hills

Lower Old Red lavas and granite.

Access:- By walking via Middleton Hall or by private transport to Middleton Hall then walking.

Maps:- Topo:- 80. Geol:- EW 3, EW 4, EW 5.

Walking distances:- 10 miles (16 km). 6 miles (9½ km) from Middleton Hall.

Itinerary:- Southwards on road to Middleton Hall on cementstones (Lower Carboniferous) with Lower Old Red lavas immediately to W., solid rock is however concealed by boulder clay. From Middleton Hall S.W. on minor road which crosses Cargy Burn and continues up Harthope Burn. Andesite-lavas outcrop on both sides of the valley for 2 miles (3 km) and then the Cheviot granite is reached. The contact is a crush-zone where it crosses the valley but by walking up the N.W. slopes for between ¼-mile (0.4 km) and ½-mile (0.8 km) a complex veining contact can be studied.

Return by same route.

Alnmouth Centre (Fig. 7.17)

Alnmouth can be reached by rail or bus from Newcastle or Durham as well as by private transport.

E7.38   Coast N. of Alnmouth

Carboniferous. Limestone and Millstone Grit Series. Permo-Carboniferous intrusions including Great Whin Sill.

Access:- By walking along the coast N. from Alnmouth, partly on roads or paths; by private transport by through road which is partly inland with stops or diversions as necessary, though some exposures may be missed. Low tide necessary.

Maps:- Topo:- 81. Geol:- EW 6.

Walking distance:- 11 miles (17½ km); reduced by using private transport.

Itinerary:- At the N. end of the golf course the foreshore at low tide shows fine sandstones, overlain by pebbly beds, of the Millstone Grit, dipping E.S.E. at 10°. Similar beds, displaced by several N.E. to E. faults, can be seen northwards to the headland of Loughoughton Steel. Near Boulmer Hall they are cut by an E.-W. quartz-dolerite dyke.

West of Loughoughton Steel a few inches of conglomerate at the base of the Millstone Grit, rest with slight unconformity, on shales over the Upper Foxton Limestone of the Upper Limestone Group.

The 1½ miles (2.4 km) shore and cliff exposure from here to E. of Howick provide the best section in Northumberland of the higher parts of the Upper Limestone Group, some 800 ft. (244 m) of sandstones, shales, thin limestone and at least one thin coal and fire clay. The Lickar Limestone which runs parallel to the coast for a mile past Sea Houses, contains a whole range of well-preserved fossils.

At the N. end of this section an E.-W. fault brings up the fossiliferous Acre Limestone of the Middle Limestone Group and, ¼-mile further on, the higher, equally fossiliferous Sandbank Limestone outcrops.

Cullernose Point where the coast juts E., marks the S. end of a 2 mile (3 km) cliff section of the Great Whin Sill. The quartz-dolerite with striking columnar jointing contains large inclusions of sandstone and grit; both they and the sediments under the sill show thermal metamorphism, including garnet formation in the grit.

A few yards S. of the sill a quartz-dolerite has been intruded along a fault. It may have been a feeder to the sill but this cannot be proved.

Return to Alnmouth by road and coastal paths or with private transport by inland roads.

E7.39   Coast S. of Alnmouth

Millstone Grit and Coal Measures.

Access:- By coastal walking. Low tide necessary.

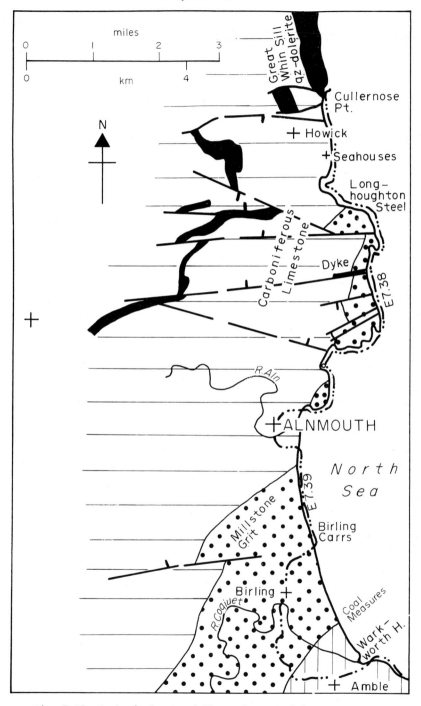

Fig. 7.17 Geological map of Alnmouth area with routes.

Maps:- Topo:- 81. Geol:- EW 6, EW 9.

Walking distances:- 10 miles (16 km). Reduced to 6 miles (9.6 km) if
private transport is available back from Warkworth Harbour.

Itinerary:- For 1¾ miles (2.8 km) S. of Alnmouth the sandy shore is fringed
with dunes, but at the Birling Scars a horizontal to gently dipping Millstone
Grit sequence is seen. This consists of sandstones, grits, shales and a
1-ft. coal underlain by sandstone with rootlets. On the low cliff at the
S. end of the Scars a glacial erratic of sandstone 60 ft. (18 m) long is
seen.

Turn inland through Birling to cross the R. Coquet, then regain the coast
at the S. pier beacon to examine a section of the Middle Coal Group. Gritty
and pebbly sandstone, dipping gently S.E., are followed by flaggy current-
bedded sandstones and greenish shales with accretions. The shales grade up
into 2-3 ft. of fire clay with carbonaceous traces, followed by 18-ft. of
sandy shales and clay ironstones with bands containing Carbonicola and
Naiadites. In the next 33 ft. (10 m) of strata there are 3 thin coals
each with underlying fire clay. Beyond a minor fault the Link House Sandstone resting on a 2 ft. (0.6 m) coal forms cliff and shore outcrops.

Return to Alnmouth by same walking route or by private transport.

Durham Centre

Durham is on the main London-Newcastle railway line.

E7.40   Haltwhistle district

Carboniferous Limestone Series. Great Whin Sill. Glacial deposits.
Roman Wall.

Access:- By train from Durham to Newcastle, then by train up Tyne valley
to Haltwhistle. By private transport from Durham via Corbridge to Haltwhistle.

Maps:- Topo:- 86. Geol:- EW 18, EW 19.

Walking distances:- 11 miles (17.6 km). If return is made from nearest
part of Whin Sill 5 miles (8 km).

Itinerary:- The run up the Tyne valley gives a number of views of fluvo-
glacial deposits and of workings for sand and gravel. From Haltwhistle
walk ½-mile (0.8 km) E. then N. on minor road to main road (following
straight Roman road) just S. of Burnhead. The minor road crosses a section
of the Carboniferous Limestone Series dipping N.; most outcrops are of sandstone. Continue N. to disused quarry in Great Whin Sill and to well-
preserved section of Roman (Hadrian's) Wall, with Milecastle. Walk E. along
Wall and crest of sill to W. end of Crag Lough where there is a high, N.-
facing cliff of the sill showing columnar jointing.

S. by minor road and by track to Melkridge over another cross-section of the
Carboniferous Limestone Series. W. back to Haltwhistle.

Those with private transport can travel E. along the Roman road before
crossing to the S. side of the Tyne at Hexham. This makes it possible to

visit on the way the well-preserved Roman Camp at Housesteads.

E7.41   Upper Teesdale

Ordovician. Carboniferous. Intrusives including Great Whin Sill. Glacial features.

Access:- For the whole excursion private transport is necessary; it is, however, possible to follow the excursion as far as High Force by taking a bus to Middleton-in-Teesdale and walking.

Maps:- Topo:- 91, 92. Geol:- EW 25.

Walking distances:- With private transport (whole excursion) 10 miles (16 km). From Middleton to High Force and back 10 miles (16 km).

Itinerary:- S.W. by road through Bishop Auckland to Staindrop. This part shows scenery typical of the Durham Coalfield, with drift mostly obscuring the solid formations. At Staindrop fork W. for Middleton, following the N. side of a structural depression which brings down Upper Carboniferous rocks between the Alston and Askrigg Blocks. Coarse Millstone Grit can be examined where overflow channels cross the road 2 miles (3 km) E. of Eggleston. Beyond Eggleston the rhythmic Yoredale facies of the Carboniferous Limestone Series forms the valley sides, and low down on the S.W. side there are large quarries in the Great Whin Sill.

Through Middleton and on to 200 yds. (183 m) W. of Newbiggin where transport could be left and a path followed to the Tees at Scoberry Bridge where the fossiliferous Cockle Shell Limestone is exposed. Upstream there is a complete cyclotherm through sandstone then shale down to the Single Post Limestone. This has been altered to a white crystalline marble and is underlain by sandstone shale beneath which the Whin Sill rises into the river bed. Thin mineral veins and replacements in the limestone carry zincblende, chalybite and pyrite. Specimens may be obtained from dumps near the S. bank, but the abandoned adits should not be entered, as like others in the orefield, they are in a dangerous condition.

From Wynch Bridge return to the road and continue to the High Force Hotel from which a path (entrance fee) leads to High Force, a picturesque waterfall over the lower part of the Whin Sill.

Above High Force the valley is wide. The Burtreeford Disturbance, an E.-facing faulted monocline roughly parallel to Langdon Beck, brings up lower beds, but the sill remains at much the same level. This part of the dale forms the Upper Teesdale Nature Reserve and permission for access should be obtained from the Nature Conservancy Officer, 2 Dents House, Middleton-in-Teesdale. Parking beside the road is usually available at Dale House, from which a path should be followed to Cronkley Bridge over the Tees, where the Whin Sill outcrops. Skiddaw Slates (L. Ordovician) of the Teesdale Inlier outcrop at Pencil Mill 8-mile (1.2 km) further upstream. Didymograptus and Diplograptus have been recorded, but are difficult to find. Several dykes cut the slates, and about ½-mile (0.8 km) further W. there are rhyolitic ashes.

From Pencil Mill there is a path to Cauldron Snout (2½ miles : 4 km), although there may be difficulty crossing the Mavie Beck tributary after rain. Falcon Clints gives a fine section of basal Carboniferous beds

beneath the Whin Sill and just above it the Cow Green Dam is sited on the
dolerite. During planning the possibility of leakage through limestone
beds was investigated (Kennard and Krill, 1919).

Return to Durham.

E7.42   North-East County Durham

Marine Permian. Glacial overflow channel.

Access:- For the whole excursion private transport is necessary; however,
by using a public bus to Sunderland, and another bus in the South Shields
direction, most of the N. part can be seen without undue walking. For the
coastal sections a low to fairly low tide is necessary.

Maps:- Topo:- 88, 93. Geol:- EW 21.

Walking distances:- Using private transport 4.5 miles (7.2 km). Using
public transport (N. of R. Wear only) 10 miles (16 km).

Itinerary:- As the Sunderland road (A690) is followed N.E. from Durham,
there is an excellent view ahead of the Permian escarpment. This is
breached at Houghton, but where about 90 ft. (27 m) of well-bedded
sparsely fossiliferous Lower Magnesian Limestone dips gently E.

On entering Sunderland follow the inner ring road for 2 miles (3 km) then
turn left at St. Luke's Road, which should be followed W. for 1 mile
(1.6 km) before turning right down an unsurfaced road. Park near a school
or near the entrance to Ford (or Hylton) Quarry. Permission to enter must
be applied for at the quarry office. On the S.E. side of the quarry there
is a magnificent exposure of Middle Magnesian Limestone reef passing
abruptly into bedded lagoonal dolomite on S.W. side. This can also be
viewed by walking down hill from the quarry entrance to the railway and
following a footpath N.E. for 600 yds. (550 m). Before leaving the area
it is well worth crossing the railway and turning right on to a footpath
downhill and bearing left to the foot of Claxhough Rock, a cliff of massive
reef rock resting on soft aeolian Yellow Sand, grey Marl Slate and bedded
Lower Magnesian Limestone. Across the R. Wear, about 43 ft. (13 m) of
Yellow Sand rest unconformably on sandstone near the top of the Middle
Coal Measures.

Rejoin the Sunderland inner ring road and drive N. to the coast at Leaburn
and along the coast road to the S.E. outskirts of South Shields where a
narrow lane leads to Trow Point.

Those without private transport should alight from the public bus as near as
possible to the lane and start the excursion on foot from there.

The section at Trow Point is:

    Collapse breccia   25 ft. (8 m) +

    Solution residue of up to 4 in. (0.10 m)

    Hartlepool Amydrite

    Cream dolomite, two beds, up to 14 in. (0.35 m)

    the upper with patchy stromatolitic

lamination at top (M. Magn. lst)

Large slumped blocks of L. Magn. lst up to 10 ft. (3.00 m)
on N.W. side passing into a single
slumped bed on S.E. side

Discordant slide plane at base

L. Magn. lst undisturbed 9.80 ft. (3.00 m) +

The shore section may be followed (starting only on a falling tide) to
Frenchman's Bay, where the slumped beds are beautifully exposed.

After returning to the Trow Point Lane the coast road should be followed,
either by private transport or on foot, towards Sunderland, with three
diversions to see the main features of the geology. At Marsden Bay the
Concretionary Limestone can again be seen, then by going inland on a minor
road opposite the bay and passing under the railway bridge, turning first
left onto Lizard Lane, the disused Marsden Hill Quarry can be reached.
Here, about 50 ft. (15 m) of spherulite Concretionary Limestone outcrops
beside a path leading up the quarry face. Further S., at Roker, about
43 ft. (13 m) of cream oolitic dolomite rests on the "Cannon Ball Rocks",
spherulitic Concretionary Limestone. A wide range of concretionary
structures is displayed beside the steps leading to the promenade.

Those on foot can finish the excursion here and walk into Sunderland for
bus or train to Durham.

With private transport it is possible to visit some exposures S. of the
Wear by following the ring road then turning S. onto Tunstall Road, ½-mile
(0.8 km) along which a rough lane (unsuitable for coaches and most cars)
leads to N. end of the Tunstall Hills. Here crags (capped by a triangulation point) of Middle Magnesian Limestone reef dolomite contain
bivalves and brachiopods; extraction is difficult.

By walking about 0.6 mile (1 km) along the ridge from the triangulation
point the reef can be followed S.E.; it is separated from lagoonal beds
of the Middle Lagoonal Limestone to the S.W. by Tunstall Hope, a spectacular glacial drainage channel.

Return to the transport and drive by coastal road to Seaham Harbour, where
the Seaham Beds, the highest in the Permian succession of the area are well-
exposed in sections round the North Dock. The Beds, partly collapse-
brecciated, consist of bedded grey limestones and dolomitic limestones
containing casts of bivalves and the stick-like alga Calcinoma (formerly
Tubulites). A walk along the beach to the N. (fairly low tide needed) below
the cliff reveals the collapse-brecciated Seaham Beds on earthy and distorted Seaham Residue left from the solution of salt and anhydrite.

Return to Durham.

E7.43   Pennine Orefield in Upper Weardale

Mineralisation in Lower Carboniferous of Alston Block. Little and Great
Whin Sills.

Access:- By private transport S.W. to Wolsingham then up Wear valley to
Stanhope. Permission required to collect from dumps of working mines.

Maps:- Topo:- 87, 92. Geol:- EW 19, EW 25.

Walking distance:- 3 miles (4.8 km).

Itinerary:- Three-quarters of a mile (1.2 km) S.E. of Stanhope the tunnel
emerges which has carried the water of the Kielder Reservoir and the Tyne
14 miles (22½ km) from Riding Mill. S. of the Wear there is the intake of
the 8½ mile (13½ km) tunnel to the Tees.

Immediately W. of Stanhope the Little Whin Sill, intruded into the Three
Yard Limestone, is seen in a roadside quarry.

Pass Eastgate, where a cement factory uses the Great Limestone and overlying shale, and the Cammock Eals Mine where fluorite and some galena are
extracted. Park transport at Westgate and take footpath N. beside Middle-
hope Burn, which gives a good section of Middle Limestone of Yoredale facies.
On reaching old Slit Lead Mine follow the opencasts up the E. side of the
valley to where the Slit Vein, carrying fluorite quartz and galena, is
exposed in the Great Limestone.

Return to transport and proceed up valley to Irehopesburn, N. of which
Blackdene Mine is operated by United Steel for fluorspar. Continue through
Wearhead to Cowshill and walk down to the R. Wear where the succession
between the Five Yard Limestone and the Tyne Bottom Limestone (Middle
Limestone Group) is exposed. The beds dip steeply E. owing to the Burtree-
ford Disturbance (see also previous excursion).

Turn N. to Allenheads. The Great Limestone, with corals and brachiopods,
is exposed in old quarries beside the road at High Greenfield. At Allenheads, where a mine has been reopened for fluorite by United Steel, turn
right onto a good secondary road to Rookhope.   Towards the village opencast iron ore workings can be seen on both sides of the valley above
Grover Lane Mine, worked for fluorite by United Steel; to the N.W. the
Redburn Mine (Weardale Lead Co. Ltd.) is worked in the Red Vein for
fluorite and lead ore.

At Rookhope a bore proved the concealed Weardale Caledonian Granite
beneath Lower Carboniferous at 1281 ft. (390 m).

Southwards to Eastgate and return to Durham.

7.5   Midland Valley of Scotland (Fig. 7.18)

The Midland Valley is essentially a graben with a broad synclinal structure,
but in detail it is complex, as its strata are affected by strong folding
of both Devonian and Hercynian dates and by numerous faults of Devonian,
Hercynian and probably Tertiary ages.

The Highland Boundary Fracture-zone (Ch. 5.4) forms the N.W. margin of the
Midland Valley and the Southern Uplands Fault (Ch. 6.3) the S.E.

The term Midland Valley is justified in so far as it forms low ground
relative to the Highlands and Southern Uplands, but the topography is
varied. It is not a valley in the ordinary sense of the term, but consists
of several low-lying areas, diversified by belts of hilly ground and
isolated hills mostly corresponding to igneous outcrops. Several of these

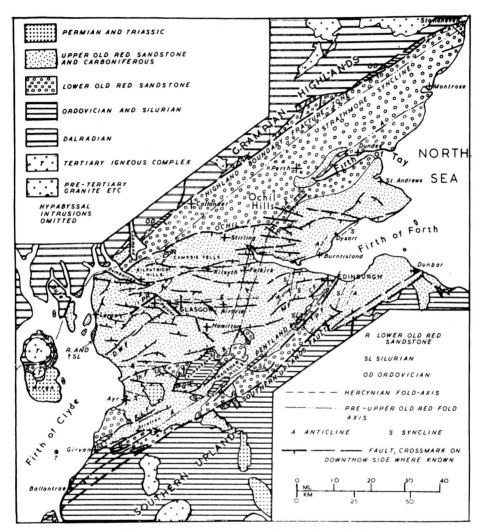

Fig. 7.18 Geological map of the Midland Valley of Scotland.
C.F. Campsie Fault. D.W.F. Dusk Water Fault. In.F. Inchgotrick
Fault. K.L.F. Kerse Loch Fault. M.F./C.F. Murieston/Colinton
Fault. P.F. Pentland Fault. P.R. Paisley Ruck.

hilly regions exceed 1000 ft. (305 m) and some 2000 ft. (610 m). The
highest point is Ben Cleugh, 2363 ft. (720 m) in the Ochil Hills.

The succession in the Midland Valley is as follows:

Triassic

Permian
    (unconformity)

Barren Red measures

Productive Coal Measures

Millstone Grit

West		East	
	⎧ U. Lst. Gr.	U. Lst. Gr.	
	⎪ Lst./Coal Gr.	Edge Coal Gr.	
Carb.	⎪ L. Lst. Gr.	L. Lst. Group	
Lst.	⎨ Upper Sediments	Oil-Shale Grps.	
Series	⎪ Clyde Plateau Lavas	Arthur's Seat Lavas	⎫ Calciferous
	⎪ Ballagan or	Ballagan or	⎬ Sandstone
	⎩ Cementstone Gr.	Cementstone Gr.	⎭ Series

Devonian ⎧ Upper Old Red Sandstone
              (unconformity)
          ⎨
          ⎩ Lower Old Red Sandstone
              (unconformity in places)

Silurian ⎧ Ludlow Series
          ⎨ Wenlock Series
          ⎪ Llandovery Series
          ⎩   (unconformity)

Ordovician ⎧ Ashgill Series
           ⎨ Caradoc Series
           ⎪   (unconformity)
           ⎩ Arenig Series

In most parts of the region there are extensive rail and public bus
services.

Four centres are suggested: Glasgow, Edinburgh, Ayr on the Firth of Clyde
and Arbroath on the North Sea.

Comparison with other areas:- Almost every series in the Midland Valley
shows differences in facies from the rocks of the same age in other parts
of the British Isles. The Lower Carboniferous, in particular, reveals
considerable variation within the region itself (see table above).

The shelf-facies with its shelly fauna of the Ordovician and Silurian of the
Girvan district (E7.56) contrasts with the thick greywacke/shale succession
of the Northern Belt of the Southern Uplands (Ch. 6.3) and, more strikingly,
with the graptolitic shale facies of the Central Belt (Ch. 6.3). Moreover
the American affinities of the fauna of the Girvan area, as opposed to the

fauna of the Lake District (Ch. 6.2) and even more of Wales (Ch. 6.1), supports the view that the Iapetus suture lies under the Solway Firth.

The ophiolite association of Girvan (E7.56) is also of significance in Plate Tectonics discussions.

The Lower Old Red Sandstone is of a strongly fluviatile/lacustrine type, similar in that respect to much of the formation N. of Devon, but the volcanics are much more widespread and thick (E7.48, E7.49, E7.55, E7.57, E7.58) than elsewhere, apart from the Cheviot (E7.37). The absence of the Middle Old Red Sandstone contrasts with its presence in the North of Scotland (E5.11 - E5.14).

The Upper Old Red Sandstone is fluviatile to lacustrine. The Lower Carboniferous, with its thick lavas (thicker in the W.),its thin limestones and its workable coals, is completely different from the carbonates of most of Central England (Ch. 7.3), of South Wales (Ch. 7.2) and of Bristol (Ch. 7.1). The Millstone Grit does contain some thick sandstones but is noteworthy for the presence in the W. (E7.54, E7.55) of basalt lavas. The lower part of the Coal Measures is similar to that elsewhere in Britain but the upper part contains the Barren Red Measures. The Permian (part of which may be in fact Stephanian) contains basic volcanics, present else-where only in the Exeter district (Ch. 7.1).

Igneous rocks are much more abundant in the Midland Valley than in any other Hercynian terrain in the British Isles, apart from Cornwall (Ch. 7.1). Volcanics range in age from Arenig to Permian, and intrusives from Arenig to Tertiary.

Glasgow Centre (Fig. 7.19)

E7.44   Fossil Grove and W. of Glasgow

Tree casts in Lower Carboniferous. Dolerite sill. Glacial and post-Glacial topography.

Access:- By bus from city centre to Victoria Park. Return by same route or, if excursion is extended (see below), from Great Western Road.

Maps:- Topo:- 64. Geol:- S 30.

Walking distances:- Victoria Park 1 mile (1.6 km). Returning from Great Western Road 2 miles (3.2 km).

Itinerary:- The route westwards is mainly over alluvial sands above marine clays and glacial deposits. These sediments hide a buried valley over 100 ft. (30.5 m) deep with a course which does not coincide with that of the modern Clyde.

The building covering the Fossil Grove is at the W. end of Victoria Park. The trees (Lepidodendron veltheimanium) are internal moulds in sandstone and were discovered in 1887 beneath a dolerite sill. Both inside and out-side the building interesting igneous intrusive features can be seen.

The excursion can be extended by leaving the Park by the N.E. gate and walking N. to Anniesland X then E. along Great Western Road. The route runs through drumlins of Boulder Clay with steeper ends towards the W.,

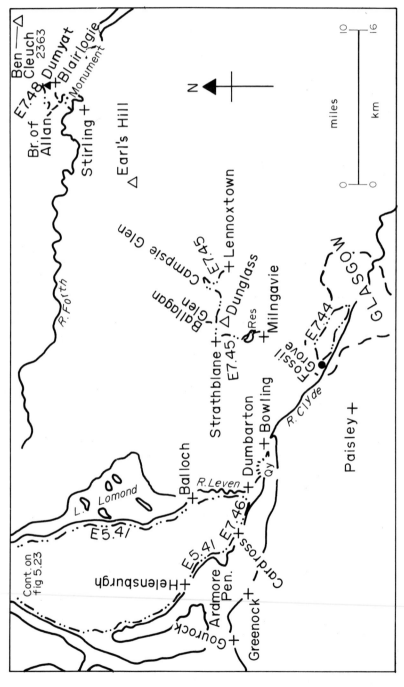

Fig. 7.19 Route map for Glasgow Centre.

the direction from which the ice travelled.

E7.45   Milngavie, Strathblane, Campsie Glen and Lennoxtown

Lower Carboniferous including Clyde Plateau Lavas.   Intrusions.

Access:- By train from Queen St. station to Milngavie;   return from
Lennoxtown by public bus.

Maps:- Topo:- 64.   Geol:- S 30, S 31.

Walking distance:- 11.5 miles (18.5 km).

Itinerary:- Turn E. from Milngavie station then N. up road leading to
Glasgow reservoirs round which the paths are open to the public.   Cross
the causeway between the two reservoirs and turn to W.   Above the path
there are exposures of a Lower Carboniferous dolerite sill intruded into
gritty sandstone, part of the Upper Sediments succession of the Scottish
Lower Carboniferous.

Join road at W. end of reservoir and climb path cutting off loop of road.
Near the top of the path there is an exposure of a basalt lava top overlain
by solid basalt of the next flow.   The lavas, stratigraphically below the
sandstone, are brought up by an E.-W. fault, one of a number downthrowing
to S. on the N. margin of the Glasgow Syncline.

Near the minor road running N. past Mugdock there are exposures of an E.-W.
quartz-dolerite dyke and of lavas of Markle-type basalt belonging to the
top of the Clyde Plateau sequence.   To the N. rises the spectacular trap-
featured scarp of the Campsie Fells.   The Markle-type basalts form the top
of the hills, brought up by the E.-W. Campsie Fault.   The view to the W.,
however, shows that the topography becomes reversed because here the lavas
form the Kilpatrick Hills, brought down by the Campsie Fault against softer
Upper Old Red Sandstone to the N.

West of the minor road, and immediately W. of the main road above Strathblane, quarried quartz-conglomerate, forming the base of the Upper Sediments,
rests on the top of the lavas.

At Strathblane turn E. along the Lennoxtown road.   A diversion to the N.
into the conspicuous Ballagan Glen makes it possible to examine the Ballagan
Beds beneath the lavas.   These consist of shales and cementstones (ferro-
dolomites) with gypsum; a thick sandstone occurs at the top of the sequence.

To the S. of the road a track leads to a disused quarry in the Dunglass
volcanic rock of Jedburgh-type basalt.   Spectacular, variably-orientated
columns suggest cooling from both the irregular walls of the pipe and a
solid cap.   The crag-and-tail form proves ice-movement from the W.

Continue along road to the E. then up a side-road to Campsie Glen.   From
the end of this road a path on the left bank of the Campsie Burn should be
taken.   Shattered basalt at the top of the sequence can be seen followed
across the Campsie Fault by disturbed Ballagan Beds, cut by a thin basalt
dyke.

The path then rises steeply alongside the stream; care should be taken in
wet or snowy conditions.   The Ballagan Beds are overlain by thin ash

followed by a succession of basalt lavas with amygdaloidal and sometimes
lateutic tops; the more massive part of each flow forms a small waterfall.
Where the slope above the left bank becomes easier an ascent should be made
to "Jamie Wright's Well" where there are exposures of fresh Markle-type
basalt.

Descend the road towards Lennoxtown. About 1¾ miles (2.8 km) from the town
an essexite intrusion forms a small crag. Half-a-mile further on a diversion
should be made to the E. along a track to long-disused quarries where there
are a few faces still exposed in the Hurlet Limestone at the base of the
Scottish Carboniferous Limestone Series downthrown by the Campsie Fault.
Fossils include productids, zaphrentids, corals and pterinopecten.

Descend to Lennoxtown.

E7.46   Bowling, Dumbarton and Cardross

Lower and Upper Old Red Sandstone. Lower Carboniferous. Volcanic rocks.
Permo-Carboniferous dyke. Glacial deposits and raised beach.

Access:- By train from Glasgow to Bowling, returning from Dumbarton or
Cardross.

Walking distances:- To Dumbarton 3.5 miles (5.6 km). To Ardmore Peninsula
and back to Cardross 11 miles (17.6 km).

Itinerary:- West of Glasgow the route passes through drumlin topography
with steeper faces to the W.N.W.

Above Bowling station road cuttings show several flows of Clyde Plateau
Lavas, mainly of the Jedburgh type. Prehnite can be found. Above Bowling
Harbour the lavas are cut by an agglomerate/basalt vent, and 9 other vents
form conspicuous features to the W.N.W. These are at the S.W. end of a line
of vents which when newly formed must have resembled the Chain of
Auvergne. A quarry (permission required) in the Drumbuck vent reveals
columnar basalt intruded, sill-like, into agglomerate.

Half-a-mile N.W. of Drum buck, Strowan's Well Road, on the E. side of the
main road, gives access to Crosslet Quarry, now disused, in an E.-W. Permo-
Carboniferous quartz-dolerite dyke cutting Upper Old Red Sandstone.

From here a road leads to the centre of Dumbarton. If the excursion is to
be continued to Cardross the R. Leven should be crossed at Dumbarton Bridge.
To the N. the Leven Valley leads to Loch Lomond, its thick fluvio-glacial
and alluvial deposits hiding a buried valley with a floor, E. of Dumbarton
Rock at least 224 ft. (68 m) below sea-level. The rock is an isolated
Lower Carboniferous neck filled mainly with basalt.

West of the R. Leven the bright-red Upper Old Red Sandstone (formerly
extensively quarried) is seen cut by another E.-W. dyke. Further W. an
E.N.E. fault brings down poorly-exposed Ballagan Beds, underlain to the
W. by Upper Old Red Sandstone which forms a cliff at the back of the well-
marked 30-ft. (raised) beach. At Cardross a N.E. fault (not exposed)
brings up the duller-red, often conglomeratic, Lower Old Red Sandstone.

North-west of Cardross on the W. shore of the Ardmore Peninsula, the Lower
Old Red Sandstone is overlain along an irregular unconformity by the Upper,

distinguished by its bright-red colour and by the absence of andesite boulders; both formations contain Highland boulders.

E7.47   Mauchline Basin (Fig. 7.21)

Upper Carboniferous/Permian sandstones and lavas. Permian teschenite sill.

Access:- By public bus to Mauchline. By train to Kilmarnock, thence by bus to Mauchline.

Maps:- Topo:- 70, 71. Geol:- S 14, S 22.

Walking distance:- 5 miles (8 km).

Itinerary:- Walk E.N.E. from Mauchline on Sorn road. Where this drops steeply at a sharp bend olivine-basalt lavas are exposed in a disused quarry. A lava-top contains large, flow-aligned amygdales.

Return to the centre of Mauchline and follow the Dumfries road S.E. North of the R. Ayr, instead of crossing a high viaduct, walk down the old road (S.W. of the main road) then descend a steep path through a wood to the R. Ayr. Bright-red, planar cross-bedded sandstones with millet-seed grains are seen in old quarries in the upper part of the succession. True dune-bedding can be seen further downstream in the R. Ayr at Stairhill, where orientation of barchan dunes indicates N.E. wind direction.

Near the base of a high cliff forming the right bank of the R. Ayr the sandstone of the old quarries pass down through an interbedded sequence into basaltic ashes which overlie basalt lavas slightly higher upstream.

Return to the road then walk to the old bridge where there are good exposures of a teschenite sill.

Return to Mauchline.

E7.48   Stirling and the W. end of the Ochil Hills

Lower Old Red Sandstone. Permo-Carboniferous quartz-dolerite sill. Glacial and post-glacial features.

Access:- By train from Queen St. station to Stirling and back. Return by public bus from Blairlogie to Stirling.

Maps:- Topo:- 57, 58. Geol:- S 39.

Walking distance:- 10 miles (16 km).

Itinerary:- From Stirling railway or bus station walk eastwards across the footbridge over the R. Forth to Cambuskenneth. The river flows in large meanders through the Carse (Flandrian) clays. To the W. Stirling Castle stands on a quartz-dolerite sill.

From Cambuskenneth N. to Stirling-Alloa road then E., along foot of scarp of columnar-joined dolerite, to path leading to highest point of sill, crowned by Wallace Monument. The sill is a faulted continuation of the Castle outcrop.

From the monument descend by another path to Causewayhead and then follow
the main road to the southern outskirts of Bridge of Allan. Here Kenilworth
Road should be followed to Mine Road which leads to a steep path through a
wood to an old adit. Further N.W. there is a more gently graded path,
signposted Coppermine.

The adit is in a fault-zone mineralised with copper and barytes striking
N.W. and cutting Lower Old Red Sandstone grits.

Return along Kenilworth Road then walk up Sherriff Muir Road. On the
steep ascent there are exposures of Lower Old Red Sandstone conglomerate
with basic volcanic boulders. Further on, where the road undulates across
moorland, andesite and basalt flows, with beds of agglomerate are seen.

North-east of the conspicuous hill of Dumyat a track branches S. across
moorland with numerous exposures of andesite and basalt. The track continues along a hillside high above Menstrie Glen, on the opposite side of
which, about a dozen andesite and basalt flows show well-marked trap-
featuring. These are underlain by tuff, with an E.-W. quartz-dolerite
dyke forming a waterfall and with a thick intercalation of platy-weathering,
flow-banded andesite.

The track descends gradually to Blairlogie below the S. face of Dumyat
where trap-featuring of the lavas is again seen; the upper part of the
hill is formed of a porphyry sill. North to north-north-west faults are
marked by gullies. Copper/barytes mineralisation has taken place along
some of the fractures, and there are several old trial adits.

The scarp face of the Ochil Hills, one of the finest in Scotland, follows
the Ochil Fault, downthrowing Coal Measures against the Lower Old Red
Sandstone. This is an active fault which gives rise to numerous minor
earthquakes.

From Blairlogie a bus can be caught to Stirling.

Edinburgh Centre (Fig. 7.20)

E7.49   Edinburgh and its neighbourhood

Lower and Upper Old Red Sandstone. Lower Carboniferous. Carboniferous
intrusions. Glacial features.

Access:- Start on foot from Princes St. Gardens, immediately W. of the
railway station. After visiting Arthur's Seat a return may be made by bus
or the excursion continued to Blackford Hill where a bus can also be caught.

Maps:- Topo:- 66. Geol:- S 32.

Walking distances:- To Arthur's Seat 3.5 miles (5.6 km). To Blackford Hill
6 miles (9.6 km).

Itinerary:- The classic crag-and-tail (ice-movement from W.) of Edinburgh
Castle Rock rises above Princes Street Gardens; the columnar-jointed
Jedburgh-type basalt can be examined from a path S. of the railway. This
path should be followed E. to an exit onto the steep road known as the
Mound which leads upwards to the High Street. This descends the "tail" of
the Castle Rock to Holyroodhouse (the Royal residence in Edinburgh) and

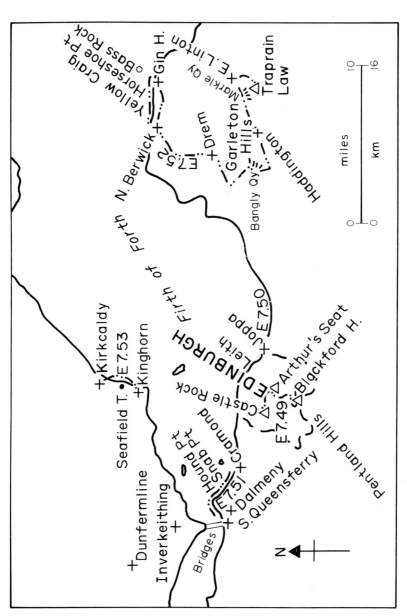

Fig. 7.20 Route map for Edinburgh Centre.

Holyrood Park (public, but hammers should not be used) where the Lower
Carboniferous, composite Arthur's Seat volcano can be studied (for details
see Mitchell et al, 1960).

Exposures are obvious; the following itinerary is suggested: From the
Holyroodhouse entrance the Queen's Drive leads to the N. end of Whinny Hill,
consisting of 13 basalt flows and some ash (well seen above the fourth
flow from the base) erupted from several vents (from petrographical
evidence).

A return should then be made to above the gate where a low crag marks the
N. end of the mid-Carboniferous Salisbury Crag teschenite (analcite-
dolerite) sill. Here the roof consists of baked shales and cementstones
(Ballagan Group). A path follows the foot of the crag which becomes higher,
then lower again, as the sill waxes and wanes. The lower contact against
baked Upper Old Red Sandstone (Holoptychius nobillisimus has been found) is
well-exposed. Where the sill bends from S.W. to S.E. it is cut by an E.-W.
quartz-dolerite dyke.

From the end of the sill the ascent of a steep slope to the top of Arthur's
Seat affords a chance to examine the coarse agglomerate and the basalt
intrusions of the Lion's Head and the Lion's Haunch vents. These are the
two main vents of the five in all which make up the composite volcano.

From the summit, on a clear day, it is possible to look across the Midland
Valley graben from the Highlands, 45 miles (72 km) away to the N.W. to the
Southern Uplands, seen beyond the Midlothian Coalfield, 12 miles (19 km)
distant to the S.E.

The type-locality of the Dunsapie-type basalt can be reached by descending
N.E. to the crag formed by the intrusion near Dunsapie Loch. A track, and
a road to the S.W., then lead past the columnar-jointed basalt intrusion of
Samson's Ribs over the low ridge formed by the St. Leonard's sill, intruded
into the Ballagan Group, to the S.W. exit from the park. A short distance
to the W. Newington Road is reached with frequent buses to the city centre.

Alternatively Newington Road then Mayfield Road (parallel to the W.) can be
followed S. across a railway bridge before turning W. into West Mains Road.
This leads past the Grant Institute of Geology of the University of
Edinburgh to Observatory Road rising to the top of Blackford Hill. This is
on the N.E. end of the Pentland Hills Anticline and consists of a single
flow of Lower Old Red Sandstone andesite.

A steep descent S.S.W. ends beside a disused quarry in the andesite with
veins of jasper and chlorite. By following a path N.E. another disused
quarry just S. of a pond is reached; here agglomerate, dipping N.E. is
exposed.

Beyond the pond, a gate opens onto a bus route.

E7.50   Joppa, S. shore of Firth of Forth

Succession from Scottish Carboniferous Limestone Series to lower part of
Productive Coal Measures.

Access:- By bus or car to Esplanade Terrace, Joppa. Low tide necessary.

Maps:- Topo:- 66. Geol:- S 32.

Walking distance:- 0.5 mile (0.8 km).

Itinerary:- The traverse starts on a massive sandstone outcropping N. of a paddling pool. About 90 yds. (82 m) further E., near the base of the sea-wall, the fossiliferous Calmy Limestone is exposed. This limestone, though thin, is remarkably persistent in the Midland Valley. It is faulted to the W. nearer low tide mark.

About 60 ft. (18 m) below the limestone grey shale marks another persistent horizon containing Edmondia punctatella and other fossils.

For the next 80 yds. (73 m) the shore is occupied by a rapidly varying succession of sandstones, shales, fire clays and then coals. Several of the shale beds contain Lingula and some, generally marked by ironstone nodules, yield marine fossils. This succession is overlain by the Castle-cary Limestone, here about 15 ft. (4.6 m) thick including shale partings. Above comes a bed of carbonaceous shale with Anthroconauta, ostracods and fish remains, overlain by sandstones and fire clays.

After an 80 ft. (24 m) gap in the succession, due to a cover of mud and boulders, the Passage Beds form a 90 ft. (27 m) wide outcrop. These consist of sandstones, shales, fire clays and marine horizons with current-bedded, sometimes pebbly sandstones dominant at the top. These were formerly known as the Roslin Sandstone Grit and are conventionally shown as Millstone Grit. The Productive Coal Measures are considered to begin at the base of the 7-ft. Coal of which only a part is seen on the shore.

The Coal Measures succession continues along the shore for about 350 yds. (320 m); 5 coal seams occur, marked only by sand-filled hollows. Current-bedded sandstones are exposed and several shaley "mussel-bands" yielding non-marine lamellibranchs of the Carbonicola communis Zone.

E7.51   Queensferry - Cramond, S. shore of Firth of Forth

Lower Carboniferous of Oil-Shale facies. Carboniferous and Permo-Carboniferous intrusions.

Access:- By train, bus or car to South Queensferry (Dalmeny station). Return by bus or car from Cramond. Many of the exposures are below half-tide level.

Maps:- Topo:- 65. Geol:- S 32.

Walking distance:- 6.5 miles (10.4 km).

Itinerary:- Queensferry has been a public ferry since about 1130. The railway bridge, opened in 1890, and the road bridge, opened in 1964, rank among the world's greatest.

Start at Queensferry Harbour, 0.5 mile (0.8 km) W. of the railway bridge. On the W. side of the harbour, sandstones, shales and cementstones of the Upper Oil-Shale Group occur. On the E. side of the harbour 8 ft. of oil-shale (Dunnet Shale) are exposed, overlain by a thin, pale-grey dolerite sill. The shale is on the W. limb of an anticline followed by a syncline. Further E. there is a steadily descending succession.

One hundred yds. (91 m) E. of the railway bridge, 3 ft. of the freshwater Burdiehouse Limestone, dividing the Lower and Upper Oil-Shale Groups, are seen at low water. The limestone contains plant and fish remains and is overlain by the Camps Shale. The succession is disturbed by a minor fault downthrowing N.W. In the thick underlying sandstone a thin, pale-grey dolerite sill occurs, and below the sandstone there is a thin sill of "white trap". This is dolerite altered to carbonate rock by gases driven from carbonaceous beds during intrusion. To the E. the Pumpherston Shales outcrop, consisting of about 86 ft. (26 m) of shaley strata with beds of oil-shale.

Between the Pumpherston Shales and two conspicuous beds of yellow-weathering cementstone the limy Pumpherston Shell-Bed contains <u>Orthoceras</u>, <u>Lingula</u>, lamellibranchs and ostracods; the fossils are mostly pyritized.

Further E., S.W. of Hound Point, a thick sill of coarse teschenite (analcite -dolerite) occurs. Hound Point itself is made of a sill of quartz-dolerite of Permo-Carboniferous age. Its chilled base over indurated sediments is seen on the E. side of the point.

In following the shore section to Cramond Ferry a diversion is necessary round an oil-terminal.

A quartz-dolerite sill (possibly the same sill as that at Hound Point brought down by faults) is seen at Snab Point. A broad development of the "25-ft. (7.6 m)" raised beach is evident.

Thin teschenite sills occur at Cramond Ferry.

<u>E7.52   Garleton Hills, Traprain Law and North Berwick</u>

Lower Carboniferous sediments. Lower Carboniferous volcanics, mainly acid. Phonolite laccolith. Volcanic necks.

<u>Access</u>:- The excursion can be most readily undertaken by car. It can also be carried out by public buses from Edinburgh on the East Linton, Haddington and North Berwick routes. In this case, unless a long day is planned, it is better to divide the excursion into two, spending one day in the Traprain Law/Haddington area and one near North Berwick.

<u>Maps</u>:- <u>Topo</u>:- 60, 67. <u>Geol</u>:- S 33.

<u>Walking distances</u>:- Whole excursion with private transport 9 miles (14.4 km). Traprain Law/Haddington area, walking 8 miles (12.8 km); near North Berwick 8 miles (12.8 km).

<u>Itinerary</u>:- By car from Edinburgh on the A1 then S. to Traprain Law Quarry (permission required). By bus to East Linton then walk to the quarry.

The Traprain Law laccolith is formed of fresh phonolite. Minerals in geodes include analcite, prehnite, natrolite and apophyllite.

Return to East Linton and drive or walk half-a-mile to the W.S.W. where the large disused Markle Quarry, 150 yds. (137 m) off the main road, is the type-locality for this type of basalt.

By car or bus back through Haddington to the junction with a side road
2 miles (3.2 km) W. of the town. Along this road 1.2 miles (1.9 km) N.N.W.
of the main road there is a large disused quarry at Bangly in trachyandesite
with aegerine.

About a mile to the E. Smithy Cottage is reached at a crossroads on the
Haddington-Aberlady road. Here a car must be left, as the route continues
along a lane to the S. side of Phantassie Hill, consisting of fine-grained
trachyte lava with conspicuous sanidine phenocrysts. To the N.W., at the
long-abandoned Garleton Haematite Mine, a vein was worked in a N.N.W.
fracture-zone in trachyte. At an old adit strings of kidney-ore can be
seen.

Travellers with private transport should return to the crossroads and
drive to a road junction close to Garleton Castle and farm. Walkers can
reach the same point by following the lane E. and then the Haddington-Drem
road N. South of the road junction the Cae Heughs scarp of the Garleton
Hills consists of trachyte-lava overlying a mugearite flow and basalt flows
to the N. Trachytes and trachyandesites also overlie the basalts to the W.
and N. and are seen in the Skid Hill roadstone quarry S.W. of the road
junction from which walkers can return S.S.E. to Haddington.

On the car journey to North Berwick, isolated exposures of trachyte are
seen around Drem, although boulder clay covers much of the solid rock.
Near North Berwick the phonolitic trachyte of North Berwick Law makes a
conspicuous feature.

The excursion along the shore E. of North Berwick, which can also be reached
by bus, should be undertaken around low tide. Between the pier and the W.
side of the paddling pool flows of Calciferous Sandstone age outcrop. A
50-ft. flow of Markle basalt is followed S.E. in ascending sequence by two
thin flows of mugearite, a 40-ft. flow of Dunsapie basalt and two 40-ft.
flows, with vesicular tops, of kulaite; this has affinities with mugearite
but contains brown hornblende and analcite.

The next 1100 yds. of shore is occupied by bedded pyroclastic and sedimentary strata penetrated by a basalt plug known as the Yellow Craig. The
sequence ends at a narrow horst bringing up tuffaceous sandstones and red
marls. To the E. of the horst there is a disturbed zone with steeply-
dipping thin-bedded tuffs. Between this zone and Quarrel Bay the bedded
pyroclastic/sedimentary sequence is cut by the two large agglomerate-filled
vents of Pattan Crag and Horseshoe Point. At Quarrel Bay a fault brings up
the sandstones and red marls again. These are cut by the Gin Head and
Tantallon vents.

From the S. side of the Gin Head vent the road can be joined at the sharp
bend S. of Castleton; 4 walking miles (6.4 km) can be saved if private
transport has been brought on here.

E7.53   Kinghorn to Kirkcaldy

Lower Carboniferous sediments and volcanics. Dolerite sills.

Access:- By rail or bus. Low tide necessary.

Maps:- Topo:- 59, 66. Geol:- S 40.

<u>Walking distance</u>:- 3 miles (4.8 km), mainly on rocky shore.

<u>Itinerary</u>:- The shore-section starts at Kinghorn Pier and runs N.N.E.
obliquely across an upward succession dipping E. at 20°-30° off the
Burntisland Anticline and into the syncline of the Fife Coalfield.
Sandstone with <u>Stigmaria</u>, at the pier, is followed by a 600 yds. (549 m)
wide outcrop of Calciferous Sandstone basalt lavas with sandy intercalations.
About 12 ft. (3.5 m) of mixed strata above the basalts is followed by the
12 ft. (3.5 m) thick First Abden Limestone. The succession is repeated by
a N.-S. fault. The limestone and associated shales are richly fossiliferous.

Above the marine limestone there is a basalt flow with pillow structure,
part of a mixed 31 ft. (10 m) volcanic/sedimentary succession (close to old
lime kilns) which intervenes between the First Abden Limestone and the
Second Abden Limestone, also richly fossiliferous. There is some doubt
which of the two forms the base of the Scottish Carboniferous Limestone
Series.

The Second Abden Limestone strikes parallel to the shore for 600 yds.
(549 m) and is then overlain by 85 ft. (26 m) of sandstones and shaley
sandstones intruded by a teschenite sill with "white trap" margins.

East of four old lime kilns, 270 yds. (80 m) S. of the conspicuous landmark
of Seafield Tower, the limestones and calcareous shales collectively mapped
as the Seafield Tower Limestone contain an abundant coral and brachiopod fauna.
The tower stands on a 50 ft. (15 m) thick current-bedded sandstone overlain
on the seaward side by 180 ft. (55 m) of sandstones, shales and thin limestones known as the Kinniny Limestones, also richly fossiliferous. These
are equivalent to the Hosie Limestones of the W. part of the Midland Valley
and therefore form the top of the Lower Limestone Group. They are succeeded
by the Limestone Coal Group; a quartz-dolerite sill altered at its margins
to "white trap" intrudes the base of the group and outcrops at the shore-
ward end of a breakwater and forms skerries offshore. The coals of the
Limestone Coal Group have weathered into sand-filled hollows; the exposures
seen N. of the breakwater towards Kirkcaldy consisting of current-bedded
sandstones with some shales containing ironstone nodules.

<u>Ayr Centre</u> (Fig. 7.21)

Ayr is easily reached by rail or road, and there are local public transport
services. There is a considerable amount of accommodation.

<u>E7.54   Clyde coast N. of Ayr and Kilwinning area</u>

Carboniferous including Millstone Grit lavas and bauxite. Permian and
Tertiary intrusions.

<u>Access</u>:- By train or car from Ayr to Troon. After coastal excursion there
by same means to Kilwinning then return from Ardrossan. Low tide necessary
for some parts of coast.

<u>Maps</u>:- <u>Topo</u>:- 70. <u>Geol</u>:- S 22.

<u>Walking distances</u>:- Troon section 3.5 miles (5.6 km). Kilwinning-Ardrossan
section 5 miles (8 km).

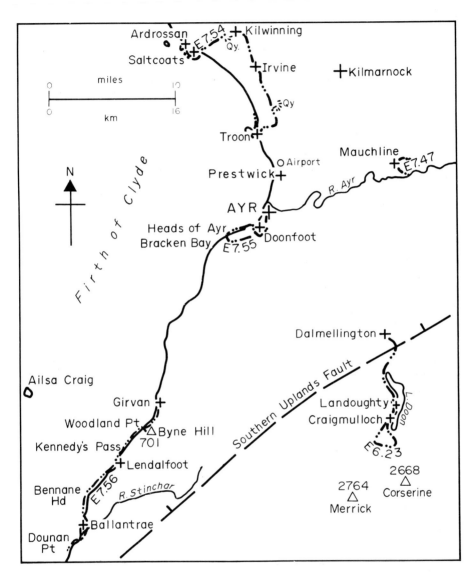

Fig. 7.21 Route map for Ayr Centre.

Itinerary:- Start at the S. side of the neck of the peninsula at Troon
where hornfelsed Coal Measures dip S. off the chilled top of the Troon
teschenite sill. The interior of the sill is of gabbroic texture in parts.

Walk along the shore N. for about 1.5 miles (2.4 km), over a thin sill of
teschenite, before reaching basalt lavas of Millstone Grit age cut by a
thick N.W. Tertiary basic dyke.

Return to Troon and travel to Kilwinning. On the way, if private transport
is available, a visit can be made to the large Hillhouse Quarry (permission
required) worked in a sill of dolerite of kylite type characterised by
analcite and abundant, fresh olivine. The columnar jointing is spectacular,
with the joint gaps infilled with late stage hydrothermal minerals.

From Kilwinning station walk W.S.W. along the Ardrossan road; two small
hills to the N. mark Permian agglomerate vents. On the S. side, Doubs
Quarry (permission required) is worked for refractory material in the
Ayrshire Bauxitic Clay. This is an impure bauxite formed from both in situ
underlying Millstone Grit lavas and alluvial decomposition products of the
lavas. At the top of the quarry, in places, the bauxite is overlain by a
thin, plant-bearing coal, taken as the base of the Productive Coal Measures.

Leave the main road on the E. outskirts of Saltcoats and walk to the shore
which should be followed in a general westerly direction. At the N. corner
of a bay, Coal Measure sediments, including two coal seams and dipping S.E.,
are exposed. A mussel-band can be easily found as it lies immediately above,
and is baked by, a conspicuous teschenite sill forming the N. end of the
bay. The mussel-band contains Anthracomaia and other lamellibranchs.
Beneath the sill the Coal Measures outcrop on a broad intertidal platform
as far W. as the bathing pool. They are cut by two thin teschenite sills
and several N.W. Tertiary dykes and also by the thick Saltcoats Main Sill
forming a point. This is composite and consists of three parts, believed
to have been intruded successively, namely flow-banded teschenite, biotite-
teschenite and hornblende-picrite. Underlying coal is burnt to columnar
coke against "white trap". At the W. end of the bathing pool about 15 ft.
(1.5 m) of Ayrshire Bauxitic Clay rests on decomposed lava. Walk N. into
the next bay (South Beach) at the S.E. end of which three fossiliferous
limestones of the Upper Limestone Group show through the sand.

Go back to the road and return to Ayr by private transport or by train from
Ardrossan (South Beach) station.

E7.55   South-West of Ayr

Lower Old Red Sandstone lavas. Upper Old Red Sandstone and Lower
Carboniferous sediments. Millstone Grit sediments, lavas and bauxitic
clay. Volcanic rocks. Tertiary dykes.

Access:- By public bus from Ayr to Doonfoot,returning from near Heads of
Ayr. Falling tide preferable.

Maps:- Topo:- 70. Geol:- S 14.

Walking distance:- 4.5 miles (7.2 km).

Itinerary:- From Doonfoot walk down to shore to examine exposures of decomposed basalt-lava of Millstone Grit age. These pass up into their

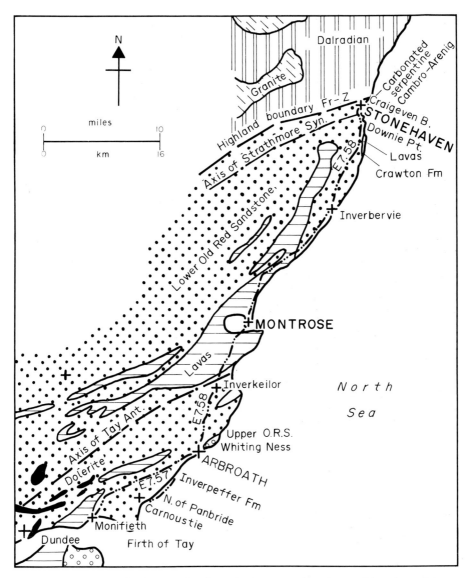

Fig. 7.22 Geological map of Arbroath-Montrose-Stonehaven District with routes.

alternation product, the Ayreshire Bauxitic Clay, and are underlain by red and green sandstones with plant remains, dipping N.E. at 10°.

To the S.W. Greenan Castle is built on a knoll of coarse pyroclastics filling a volcanic neck of probable Lower Carboniferous age.

The Millstone Grit overlaps onto a low horizon in the Lower Carboniferous, and for the next $1\frac{1}{4}$ miles there are fairly frequent exposures of cementstones, marls and sandstones of the Calciferous Sandstone Series, dipping N.E. and cut by N.W. Tertiary dykes of the Arran swarm. Quarter of a mile E. of the exposed E. edge of the Lower Carboniferous pyroclastic vent which forms the Heads of Ayr there is a thin bed of fossiliferous shale. The fossils include small mussels, <u>Lingula mytiloides</u>, <u>Spirorbis</u> and fish remains.

The Heads of Ayr vent is composite, with a S.W. vent, showing sag and collapse structures, faulted against a N.E. vent filled with intrusive breccia and monchiquite-basalt.

W. of the vent there are further exposures of Calciferous Sandstone strata brought down by a fault at the S. end of Bracken Bay against pink sandstone of the Upper Old Red Sandstone. Where the cliffs become steeper the Upper Old Red Sandstone is faulted against Lower Old Red andesite-lavas; a volcanic neck considered to be of the later age complicates the junction. The lavas have sedimentary intercalations and contain agates in vesicles.

From the S. end of Bracken Bay walk up to the main road across a boulder clay surface.

E7.56   Girvan and Ballantrae

Ordovician and Silurian sediments. Ordovician igneous rocks of ophiolite association. New Red Sandstone. Tertiary dykes. Raised beach.

Further details are given, for example, by Lawson and Lawson, 1976, 102-104.

Access:- By train or private transport from Ayr to Girvan, thence by local bus or private transport along parts of coastal road to Ballantrae.

Maps:- Topo:- 76. Geol:- S 7.

Walking distances:- 4 to 12 miles (6.4 to 19.2 km) depending to what extent transport is used along the coastal road. Low tide necessary for most exposures.

Itinerary:- Start at Woodland Point; parking is possible close to Milestone 3 S. from Girvan.

Near low tide mark at Woodland Point dark shales yield <u>Monograptus</u> and are followed southwards by impure limestones containing pentamerids and other typical shelly fauna fossils of Silurian age. Current-bedding in sandy and pebbly basal Silurian beds shows that the dip to the S.E. at 70° is inverted. The Silurian beds unconformably truncate asymmetrically folded Upper Ordovician grit. These grits, with siltstones (Caradoc), containing a sparse shelly fauna, continue to Kennedy's "Pass" where there are excellent exposures of basal Caradoc conglomerate with clasts of Arenig igneous rocks. The unconformity cuts out the Llanvirn and Llandeilo Series; the contact on the shore-section is, however, a fault.

About 10 miles (16 km) out to sea Tertiary riebeckite-granite forms the
island of Ailsa Craig.

Arenig agglomerates and spilite-lavas (some with pillow structure), black
shales (with scarce Arenig graptolites) and radiolarian cherts are well
exposed along the coast from Kennedy's Pass to Bennane Head. Tight folding
with vertical axes is conspicuous in the raised beach cliff. A more
northerly serpentinite between Pinbain Burn and Lendalfoot is thrust over
the volcanics with a sheared zone containing tectonic melange exposed at
Pinbain Burn   The dyke-like intrusions into the serpentinite are coarse-
grained cumulate gabbros and pyroxenites in the serpentinised peridotite,
tilted up during later Caledonian deformation. Dolerite dykes are conspicuous as elongated former sea-stacks on the raised beach platform. A
more southerly serpentinite is not well exposed along the coast.

Between Bennane Head and Ballantrae the Lower Palaeozoic is overlain by red
breccias and sandstones of New Red Sandstone facies. Both these and the
older rocks along the coast are cut by basic Tertiary dykes of the Arran
swarm.

At Ballantrae the R. Stinchar is diverted southwards by a striking shingle
bar. South of the bridge over this river the shore should be regained and
closely followed to Downan Point, where the spilitic pillow-lavas are
perfectly displayed and can be easily used to show that the sequence here
is inverted.

To the S.E. the Stinchar Fault brings down an Upper Ordovician succession
which is thicker and of deeper water facies than that seen S. of Girvan.

Geologists with their own transport may continue the excursion from Ballantrae to the Southern Uplands Fault at Glen App and beyond (E6.24).

Alternatively a stop may be made on the way back to Girvan to climb Byne
Hill. The summit consists of trondhjemite which grades down through diorite
to gabbro. On the S.E. flank of the hill the gabbro is chilled against
serpentinite. The contact could be interpreted as the Ordovician Moho.
The whole complex is overlain by Caradocian conglomerate, well seen in crags
on the N. slopes.

Arbroath Centre (Fig. 7.22)

Arbroath can be easily reached by train or by private transport. There are
local public bus services and a fair amount of accommodation.

E7.57   Arbroath to Monifieth

Lower Old Red Sandstone. Upper Old Red Sandstone. Raised beach.

Access:- By public bus or by private transport. Road and coastal walking.
Fairly low tide necessary.

Maps:- Topo:- 54. Geol:- S 49.

Walking distances:- 8 miles (12.8 km).

Itinerary:- Walk E.N.E. to the end of Arbroath promenade at Whiting Ness.
On the cliff nearly flat red sandstones and conglomerates of the Upper Old

Red Sandstone unconformably overlie red sandstones of the Lower Old Red Sandstone, dipping S.E. at 15°-20°. Further N.E. the Upper Old Red Sandstone is banked against a cliff of lava. The Upper Old Red Sandstone shows current-bedding and washout structures and contains cornstones (replacement limestones which are probably fossil calcareous soils); one of these penetrates downwards through the unconformity and has developed carbonate veins in the Lower Old Red Sandstone.

Return to the centre of Arbroath then travel by public bus or private transport S.W. for 3 miles (4.8 km) on the main Dundee road, then turn S.E. onto a minor road leading to Inverpeffer Farm. Walk to the coast and along the shore to Newton of Panbride, if a car can be sent on, or if a bus is being caught; if this cannot be arranged, walk as far as desired along the shore then return to the car at Inverpeffer Farm.

The shore section reveals a continuous outcrop of Lower Old Red Sandstone sandy and conglomerate sediments dipping S.E. at 15°-20° on the S.E. limb of the Tay Anticline. Sedimentary structures are well seen, and there are several N.E.-striking faults.

Continue by car or bus along the main Dundee road. Around Monifieth there is a broad raised beach backed by a feature at 50 ft. (15 m) O.D. The shore section on the S.W. side of Monifieth shows Lower Old Red Sandstone sediments overlying andesite lavas.

Return to Arbroath.

E7.58   Arbroath to Stonehaven

Dalradian. Cambro-Ordovician. Carbonated serpentinite along Highland Boundary Fracture-zone. Lower and Upper Old Red Sandstone.

Access:- By private transport or public bus with walking diversions. Those without private transport might find it more convenient to stay in Montrose or Stonehaven or to divide the excursion into two. Fairly low tide necessary.

Maps:- Topo:- 45, 54. Geol:- S 49, S 57, S 67.

Walking distance:- 9 miles (14.4 km).

Itinerary:- Travel N. from Arbroath on the Aberdeen road by private transport or public bus. N. of Inverbervie the road rises steeply onto Lower Old Red Sandstone lavas. Further on it drops to the Montrose Basin, occupied by Quaternary clays.

The car or bus should be left at the bridge over the Catterline Burn, 6 miles (9.6 km) N. of Inverbervie. An unclassified road leads to the coast at Crawton Farm. On the W. side of the bay a coarse conglomerate of Lower Old Red Sandstone age forms a 100-ft. cliff. Derivation from both a local source and from the Highlands is shown by boulders both of the nearby basalts and of quartzite and of granite. Below the cliff there is a spectacular storm beach.

The dip of the Lower Old Red Sandstone is westwards into the Strathmore Syncline. A traverse eastwards along the N. shore of the bay makes it possible to examine four basalt lava flows with intervening sediments,

mainly conglomerates with a high proportion of Highland rocks. The centres
of the flows are massive with columnar jointing. The lowest flow contains
exceptionally large amygdales with chalcedony, clear quartz, amethyst and
calcite.

After walking back to the main road continue N. to Stonehaven, crossing on
the way the axis of the Strathmore Syncline. Leave private transport or
the public bus at Stonehaven Harbour and walk E. to Downie Point. This
headland is composed of about 600 ft. (183 m) of conglomerate belonging to
the Dunnottar Group of the Lower Old Red Sandstone and containing boulders
of Highland origin and also some serpentinite and chert boulders from the
Highland Border.

The vertical, N.E.-striking beds are on the steep, N.W. limb of the
Strathmore Syncline and are cut by a 40-ft. (12 m) quartz-dolerite dyke.

Walk through Stonehaven then go down to the shore at the little inlet of
Cowie Harbour where there is a spectacular intertidal platform eroded across
vertical Downtonian striking N.E. On the S. side of the inlet there is a
bed of tuff and to the S.E. of this bed shale with Dictyocaris. Walk along
the shore N.E., crossing two felsite dykes, to Craigeven Bay. The conspicuous feature here is a steep band of orange-weathering dolomitic rock-
carbonated serpentinite along the Highland Boundary Fracture-zone. On its
S.E. side spilite lavas (pillows can just be made out), black shales and
irony cherts are seen. Fossils, which are difficult to find, indicate an
Arenig, or possibly late Cambrian age.

North of the Fracture-zone Upper Dalradian gritty metagreywackes and slates
are well exposed. In spite of the cleavage, graded bedding can be made out
which shows the psammites to be younger than the pelites.

Return to Stonehaven Harbour, then to Arbroath.

7.6   Central Ireland (E7.24)

Central Ireland is regarded as extending to the Highland Boundary Fracture-
zone on its (partially hidden) course across Ireland, excluding the Irish
"Southern Uplands" and Connemara. In the S. it extends to the Hercynian
front. Apart from a coastal stretch near Dublin, on the E. the region
comes against the Lower Palaeozoic and granite of South-East Ireland.

In the W. fine coastal exposures overlook the Atlantic. Much of the region
is between only 200 and 400 ft. (60-120 m) above sea-level (the Central
Irish Plain). Hills and mountains, mainly eroded in anticlinal inliers of
Old Red Sandstone and Silurian, are more common in the S. and rise to
3127 ft. (952 m) at Brandon Mountain, N. of Dingle Bay.

The succession is as follows:-

> ?Oligocene
> Eocene (lavas)
> Cretaceous
> Jurassic (Lias)
> Triassic
> Permian
> Coal Measures

Millstone Grit
Carboniferous Limestone
Old Red Sandstone
Ordovician
?Dalradian

Nearly the whole of the Central Plain is underlain by the Carboniferous
Limestone, covered by a thick mantle of glacial deposits which greatly
limits the exposures in a number of districts. In the W., however, in
Co. Clare and Fermanagh, there are extensive, bare limestone pavements
with cave systems.

Central Ireland is crossed by a number of roads and by a few railways.
Two centres are suggested, Dublin (also a centre for part of Eastern
Ireland, Ch. 6.5) and Limerick.

Comparison with other areas:- The oldest rocks occur in the N.E. of the
region, in the Tyrone Inlier. These biotite-schists could be Dalradian but
they are of higher regional grade than Dalradian rocks exposed a short distance away N. of the Highland Boundary Fracture-zone (Ch. 5.4) and may,
therefore, be older.

The Lower Palaeozoic strata broadly resemble those of Eastern Ireland
(Ch. 6.5) and Wales (Ch. 6.1) and have similar faunas.

The Old Red Sandstone is of the fluviatile/lacustrine type seen in much of
Great Britain (Ch. 7.2, 7.4, 7.5) but differs in the absence of the Lower
division in most of the region; in fact the Dingle Beds in the S.W. may be
the only Lower Old Red Sandstone in the southern part of Central Ireland.

The Carboniferous Limestone is mostly of carbonate facies similar to that
of much of England (Ch. 7.3) and Wales (Ch. 7.2). Its thickness in the centre
of the region is about 3000 ft. (915 m). thinning towards the margins of
Leinster and Longford-Down massifs but swelling to almost 4000 ft. (1220 m)
in the intervening Dublin Basin. The succession is noteworthy for reef
limestones (E7.60)including the great sheet reef, which extends as far N.
as County Clare in the W. and County Kildare in the E., an area of nearly
3000 square miles. This massive reef, nearly 2000 ft. (610 m) thick in the
Cork district, belongs to the lowest Visean. In the Limerick district the
Dinantian contains two basaltic horizons recalling the Peak district (E7.33,
E7.34). At the top of the Dinantian sequences occur black Pendleside-type
shales and thin limestones, and on the S. side of the Balbriggan Massif these
are followed by sandstones and shales (cf. E7.41). In South Clare, in the
vicinity of the Shannon estuary, the Namurian is 3500 ft. (1065 m) thick,
comprising shales, flags, siltstones and sandstones and with many slump
sheets and sand volcanoes (E7.61).

There are limited areas of Coal Measures but unlike those of Great Britain
these contain few workable seams, which are mostly anthracitic. In the S.,
near the Hercynian Front, the structure is complex.

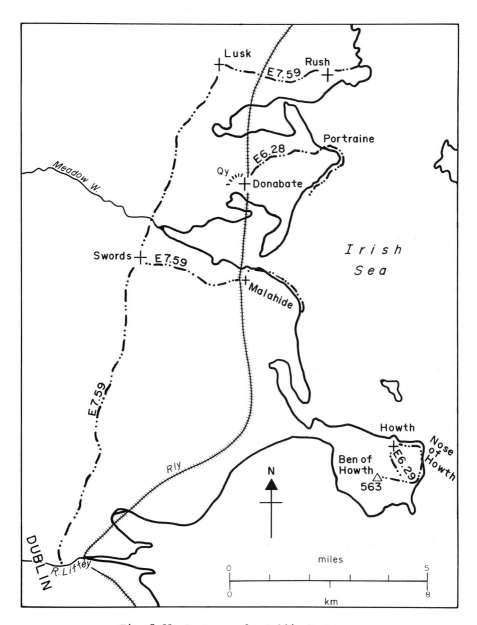

Fig. 7.23 Route map for Dublin Centre.

Dublin Centre (E7.23)

E7.59   N. of Dublin

Carboniferous Limestone including reef-knolls.

Access:- By private transport. By public bus to Swords, returning from
Malahide (missing out Rush Harbour).

Maps:- Topo:- ½-inch Ireland 13. Geol:- 1-inch Ireland 102.

Walking distances:- With private transport 3 miles (4.8 km). Using public
bus 6 miles (9.6 km).

Itinerary:- Travel N. on Belfast road by private transport or public bus
to Swords. Visit quarry (permission required) where the Feltrim reef-knoll
(Lower Visean) is well exposed. On the S. face of the E. quarry the reef
can be seen in contact with non-reef rocks. The reef is a steep antiformal
structure, and most of the dip is probably depositional. The fossils
include the goniatite Nautellipsites.

Those with private transport continue N. then N.E. to Rush Harbour, those
without walk E. to Malahide.

At Rush Harbour the Lower Visean Rush Conglomerates are seen. The conglomerates are coarse and graded with interbedded shales and limestones
and are of turbidite origin. The Lambay-Portrane Caledonoid Axis (E6.27)
is crossed between Swords and Rush Harbour. This is concluded to have
played a part in sedimentation as the conglomerates are not found S. of the
axis.

Return to Malahide where coastal sections show dark Lower Visean limestones.

Limerick Centre (E7.25)

Limerick can be reached by road or rail, and there are some local bus
services. Accommodation should be readily found. Limerick can also be
quickly reached from Shannon Airport.

E7.60   Limerick area

Silurian. Old Red Sandstone. Lower Carboniferous sediments, volcanics and
sill.

Access:- Best done by private transport. It is possible to see the
Carboniferous rocks as far as Nicker by using the Limerick-Tipperary bus
service. Road and hillside walking.

Maps:- Topo:- ½-inch Ireland 17, 18. Geol:- 1-inch Ireland 144.

Walking distances:- With private transport 3 miles (4.8 km). Without
private transport (as far as Nicker) 10 miles (16 km).

Itinerary:- By private transport S.E. for 7.5 miles (12 km) on Tipperary
road then turn off S. on road to Caherconlish or by public bus to same road
junction. Along Caherconlish road to quarry on E. side N. of village in a

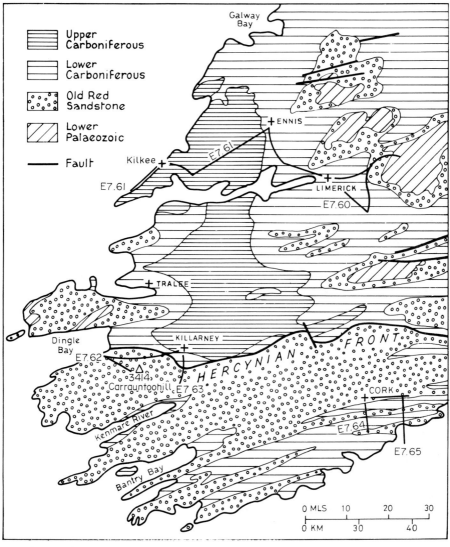

Fig. 7.24 Geological map of South Central and Southern Ireland with
          routes.

trachyte sill with columnar jointing. Continue to Quirke's Quarry, a mile
(1.6 km) S.W. of the village in tuffs of the Lower Volcanic Group (early
S_1). The tuffs contain fragments of Old Red Sandstone and Carboniferous
Limestone.

Return to main road then by private transport or public bus to Linfield
Quarry N.W. of Nicker. Here a phacoidal sill of analcite-basalt shows
excellent columnar jointing. At the S. end of the quarry there is a radial
grouping of small and large columns about a pyroclastic core.

On to Nicker village where a quarry shows limestones (S_1/D_1) overlain,
higher up the hill, by weathered basalts of the Upper Volcanic Group (D_1).

A return to Limerick can then be made by public bus. However, those with
private transport should travel N. to the crossroads at Newport (part of
route not suitable for coaches) then E. to Rear Cross. This enables Old
Red Sandstone (probably Upper) underlain by Upper Silurian greywacke to be
seen in one of the anticlinal inliers which come up in Central Ireland.

Return to Limerick. Lower Carboniferous pyroclastics can be seen close to
the ruins of New Castle, 2.5 miles (4 km) E. of the town.

E7.61   Shore near Kilkee

Carboniferous Limestone. Namurian.

Access:- To carry out the excursion from Limerick in a day private
transport is needed. By travelling to Kilkee by public bus and possibly
staying overnight the sections as far as Goleen Bay and back can be seen
by walking. Fairly low tide necessary.

Maps:- Topo:- $\frac{1}{2}$-inch Ireland 17. Geol:- 1-inch Ireland 133, 140, 141,
                                                        142, 143.

Walking distances:- With private transport 4 miles (6.4 km). Without
private transport 10 miles (16 km).

Itinerary:- Travel from Limerick to Ennis, mainly across the Carboniferous
Limestone. There is a descending sequence to $6\frac{1}{4}$ miles (10 km) out of
Limerick, where a short diversion should be made to the village of Cratloe.
On the hillside to the N.E. Old Red Sandstone (probably Upper), dipping S.
by W. is exposed at the S.W. end of the Slieve Bernagh Anticline. Towards
Ennis the higher part of the Carboniferous Limestone comes on again and
further W. the Namurian.

At Kilkee the section in Moore Bay close to the town in this formation
should first be examined. On the N. side where the stream enters the bay
the Reticuloceras stubblefieldi Marine Band can be seen as can slump folds.
At the top of the highest cliff large-scale "rafting" of about 20 ft. (6 m)
of flagstone overlying 20 ft. (6 m) of shale is seen. Brecciated shale
rises at least 15 ft. (4.6 m) between the rafts. On the S. side of Moore
Bay thin balled-up sandstone sheets are seen at the "amphitheatre".

S.W. along coast road to Goleen Bay where about 300 ft. (91.5 m) of beds
are exposed on a broad rock platform. The striking sedimentary structures
include:- Thin slumped sheets in which the fabric consists of slabs about
6 ins. (15 cm) across, all dipping at 50° to 60° in what are considered to be

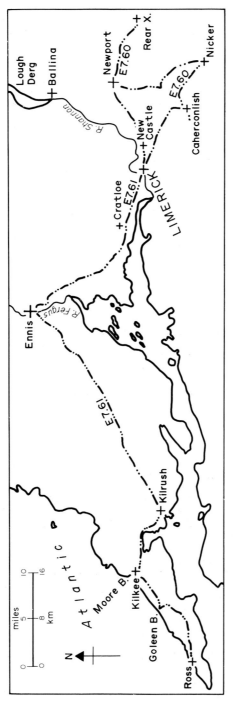

Fig. 7.25 Route map for Limerick Centre.

two directions of movement.

A channel slump at least 20 ft. (6 m) deep.

The largest sand volcanoes in Co. Clare occur on top of this channel; they
are from 8-10 ft. (2.4-3 m) in height and up to 35 ft. (10.7 m) in diameter.

At the northern cliff margin of the deep cove at Goleen Bay, at the seaward
end, a sand volcano, about 15 ft. (4.6 m) across and 8 ft. (2.4 m) high,
has slumped along arcuate shear planes into the substratum. Half the
volcano has broken away along a joint to reveal the structure of the vent
and crater.

Those with private transport should continue S.W. to exposures of slump
sheets and sand volcanoes near the Natural Bridge of Ross.

Return to Kilkee and Limerick.

7.7   Southern Ireland (E7.24)

Much of Southern Ireland consists of rugged ground, culminating in
Carrauntoohil, 3414 ft. (1035 m), the highest point in Ireland, formed of
hard, fine-grained Devonian sandstone. The whole region has undergone
Pleistocene glaciation which left many rock outcrops on the higher ground
but covered the lower areas with thick drift deposits. There are good
exposures, however, along the coast, which, in the S.W., is deeply indented
by strike-controlled ria-inlets.

The succession consists of:-

                    Carboniferous
                    Devonian

These rocks are highly folded and are separated by the Hercynian Front,
extending from Dungarvan Harbour to Dingle Bay, from a structurally simpler
region further N. where, moreover, the Devonian and Carboniferous sedimentary
piles show rapid thinning. Near Killarney the front is marked by powerful,
mainly reversed, faults. Further E. the dislocation is for the most part
along a single thrust dipping S. at 45°. Further E. still, towards
Dungarvan, Philcox (1963) believes that the front splits up into echelon
zones and that this "fanning out" may be related to Caledonoid influences.

There are a fair number of roads. The Dublin-Cork line crosses the region
from N. to S. but otherwise there are no railways.

Two centres are suggested, Killarney and Cork. See also E10.31.

Comparison with other areas:- Southern Ireland continues the Armorican Arc
from South-West England (Ch. 7.1) and structurally consists of long westerly
folds curving W.S.W. Pelitic beds are cleaved but the folding, on the whole,
is not as tight as in Devon and Cornwall. Granites are absent, and
mineralisation on a comparatively small scale.

The Devonian differs from that of South-West England in the probable absence
of the Lower and Middle divisions and in the dominantly sandy fluviatile

nature of the Upper.

The Lower Carboniferous contains a Culm facies similar to that of Devon and Cornwall (Ch. 7.1) and contrasting with the succession in Central Ireland (Ch. 7.6). Carbonate rocks, however, are also present (E7.64) in the Lower Carboniferous. The Upper Carboniferous is absent.

Killarney Centre

Killarney can be reached by rail or road. There is plentiful accommodation but booking is advisable in the tourist season. The town lies just N. of the Hercynian Front.

E7.62   Carrauntoohil (Macgillycuddy's Reeks)

Old Red Sandstone (probably Upper). Glacial features.

Access:- By car to N. end of Lough Acoose. It is possible to stay at Glencar Hotel, 2 miles W. of Lough Acoose, and to walk from there. Rough mountain walking. The excursion should not be undertaken in bad weather.

Maps:- Topo:- ½-inch Ireland 20. Geol:- 1-inch Ireland 173, 184.

Walking distances:- 10 miles (16 km). From Glencar Hotel 14 miles (22.4 km).

Itinerary:- From the N. end of Lough Acoose walk along a rough track on E. side of lough then up steep grass slope to the summit of Caher, 3200 ft. (975 m). Here there are grey and pale pinkish sandstones and pebbly sandstones with subordinate sandy shales dipping S. by E. Current bedding is seen, and the palaeocurrent direction was clearly variable.

E. along rough and narrow ridge to col at about 3000 ft. (915 m) where hard, fine, greyish sandstones show sharp folding along E. by N. axes and rough cleavage dipping S. by E. at about 60°. Up slope N.E. to summit of Carrauntoohil, 3414 ft. (1035 m) consisting of hard, fine sandstone.

Walk back about 50 yds. (46 m) S.W. from summit then N. along rough and in places narrow ridge to top of Beenkeragh, 3314 ft. (1010 m), where conglomerate is seen. There is a good view S.W. into a corrie with two moraine -dammed lakes.

Descend easy slope W. by N. to lower lake then walk W. to Lough Acoose.

E7.63   Lough Leane (S. of Killarney)

Old Red Sandstone (probably Upper). Carboniferous Limestone. Glacial features.

Access:- By walking on road S. from Killarney.

Maps:- Topo:- ½-inch Ireland 20. Geol:- 173, 184.

Walking distance:- 10 miles (16 km).

<u>Itinerary</u>:- Walk S. from Killarney. From the bridge over the R. Flesk there is a fine view of the glaciated Old Red Sandstone mountains of the Macgillicuddy's Reeks. Continue S. to Castlelough Bay N. of Muckross Abbey then walk along shores of lough. Here there are numerous exposures of strongly sheared Carboniferous Limestone dipping S. and cut by reversed faults. Solution effects are seen.

The limestone is overridden from the S. (the actual thrust is not seen on this traverse) by the Old Red Sandstone. This formation can be studied by branching off the main road and walking some distance up the narrow road in the spectacular glaciated valley N.W. of the Devil's Punch Bowl. The Old Red Sandstone strikes E. by N., and cleavage in the fine beds dips S. by E.

Return to Killarney by same route.

Cork Centre

Cork is the terminus of the main line railway from Dublin and can readily be reached by road. There is a boat service from Pembroke Dock in Wales. There are local bus services and plentiful accommodation.

E7.64   W. side of Cork Harbour

Old Red Sandstone. Lower Carboniferous. Ria estuary.

<u>Access</u>:- By private transport to Myrtleville or by public bus to Crosshaven and walking S. from there to Myrtleville. Return from Monkstown. Shore and road walking. Fairly low tide an advantage.

<u>Maps</u>:- <u>Topo</u>:- ½-inch Ireland 25. <u>Geol</u>:- 1-inch Ireland 187, 195.

<u>Walking distances</u>:- With private transport 3 miles (4.8 km). With public transport 7 miles (11.2 km).

<u>Itinerary</u>:- At Myrtleville the shore section shows pale sandstones (probably K zone) at the base of the Carboniferous on the S. limb of an E. by N. striking anticline. To the S. these pass upwards into the Ringabella Series of grey slates (Z zone).

Next walk N. across the pale sandstones and underlying purple slates into the Old Red Sandstone (Kiltorcan Beds) which continues to Crosshaven and Corrigaline. From here to Monkstown the Cloyne Syncline is crossed diagonally in which the Lower Carboniferous contains limestones. The ria form of Cork Harbour is well seen.

Return by private transport or public bus to Cork, crossing the Great Island Anticline and the Cork Syncline.

E7.65   E. of Cork

Old Red Sandstone. Carboniferous.

<u>Access</u>:- By private transport or public bus on Cobh road to Glounthaune, returning from Black Rock. Mainly road walking.

<u>Maps</u>:- <u>Topo</u>:- ½-inch Ireland 25. <u>Geol</u>:- 1-inch Ireland 187.

<u>Walking distance</u>:- 4 miles (6.4 km).

<u>Itinerary</u>:- At Glounthaune a road cutting near the church shows the top of the Old Red Sandstone (Kiltorcan Beds or Yellow Sandstone) on the N. limb of the long E.-W. Cork Syncline. Small lamellibranchs and plant fragments are common, and poorly preserved specimens of <u>Archanadon</u> can also be found.

Walk E. to Little Island where limestones of the C_2-S_1 to D_1 zones in the Cork Syncline are seen in the Rock Farm Quarries. By crossing part of a golf course the Cork Red "Marble" can be seen in a sunken quarry near the farmhouse.

Continue E. then cross to Black Rock where reef limestone is seen in the large disused Temple Hill Quarries. (Underlying Cork Red "Marble" is no longer exposed.) Calcite infilling at the S. end show something of the diagenetic history. Many cavities are filled in part by dark sediment and in part by a white mosaic of calcite. The original flat floors of the calcite are now vertical and are parallel, except where dragged by trans-current shears.

Return by private transport or public bus to city centre.

References

Anderson, J. C. C. 1977. Geology around the University Towns: The Cardiff District. <u>Geol. Assoc.</u> Guide No. 16. 2nd ed.

Bassett, M. G. (ed.) 1982. Geological Excursions in Dyfed, South-West Wales. Nat. Mus. Wales.

Cope, F. W. and others. 1972. The Peak District. <u>Geol. Assoc.</u> Guide No. 16. 2nd ed.

Dearman, W. R. and others. 1970. The North Coast of Cornwall from Bude to Tintagel. <u>Geol. Assoc.</u> Guide No. 10.

Downie, C. 1960. The Area around Sheffield. <u>Geol. Assoc.</u> Guide No. 9.

Durrance, E. M. and Lanning, D. J. C. (eds.) 1982. The Geology of Devon. University of Exeter.

Edmonds, E. A. and others. 1975. British Regional Geology: South-West England. <u>M.G.S.</u> 4th ed.

Edwards, W. and Trotter, F. M. 1954. British Regional Geology: The Pennines and Adjacent Areas. <u>M.G.S.</u> 3rd ed.

George, T. N. 1970. British Regional Geology: South Wales. <u>M.G.S.</u> 3rd ed.

Haines, B. A. and Horton, A. 1969. British Regional Geology: Central England. <u>M.G.S.</u> 3rd ed.

Hall, A. 1974. West Cornwall. <u>Geol. Assoc.</u> Guide No. 19.

Johnson, G. A. L. 1973. The Durham Area. <u>Geol. Assoc.</u> Guide No. 15.

Kennard, M. F. and Knill, J. L. 1969. Reservoirs on Limestones with particular reference to the Cow Green Scheme. <u>J. Inst. Water Eng.</u>, 23, 87-136.

Lawson, J. H. and Lawson, J. D. 1976. Geology explained around Glasgow and South West Scotland, including Arran. David and Charles.

MacGregor, M. and MacGregor, A. G. 1978. British Regional Geology: The Midland Valley of Scotland. 2nd ed.

Mitchell, G. H. and others (eds.) 1960. Edinburgh Geology. An Excursion Guide. Edinburgh.

Owen, T. R. and Rhodes, F. H. T. 1960. Geology around the University
     Towns: Swansea. Geol. Assoc. Guide No. 17.
Owen, T. R. 1954. The structure of the Neath Disturbance between Bryniau
     Glesion and Glynneath, South Wales. Q.J.G.S., 109, 333-365.
Philcox, M. E. 1963. Compartment deformation near Buttevant, Co. Cork,
     Ireland, and its relation to the Variscan thrust front. Sci. Proc.
     Roy. Dublin Soc., A2, 1-11.
Savage, R. J. G. 1977. Geological Excursions in the Bristol District.
     University of Bristol.
Taylor, B. J. 1971. British Regional Geology. Northern England. M.G.S.
     4th ed.

CHAPTER 8

Alpine Terrains
(Mesozoic and Tertiary Districts)

8.1   Eastern England (Fig. 8.1)

The western margin of this area extends along the E. margins of North-East
England ((7.4) and Central England (7.3) to Gloucester, whilst the southern
margin runs from Gloucester to Swindon and then eastwards to the Thames
estuary. The area has a long eastern coast formed for the major part of
mainly Chalk or the overlying Tertiary-Pleistocene sediments. Jurassic
rocks form the coast around the Wash and again along the Yorkshire coast
from Filey to Redcar.

Topographically, the region is one of diverse relief with long hill scarps
and intervening clay vales, but apart from the Cleveland Hills in the north,
the surface height rarely exceeds 1000 ft. (305 m) O.D. The higher ground
tends to follow two main belts, corresponding to the Middle Jurassic limestones and the Chalk respectively.

The succession is as follows:-

> Pliocene- Pleistocene
>   (unconformity)
>
> Eocene
>   (unconformity)
>
> Cretaceous
>   (unconformity in most places)
>
> Jurassic

There are numerous roads and many public transport routes.

Five centres are suggested: Scarborough, Hull, Ipswich, Oxford and London.

Comparison with other areas:- The estuarine and deltaic beds in the
Jurassic (E8.1, E8.2) of the N. part of the region contrast with those of
the S. part of the region and of Southern England (8.2) but resemble those
of Western Scotland (E9.1). The region includes the N. part of the London

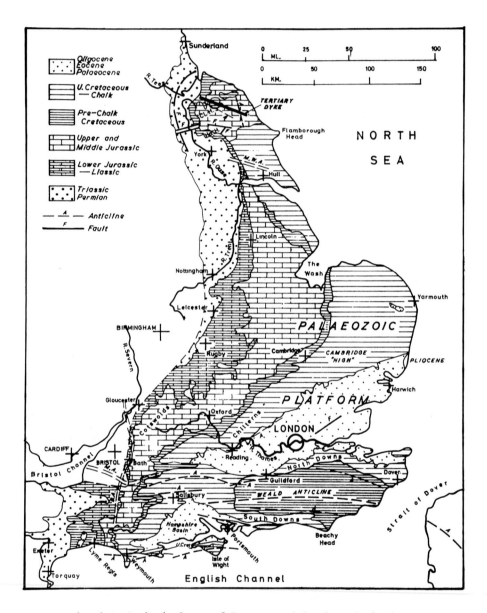

Fig. 8.1 Geological map of Eastern and Southern England.

Basin with widespread Lower Tertiary (E8.9, E8.10) and also contains the
most extensive Pliocene deposits (E8.6) in Great Britain.

Structurally the region is relatively simple, Alpine folding having had much
less effect than in Southern England (8.2)

Scarborough Centre (Fig. 8.2)

Scarborough can be easily reached by road or rail, and there is plentiful
accommodation, although much of it may be taken up in the holiday season.
For other excursions in the district reference should be made to Hemingway
and others (1968).

E8.1   Coast near Scarborough

Middle and Upper Jurassic Glacial deposits.

Access:- Road and coastal walking. Low tide necessary for some exposures.

Maps:- Topo:- 101. Geol:- EW 35 with 44, EW 54.

Walking distance:- 10 miles (16 km).

Itinerary:- Walk to Castle Hill, on N. side of town, which consists of
Oxford Clay (seen along Marine Drive) overlain by the Lower Calcareous
Grit and the Hambleton Oolite of the Corallian. Fossils may be collected
from limestones in these formations on the N. side of the hill and at a
higher level just under the walls of the castle.

Walk S.E. along foreshore of South Bay. The sandstones of the Upper Deltaic
Series outcrop just S. of the bathing-pool, and, further S.E., the Grey
Limestone in the Scarborough Formation forms a low anticline visible at low
tide. At the back of the next bay, Carnelian Bay, the Boulder Clay forming
the cliffs contains many erratics of Scandinavian origin.

Continue along coast to Carnelian Bay. This is enclosed by slipping Boulder
Clay cliffs but at its S. end High Red Cliff. The lower massive beds of the
Lower Calcareous Grit form the uppermost part of the cliff, underlain by the
Oxford Clay which makes a very steep, unstable slope down to the sandstones
of the Hackness Rock and Kellaways Rock. At the E. end of High Red Cliff a
N.-S. fault brings up the Deltaic Series; generally not well exposed.

From Carnelian Bay go up to road and return to Scarborough.

E8.2   Coast near Whitby

Lower and Middle Jurassic.

Access:- By public bus or private transport to Whitby. Coastal walking.
The itinerary should be followed only during an ebbing tide, and the foot
of cliffs should be avoided because of the danger of falling blocks.

Maps:- Topo:- 94. Geol:- EW 35 with 44.

Walking distance:- 7 miles (11.2 km).

Itinerary:- The East Cliff at Whitby shows some 40 ft. (12 m) of Alum Shales (Upper Lias) followed by 2½ ft. (0.75 m) of Dogger (basal Middle Jurassic) followed by the Lower Deltaic Beds.

From the East Pier descend to shore level and cross the breakwater below the Spa Ladder onto the Scaur. From here to Saltwick Bay and beyond, the shore is made up entirely of gently folded Upper Lias. At Land Bight the upper part of the Alum Shales is seen with ammonites of the genus Hildoceras and belemnites. Here a large rock-fall (c. 1912) hides the base of the cliff. On this part of the cliff, and also beyond the rock-fall, wedging sandstone cuts the silts and bedded sandstones of the Lower Deltaic Beds.

Further E. at Saltwick Nab, a five-inch bed of sideritic mudstone separates the Alum Shale Series from the underlying Hard Shales. Beneath these come the Bituminous Shales and the Jet Rock, a finely laminated black or black-brown shale, with a conifer allied to Araucaria, which is accessible only at low spring tides.

At Saltwick Bay disused workings for alum shale are seen. On the upper part of the cliff the Lower Deltaic Beds are dominated by massive sandstones.

Ascend the cliff by the track above the sandy beach and follow the cliff path back to Whitby.

E8.3   Reighton Gap to Speeton Cliffs

Cretaceous.

Access:- By public bus or private transport to Reighton village. Walking mainly on shore. Low (preferably falling) tide necessary.

Maps:- Topo:- 101. Geol:- EW 55.

Walking distance:- 6 miles (9.6 km).

Itinerary:- Walk from the bus-stop on the Bridlington road at Reighton village to Reighton Gap and descend to the shore. Along the beach S.E. for ½-mile (0.8 km) past Boulder Clay cliffs until grey clay outcrops at the cliff-foot. From here to the mouth of Speeton Beck the Speeton Clay (Lower Cretaceous) is exposed more or less continuously, but the occurrence of clean exposures depends on the effects of recent tides and storms. Belemnites are the commonest fossils. Ammonites are plentiful but are often crushed. Many species of lamellibranchs and gastropods are present.

S.E. of the mouth of Speeton Beck slipped masses of Chalk (Upper Cretaceous) and Red Chalk (Lower Cretaceous) appear, but good exposures of Red Chalk do not begin until about ¾-mile (1.2 km) past the beck. A Lower Red Chalk, a Grey Band and an Upper Red Chalk can be recognised. Belemnites are abundant up to the Grey Band and brachiopods throughout.

The Cenomanian Lower Chalk can be examined in clear patches between talus slopes. In a Lower Pink Bank echinoids can be found.

Return to Reighton and Scarborough.

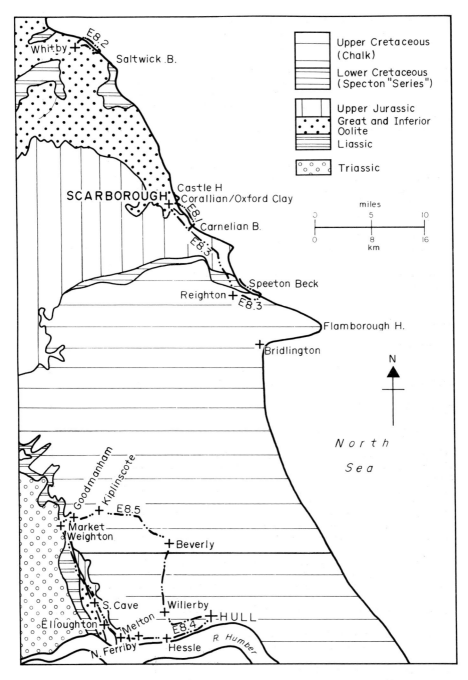

Fig. 8.2 Geological map, with routes, for Scarborough and Hull Centres.

Hull Centre (Fig. 8.2)

Hull can be reached by road or rail, and there are local public transport services in a number of directions. Some excursions are described by Bisat and others.

E8.4   North side of R. Humber from Elloughton to Hessle

Jurassic. Cretaceous. Glacial deposits.

Access:- By public bus. Walking.

Maps:- Topo:- 106. Geol:- EW 80.

Walking distance:- 9 miles (14.4 km).

Itinerary:- From the Central Bus Station take a South Cave, Goole, Selby or Brough bus to the roundabout ⅝-mile (1 km) W. of Elloughton village. Walk N. on Boothferry Road to Elloughton South Quarry (permission from W. Clifford Watts, Gransmoor, Burton Agnes). The Cave Oolite (M. Jurassic) is overlain by drift deposits consisting of sands and gravels containing Cave Oolite cobbles and lenses of pebbles. A storm beach or shoal deposit origin at the end of the glaciation has been suggested.

Take a bus to Melton and alight at the Beverley Turn on Gibson Lane. Walk along Beverley road to the Melton Clay Pit (permission from Works Manager, G. and T. Earle Ltd., Melton Works). About 60 ft. (18 m) of Oxford Clay are exposed with fossils which include Perisphinchs, Amoeboceras, Gryphaea and Chlamys. A 2-inch band of small grey nodules marks the top of the Jurassic which can be seen to dip gently E. at a greater angle than the overlying Cretaceous. The latter consists of the Greensand (M. to U. Albian), the Red Chalk (about 5 ft. , 1.5 m) and the Cenomanian White Chalk with Inoceramus and terebratulids.

Return to road and continue E. towards Ferriby for about 300 yds. (275 m) to Melton Chalk Pit, with a working face ½-mile (0.8 km) long.

Return to road and retrace steps to the T-junction at Gibson Lane where a turn to the left leads to the A63. Follow this for 100 yds. (91 m) then turn right and continue for 1 mile (1.6 km) down to the river which should be followed E. to the Ferriby Red Cliff. This section is in a morainic ridge, the deposits of which are well worth detailed study.

Alongside river to Ferriby Landing and follow road N. to North Ferriby. About 1½ miles (2.4 km) along the A63 the Humberfield Quarry (permission from W. Marshall (Hessle) Ltd., The Cliff, Hessle, East Yorkshire) is worked in the Middle Chalk.

Continue to Hessle Quarry (permission as above) where some 40 ft. (12 m) of Chalk are overlain by 20 ft. (6 m) of drift. About 140 yds. (127 m) W. of the entrance the drift consists of orange sandy clays (Lower Hessle Clay) with abundant flint fragments and some far-travelled erratics of the Whin Sill etc. The Whin Sill outcrops are in N.E. England (See Ch. 7.4). The Lower Hessle Clay is overlain by well-bedded gravels covered by red Upper Hessle Clay.

Back to main road and return to Hull by public bus.

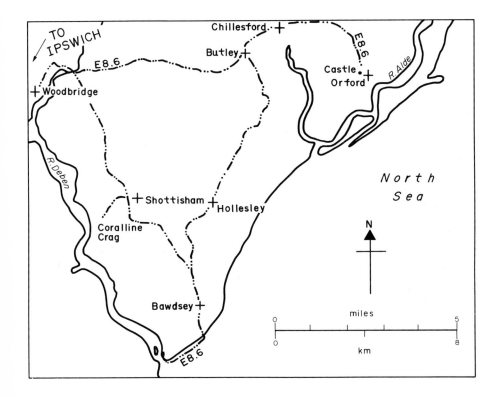

Fig. 8.3   Route map for Ipswich Centre.

E8.5   Market Weighton area

Jurassic. Cretaceous. Glacial features.

Access:- Best done by private transport. If public transport is used two
or more days are necessary.

Maps:- Topo:- 106. Geol:- EW 71, 72.

Walking distance:- 1 mile (1.6 km).

Itinerary:- Take Boothferry road (A63) out of Hull, crossing at first a
flat expanse of post-glacial estuarine muds (Humber "warp") and then rising
sharply onto the Chalk of the Wolds. Fork right from the A63 onto the A1034
about 1½ miles (2.4 km) beyond a roundabout near Elloughton Quarries (E8.4)
and travel through South Cave village to North Newbald Quarry where there
is a good section in the Cave Oolite (M. Jurassic).

Continue into Market Weighton, turn left on the York road, right on the
Driffield road (A163) and right to Goodmanham. Turn right again through
the village, leaving the church on the left. 100 yds. beyond the church
turn right through a gate and follow a narrow metalled road for 0.6 mile
(1 km) to the Rifle Butts section. This shows the Upper Cretaceous resting
on the Lower Lias above the Market Weighton Axis or Upwarp.

Continue in the same direction to a T-junction and follow the road E. The
route runs along the Market Weighton spillway which carried drainage into
the Vale of York during the later part of the glaciation. A number of dry
valleys hang above the general level of the spillway. About a mile (1.6 km)
beyond the junction the Kiplingscote Railwayside Pit shows the Holasterplanus
Zone of the Chalk. Continue to Kiplingscote station where the same zone is
seen in another pit. The Chalk contains highly fossiliferous lenticles of
iron-stained nodules.

On to Beverley, then fork right onto the Cottingham road (A164) which should
be followed to Willerby, where the Willerby Railwayside Pit should be
examined. This is in the M. cortestudinarium Zone (Senonian) of the Chalk.
The zone fossil is rare but M. praecursor is common as are large specimens
of Inoceramus.

Return to Hull.

Ipswich Centre (Fig. 8.3)

The Pliocene and Pleistocene beds of East Anglia occur over an extensive
area but good exposures are limited and variable in response to coastal
conditions and the state of pits. An excursion in the neighbourhood of
Orford is suggested. Ipswich is the nearest big centre, but accommodation
may be available in Woodbridge or Orford itself. Other excursions are
described by Greensmith and others (1973).

E8.6   Orford and Bawdsey

Eocene. Pliocene. Pleistocene. Shingle spit.

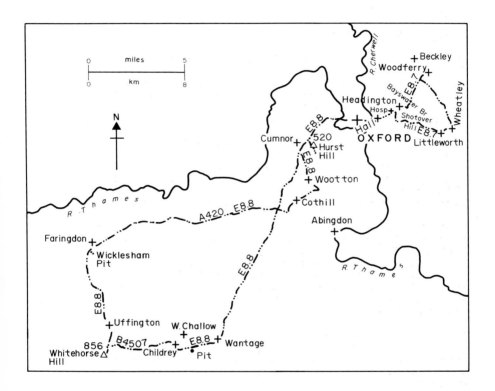

Fig. 8.4 Route map for Oxford Centre.

Access:- For the whole excursion private transport is necessary. As far as Orford it may be possible to use public buses to the Butley road-end and return from Orford.

Maps:- Topo:- 156, 169. Geol:- EW 208, EW 205.

Walking distances:- With private transport 3 miles (4.8 km). Butley to Orford 6 miles (9.6 km).

Itinerary:- From Woodbridge or Chillesford to Butley where a pit (permission from Estate Office, Sudbourne Hall, Orford, Woodbridge, Suffolk) shows Red Crag (early Pleistocene). Marine bivalves are common; occasionally the freshwater gastropod Planorbis can be found.

On to Chillesford Pit (permission from Church Farm opposite the Church), on N. side of road.

The succession is:-

> Chalky Till
> Chillesford Clay
> Chillesford Crag
> Scrobicularia Crag
> Red Crag

Continue to Orford Castle (entrance fee) from which there is a fine view of the spit which has diverted the R. Alde or Ore at Aldeburgh for 7 miles (11 km). Within the grounds a visit (inform Custodian) may be made to the upper "Bryozoan rock-bed" of the Coralline Crag (Pliocene). This soft yellow limestone contains comminuted fragments of shells and bryozoa, the latter not very common.

S.W. to Bawdsey via Butley and Hollesley. The cliffs and shore at Bawdsey are M.O.D. property and prior written authority must be obtained from the C.O., R.A.F. Bawdsey, Woodbridge, Suffolk. From the ferry quay walk S. along the beach to the point then N. to the cliff which shows a fine section of Red Crag resting on London Clay. The Red Crag is packed with brittle, often broken molluscs.

N.W. to Shottisham. Enter Sutton Farms Estate (prior permission from Estate Office). Walk to a small wooded hill, Rockhall Wood, which is an inlier of Coralline Crag surrounded by younger Red Crag. A small pit shows the lower part of the Coralline Crag consisting of light yellow, fine sand with small molluscs and bryozoa. The very top of the section is probably in the "Bryozoan rock-bed".

Return to Woodbridge and Ipswich.

Oxford Centre (Fig. 8.4)

Oxford can be easily reached by road and rail; accommodation at various levels is plentiful.

E8.7 East of Oxford

Upper Jurassic and Lower Cretaceous.

Access:- By public bus to Headington and back or by private transport.
Road walking.

Maps:- Topo:- 164. Geol:- EW 236, EW 237.

Walking distances:- 10 miles (16 km). With private transport 1 mile
(1.6 km).

Itinerary:- At Headington visit the disused Cross Roads Quarry, at the
junction of Windmill Road and Old Road beyond the Nuffield Hospital. This
shows interdigitation of the Wheatley Limestone facies and the Coral Rag
facies of the Corallian Group. Isastrea and Thecosimilia may be found.

Through Headington then follow a minor road N. by E. signposted Beckley.
Descend to Bayswater Brook, which flows on Oxford Clay, then ascend to
Woodferry Quarry S.E. of Beckley. This shows further facies of the
Corallian. The lower beds are the Beckley Sands, overlain by rubbly and
marly limestones, a variant of the Wheatley Limestone. At the base of the
latter is the richly fossiliferous Shell Pebble Bed.

S.E. to Wheatley then W.S.W. to Littleworth. At the disused brickpit there
were formerly good exposures but all that is likely to be seen are the top
23 ft. (7 m) of sandy Kimmeridgian overlain by 16 ft. (5 m) of Portland Beds
on top of which are remnants of Shotover Ironsands (Wealden).

Return over Shotover Hill to Headington and Oxford.

E8.8   Vale of White Horse

Jurassic and Cretaceous.

Access:- The excursion as a whole has to be done by private transport but
the first part as far as Cothill Quarry can be seen by using public bus to
Cumnor and back.

Maps:- Topo:- 163, 164, 174. Geol:- EW 235, 252, 253.

Walking distances:- With private transport 4 miles (6.4 km). From Cumnor
to Cothill Quarry and back 8 miles (12.8 km).

Itinerary:- S.W. from Oxford towards Cumnor. Stop short of village and
walk up N.W. slopes of Cumnor Hurst. A brickpit was formerly worked in the
Kimmeridge Clay but this is now poorly exposed. On the S. side of the hill,
however, the succeeding beds are well exposed, consisting of coarse and
pebbly red sands. As these are considered to be Lower Cretaceous there is
considerable overlap. From the summit of the Hurst, 520 ft. (158 m) there
is a fine view over a wide area.

Return to main road then S.E. of Cumnor turn left through Wooton, ½-mile
(0.8 km) after which a road should be followed to Cothill Quarry, 100 m
beyond the W. boundary of Abingdon Aerodrome. This shows the main divisions
of the Corallian, only the Lower part of the Lower Calcareous Grit, the top
of the Coral Rag not being visible.

Return to main road (A420) and travel W. to Faringdon to visit the well-known
Lower Greensand (Lower Cretaceous) outlier S. and E. of the town. The
succession is:-

Sands with chert and ironstone
Sandy clays
Red Gravels
        pebble bed ⎫
Yellow Gravels ⎬ Sponge Gravels
                ⎭

There are a number of pits; a suitable one to visit is Wicklesham Pit,
reached by following road A420 out of Faringdon then turning left alongside
a disused railway and in ½-mile (0.8 km) turning right. This pit is in the
Red Gravels with abundant sponges, brachiopods, bivalves and other fossils.

From Faringdon S. through Uffington to the B4507. Ascend the hill in which
the famous White Horse is carved, best done on foot as the road is steep
and narrow. There are numerous small exposures in the Lower and Middle
Chalk. At the foot of the escarpment the "normal" valleys in the Gault
contrast with the absence of surface drainage in the Chalk.

Along B4507 eastwards for 5 miles (8 km) and turn right, just beyond
Childrey to a disused pit S. of West Challow in fossiliferous Lower Chalk
(Lower Cenomanian).

Return via Wantage to Oxford.

London Centre (Fig. 8.5)

To undertake excursions from London obviously requires considerable travel
and exposures are limited. However, two excursions N. of the Thames are
suggested here and two S. of the river in the next section (E8.11, E8.12).
Other excursions from London are described by Pitcher and others (1967).

E8.9   Rickmansworth to Denham

Cretaceous. Eocene.

Access:- By rail from Baker Street or Marylebone to Rickmansworth, returning from Denham. Road and path walking.

Maps:- Topo:- 176. Geol:- EW 256.

Walking distance:- 8 miles (12.8 km).

Itinerary:- From Rickmansworth station follow the main Uxbridge Road
(A412) then turn left along Springwell Lane and after crossing the river
and the canal visit (if permission is granted at office) a quarry in Upper
Chalk.

Return to canal and walk along E. side to disused pit opposite Springwell
Farm which shows "piped" Upper Chalk. Follow a footpath which leaves
Springwell Lane immediately E. of Springwell Farm S. for two-thirds of a
mile (1 km) to another disused pit in the Upper Chalk with "pipes" infilled
with gravel and some Eocene material.

Continue along path and road to approach to Harefield Church. A pit
(permission from W. W. Drinkwater (Willesden) Ltd., Dudden Hill Lane,
Willesden, N.W.10) opposite shows

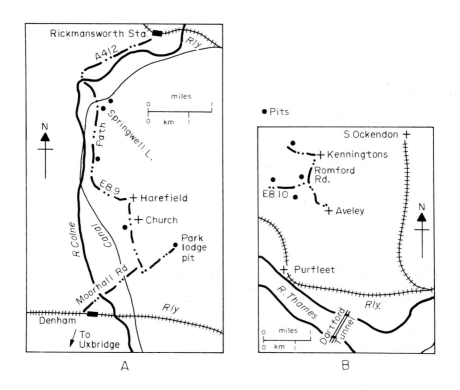

Fig. 8.5 Route maps for North London Centre. A. Rickmansworth area.
B. Aveley area.

London Clay ⎱ (Eocene)
Reading Beds ⎰

Upper Chalk   (Upper Cretaceous)

The Chalk contains the echinoid <u>Micraster coranguinum</u>. A section showing
the unconformity with the Reading Beds is preserved (Site of Special
Scientific Interest).

Along Uxbridge Road S. to about ¼-mile (0.4 km) beyond Moorhall Road corner
and turn off left on approach road to sand pits in the Reading Beds, the
lower part of which contains glauconitic material. The sections are
variable according to the state of development, but the overlying London
clay can generally be seen and sometimes the underlying Upper Chalk.

Along Moorhall Road S.W. to Denham for return to London.

<u>E8.10   Aveley</u>

Eocene. Interglacial and Terrace deposits.

<u>Access</u>:- By rail from Fenchurch Street to Rainham then by London Transport
bus to Aveley.

<u>Maps</u>:- <u>Topo</u>:- 177. <u>Geol</u>:- EW 257.

<u>Walking distance</u>:- 4 miles (6.4 km).

<u>Itinerary</u>:- On the N.W. side of Aveley a clay pit (permission from Tunnel
Portland Cement Co.) is worked in the London Clay (Eocene), and from this
pit there is access to another pit to the W. (same owner) where channels in
the London Clay contain freshwater sediments of the Ipswichian (Eemian)
interglacial. The London Clay contains <u>Balanocrinus subbasaltiformis</u> and
<u>Thyasira angulata</u> together with <u>Terebratulina waretenensis</u> and teeth of
Odontaspid and Lamnidfish.

To the E.,where the road past these pits joins Romford Road, a pit (permission from Hall and Co. Ltd., Victoria Wharf, Cherry Orchard Road, Croydon,
Surrey) is worked in the Swanscombe Gravels which have yielded <u>Corbicula
fluminalis</u>, and <u>Mammuthus</u> sp.

N. along Romford Road then left at Kenningtons to another pit (permission
from Associated Portland Cement Co.) in the London Clay.

Return to Aveley and London.

<u>E8.2   Southern England</u> (Fig. 8.1)

This is the region S. of a line from Gloucester through Swindon to the
Thames estuary and E. of the Jurassic southwards to Bridport and Lyme Regis.

There are excellent exposures along much of the south coast. No part
reaches 1000 ft. (305 m); the Chalk in places rises to over 800 ft. (244 m).

The succession is as follows:-

> Oligocene
>
> Eocene
>   (unconformity)
>
> Cretaceous
>   (unconformity in places)
>
> Jurassic

Dominant structures are the London and Hampshire Basins separated by the Weald Anticlinorium (the continuation of that of the Boulonnais in France).

The region is densely populated with a network of roads and railways and of public transport.

Five centres are suggested: London, Folkestone, Newport (Isle of Wight), Bristol and Weymouth.

Comparison with other areas:- The Jurassic forms the "typical" succession for the British Isles and it was in the region, near Bath, that William Smith laid some of the foundations of stratigraphy. The Wealden facies (Lower Cretaceous) of the southern part of the region has distinctive features; the region shares with Eastern England the main development of the Chalk. The London Basin (partly in Eastern England, 8.1) and the Hampshire Basin contain the largest outcrops of sedimentary Tertiary in the British Isles.

The region is also the only part of the British Isles affected by powerful Alpine folding with vertical strata in the extreme S. (E8.16, E8.17).

London Centre (Fig. 8.6)

See remarks under London Centre in previous section.

E8.11   Upnor and Strood

Cretaceous. Tertiary.

Access:- By rail to Strood. Road and path walking.

Maps:- Topo:- 178. Geol:- EW 272.

Walking distance:- 5.5 miles (8.8 km).

Itinerary:- Follow Station Road, Strood, N., then turn right up Frindsbury Road (A228) and through Frindsbury before taking a turning on the right signposted Upnor. After a sharp left bend leave the road for a footpath on the right going up Tower Hill. Quarries show Chalk of the zone of Micraster coranguinum which is overlain by the Thanet Beds (Eocene).

Follow the path and road N. to Lower Upnor on the Medway and enter a sandpit near the Ship Inn. Here the top part of the Thanet Beds is overlain about half-way up the face by a pebbly layer at the base of the Woolwich Beds (Eocene), above which are current-bedded sands, striped clays of the Shell

Beds and yellowish-brown sands. Above comes the London Clay which has
extensively slipped. Shark teeth can be picked up from the weathered surfaces of mixed sand and clay.

Return to Strood.

E8.12   Herne Bay

Tertiary.

Access:- By rail to Herne Bay. Walking mainly on shore. Low tide
necessary.

Maps:- Topo:- 179. Geol:- EW 273.

Walking distance:- 5 miles (8 km).

Itinerary:- Walk along the promenade, the eastern end of which is on large
slips of London Clay. Continue E. along the shore; due to the dip to the
W. older beds appear. A thin, discontinuous layer of flint pebbles occurs
under the current-bedded Oldhaven Sands where the latter rise above the
beach. Shell seams occur in the sands but the fossils are fragile and are
best collected from ferruginous tabular concretions. Rosettes of barytes
also occur in the sands. The base of the Oldhaven Beds is marked by a pebble
bed with black flint pebbles in a sand matrix which contains shark teeth.

The glauconitic Woolwich Sands are best seen at the opening of Bishopstone
on the side opposite the wooden steps. The junction with the Thanet Beds
is taken at a layer of tabular sandy concretions. Exposures of the Thanet
Beds are most likely to be found at the terminations of groynes to the E.
The stiff glauconitic sandy clays contain fragile specimens of Cyprina,
Thracia etc.

Ascend cliff by the steps, noting at top Palaeolithic gravels in which
implements may be found. Return to Herne Bay.

Folkestone Centre (Fig. 8.6)

The pleasant town and port of Folkestone is easily reached by rail and road
and has plentiful accommodation. For very many visitors arriving by sea
from continental Europe it is their first sight of England, and some foreign
geologists may find it useful to stop over and view the coastal geology.
For other excursions in the Weald see Kirkaldy (1967) and Gibbons (1981).

E8.13   Coast W. of Folkestone

Cretaceous. Landslips.

Access:- By road and coastal walking.

Maps:- Topo:- 189. Geol:- EW 305 and 306.

Walking distance:- 6.5 miles (10.4 km).

Itinerary:- For those with limited time in Folkestone this itinerary
provides a short excursion.

Walk along the shore road W.S.W. from the town. Exposures occur of brown-weathering Lower Greensand (Folkestone Beds) both in the cliff and on the shore (at low tide). These beds dip gently N. on the N. flank of the Weald Anticlinorium. Considerable slipping has occurred from the cliff. Fossils are rather scarce.

After about 1½ miles (2.4 km) ascend the cliff by a zig-zag path and return to town by a higher road.

E8.14   Coast E. of Folkestone to Dover

Cretaceous. Major landslips.

Access:- By road and coastal walking. Low tide necessary for sections below cliffs.

Maps:- Topo:- 179, 189. Geol:- EW 305 and 306.

Walking distance:- 7 miles (11.2 km).

Itinerary:- Walk E.N.E. along road following sea front. The cliff consists of gritty Lower Greensand (Folkestone Beds) overlain by Gault clay which has slipped over the Greensand in places. Ascend steps at the end of the road to cliff path leading N.E. then N. Below the first Martello Tower Glauconitic Marl at the base of the Lower Chalk overlies the Gault in a slipped mass. About 50 yds. (46 m) further on another slipped mass of the Glauconitic Marl contains harder beds which yield ammonites, rhynchonellids and other fossils.

The slipped masses are part of a major landslip area known as The Warren. Rough paths make it possible to descend to the shore where there are seaweed-covered reefs of Chalk Marl (Schloenbachia varians Zone). Fallen blocks belong both to this zone and to the overlying Grey Chalk.

Continue along shore examining sections in Grey Chalk and overlying Belemnite or plenus Marls (both Lower Chalk). At the E. end of Shakespeare Cliff the abrupt junction of the Lower Chalk with the purer and whiter Middle Chalk is easily seen.

Leave the beach by the Ropewalk Steps at the W. end of Dover and return to Folkestone by train.

E8.15   Tonbridge and Sevenoaks

Cretaceous.

Access:- By train or by private transport. The excursion can also be carried out from London.

Maps:- Topo:- 188. Geol:- EW 287.

Walking distances:- Without private transport 15 miles (24 km). With private transport 2 miles (1.4 km).

Itinerary:- If a start is made at Tonbridge station, walk S.W. on the A26 road then turn S.E. to the Quarry Hill Brickworks (permission required from

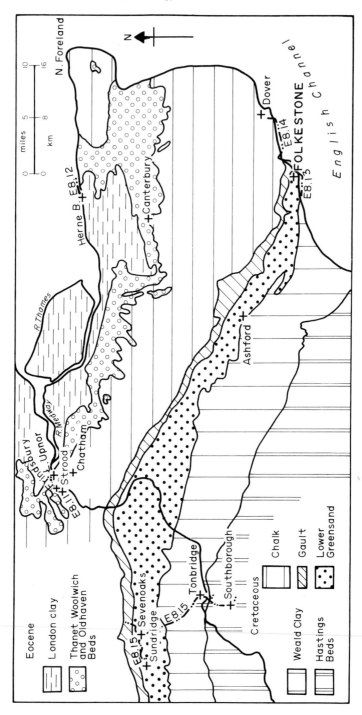

Fig. 8.6 Geological map, with routes, for South London and Folkestone Centres.

Sussex and Dorking United Brick Companies Ltd.). From the road to the
quarry there is a good view to the N. across the Weald to the scarp face
of Chalk of the North Down. In the quarry the dark brown Wadhurst Clay
(Hasting Beds of Lower Cretaceous age) is worked. Calcareous layers yield
ostracods, poorly preserved lamellibranchs (Neomiodon) and fragments of
plants and ganoid fish. At the E. end of the workings the Wadhurst Clay is
faulted against massive white sandstones (Ashdown Sands - lowest Cretaceous
beds of area).

Continue S.W. along main road (A26) to Southborough then turn left to the
High Broom Brick and Tile Co.'s Pit N. of the station. A section nearly
100 ft. (30 m) high shows Lower Tunbridge Wells Sands resting on Wadhurst
Clay.

Those without private transport should return to Tonbridge station and
travel to Sevenoaks. Just N. of Tonbridge station there is a cutting
through sandstone (Tunbridge Wells Sands) beyond which flat country marks
the Weald Clay.

From Sevenoaks station W. on A25 to about a mile E. of Sundridge then turn
S.,by a nursery garden, to the disused Dryhill Quarries in folded Hythe
Beds (Lower Greensand) consisting of alternations of sandy and glauconitic
limestone (Kentish Rag) and loamy and glauconitic sand ("hassock") with
chert developed in the synclines. Fossils include Exogyra, Trigonia, other
lamellibranchs and brachiopods, belemnites, nautiloids and ammonites. The
quarry lies in the middle of a belt of short sharp folds, probably of a
superficial nature.

Return E. along A25 to N.E. of Sevenoaks then turn N.E. along Greatness Lane
to a pit worked by the Sevenoaks Brick Co. Ltd. in Lower and Upper Gault.
A conspicuous seam, a few inches thick, of phosphatic nodules contains
Anahaplites splendens, Euhoplites lautus, Hamites sp. and other ammonites
as well as lamellibranchs, gastropods etc. Beneath this bed are dark clays
with lenticles of reddish clay-ironstone. Above the bed are 4 ft. (1 m)
of dark clays overlain by about 5 ft. (1.5 m) of light grey clays of the
Upper Gault. To the N. rises the scarp of the Lower Chalk.

Return to Sevenoaks station.

Newport Centre (Isle of Wight) (Fig. 8.7)

The Isle of Wight can be easily reached by travelling by rail or road to
Portsmouth and crossing by ferry (vehicles carried) to Ryde, from which
there are public buses to various parts of the island. Newport is the best
centre for travel. There is plentiful accommodation here and elsewhere but
there may be difficulties in the holiday season.

E8.16   Whitecliff Bay

Cretaceous. Eocene. Oligocene.

Access:- By public bus or private transport to Whitecliff Bay Hotel or to
Brading station. Path, road and shore walking. A falling tide is necessary
to examine the Chalk but is not essential for a study of the Tertiary beds.

Maps:-   Topo:- 196. Geol:- Special 1:50,000 Sheet Isle of Wight.

Walking distance:- From Whitecliff Bay and back 1.5 miles (2.4 km). From Brading station and back 5.5 miles (8.8 km).

Itinerary:- Descend to Whitecliff Bay and walk to S. end where the Upper Chalk forms Culver Point. The Belemnitella macronata Zone is seen at this end of the bay; lower zones can be examined by walking round the point if the tide is suitable. The dip is N. at 65°.

On a traverse N. the following Tertiary succession is seen:-

Reading Beds (resting on Chalk) - brick-red, purple and mottled clays.

London Clay - clays with sand; crushed molluscs.

Bagshot Sands - yellow, grey and red sands with clays.

Bracklesham Beds (580 ft., 177 m) - basal bed of rolled flint pebbles over-lain by dark grey-green marine clays. Fossils at several horizons include Nummulites variolarius.

Barton Beds - clays followed by pale marine sands; fossils source here and most of section hidden by slips.

Lower, Middle and Upper Headon Beds (Oligocene) - clays, sandy clays and sands. Freshwater molluscs (Potomomya and Viviparus common near top.

Osborne Beds - badly exposed clays and limestones with a few Unio and Viviparus.

Bembridge Limestone (28ft., 8.5 m) - cream and pale brown limestones with Chara "fruits" and moulds of freshwater (and, more rarely, land) snails.

Bembridge Marls - silts, clays and marls with occasional thin limestones and sands. Bright green clay with brackish-water fossils at base followed by a shelly sand - the Bembridge Oyster Bed.

Up to the lowest part of the Oligocene the beds are vertical or dip steeply N. in the steep limb of the Miocene monocline which traverses the island, but they flatten rapidly to the N. in the Bembridge horizons.

Return to Newport.

E8.17   Alum Bay and Headon Hill

Cretaceous. Eocene. Oligocene.

Access:- By public bus via Yarmouth or by private transport to Needles Hotel. In the winter months the bus goes only as far as the Totland Bay War Memorial. Tide not critical. Path, road and shore walking.

Maps:- Topo:- 196. Geol:- Special 1:50,000 Sheet Isle of Wight.

Walking distances:- From Needles Hotel and back 3 miles (4.8 km). From Totland Bay War Memorial and back 5 miles (8 km).

Itinerary:- From the hotel car park follow a path down to Alum Bay and walk to the S. end. Here the Upper Chalk dips steeply N. and forms High

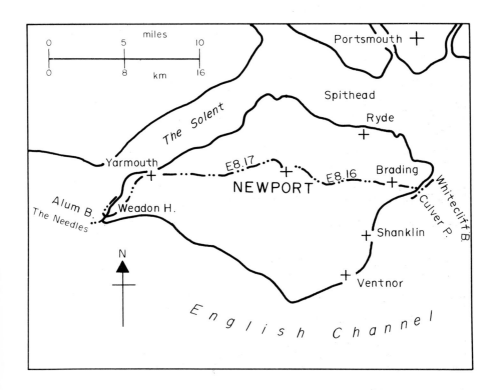

Fig. 8.7 Route map for Newport Centre (Isle of Wight).

Down. It terminates in the spectacular stacks of The Needles.

In a traverse N. and N.E. the following Tertiary succession is seen:-

Reading Beds (resting on Chalk, see previous excursions E8.15).

London Clay (40 ft., 124 km) - sands and clays. Molluscs in lower part, glauconitic sand at the base contains Ditrupa and shark teeth.

Bagshot Sands - white, yellow and grey sands with 6 ft. (1.8 m) of white pipe-clay (the Alum Leaf-Bed) near the middle.

Bracklesham Beds (575 ft., 175 m) - sands and clays.

Barton Beds - clays and sandy clays; rich fauna in middle part.

Lower, Middle and Upper Headon Beds (Oligocene) - clays and sands with variable limestones; fossils fairly common.

Osborne Beds - mainly poorly fossiliferous greenish and mottled clays with calcareous sandstone and hard limestone near base. Slips partly obscure the last two formations on shore below Headon Hill. The Bembridge Limestone is mostly hidden.

Structure as at Whitecliff Bay (E8.16).

Return to Newport.

Bristol Centre (Fig. 7.5)

Bristol is also a centre for excursions E7.15, E7.14 and E7.15. The Mesozoic excursions described below can also be carried out from Bath.

E8.18   Wotton under Edge district

Jurassic.

Access:- By public bus to Wotton under Edge or by private transport. Road walking.

Maps:- Topo:- 162. Geol:- EW 251.

Walking distance:- 6 miles (9.6 km).

Itinerary:- From Wotton under Edge N.W. to Bradley then W. and N. on minor roads to road junction ¼-mile (0.4 km) N.W. of Hawley, below terrace of the Middle Lias, with the Lower Lias forming lower ground to the W. Turn up narrow road N.E. along which a cutting shows fossiliferous ferruginous marlstone of the Middle Lias. At the road junction above sandy clays (Cotswold Sands) contain abundant Upper Lias ammonites and other fossils.

S.S.E. to W. outskirts of Wotton under Edge then turn N.N.E. up Wotton Hill, where rather poor exposures are seen of the Cotswold Sands. Higher up a disused quarry on the W. side is an excellent place for collecting. On a bored and planed surface of oolitic "freestone" of the Lower Inferior Oolite rests the Upper Inferior Oolite with Trigonia and many other fossils. Continue N.E. to main road where there are exposures of fissile limestone known as the Stonesfield Slate. This belongs to the Great Oolite, but the oolite itself is not exposed here.

Descend main road to Wotton under Edge; several quarries are seen in the Inferior Oolite.

E8.19   South of Bath

Jurassic.

Access:- By train to Bath. Walking on roads and disused railways.

Walking distance:- 11 miles (17.6 km).

Itinerary:- Southwards from Bath on Radstock road (A367) then at Odd Down turn S.E. to Midford. Cross valley and walk N.E. along dismantled railway to examine Midford Sands (Upper Lias) overlain by the Trigonia Grit (Inferior Oolite). The sandy facies is due to the influence of the Bath Axis. Landslipping greatly confuses the sections in the Midford area.

Next walk along road N.E. from Midford towards Monkton Combe. This road is too narrow for coaches. The Midford Sands are again seen overlain by disturbed Trigonia Grit. At Tucking Mill, nearby, there is the house which William Smith owned from 1798 to 1819. In fact, a plaque has been placed on the wrong house. The correct house is 55 yds. (50 m) to the N.E.

Return to Midford and walk W. along dismantled railway. Exposures of the Inferior Oolite provide good opportunities for collecting.

Leave the railway near Combe Hay and walk N. up minor roads to main road (Roman Fosse Way) examining exposure of Great Oolite Limestone on way. The underlying Fullers Earth is mined S.E. of the main road. E. of Odd Down the Great Oolite Limestone is quarried.

Down road to Bath.

Weymouth Centre (Fig. 8.8)

The pleasant coastal town of Weymouth is easily reached by rail or road. Like Folkestone, it is a port for boat services to France. Accommodation is plentiful but the town is very busy in the holiday season.

E8.20   Swanage

Upper Jurassic. Cretaceous. Tertiary.

Access:- By bus or private transport to Swanage. Road and coastal walking. Low tide necessary for some exposures.

Maps:- Topo:- 195. Geol:- EW 343.

Walking distance:- 10.5 miles (16.8 km).

Itinerary:- S. along cliff path to Tilly Whim Inn then descend zig-zag path to shore about the middle of Durlston Bay. The path roughly follows a fault zone.

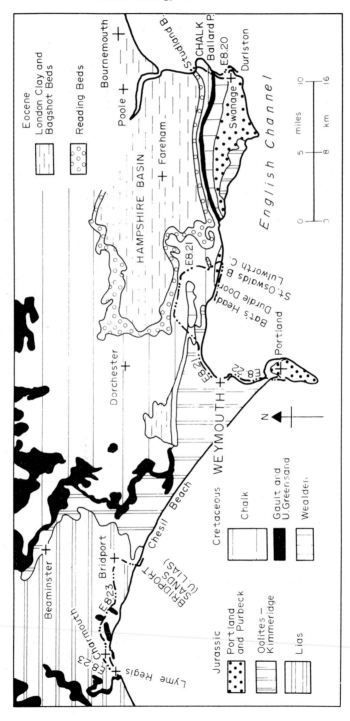

Fig. 8.8 Geological map, with routes, for Weymouth Centre.

Walk N. along the shore over a Purbeck succession dipping N. at around 12°.
The Lower Purbeck consists of thick shales with gypsum. The Middle Purbeck
begins with the Lower Building Stones at the base of which mammals and reptiles were formerly obtained. 15 ft. (4.6 m) further up the Cherty Freshwater Bed yields well-preserved gastropods including <u>Valvata</u> and <u>Planorbis</u>.
A ledge further N. marks the "Cinder Bed", with <u>Ostrea distorta</u>, followed
by the Upper Building Stones, containing <u>Corbula</u> and <u>Viviparus</u>. The Upper
Purbeck starts with the Broken Shell Limestone followed by the <u>Unio</u> Beds
and the Marble Beds, full of <u>Viviparus</u> and sharply folded as seen round
Peveril Point.

Walk along the front at Swanage to the beginning of the cliffs at the Ocean
Bay Hotel, then go down to the beach to examine the Wealden Beds (clays with
thin sandstones and a grit band) above the beach huts. Continue beyond the
sea-wall then turn up a small ravine to join the cliff path. Along path to
the S. edge of Ballard Down then descend a large recess to outcrop of Lower
Greensand. The Gault is usually obscured but the Upper Greensand, glauconitic
in places, can be examined along the shore to the N.E., as can the Lower and
Middle Chalk. Most of the succession is moderately fossiliferous, and
abundant fossils of the Zone of <u>Inoceramus labriatus</u> can be collected from
fallen blocks.

Further N.E. the cliffs become inaccessible, and it is necessary to follow
the cliff path. The beds become vertical and the Upper Chalk is cut by the
Ballard Down Fault which can be examined only from the sea.

Continue to the S. end of Studland Bay where the Reading Beds (Tertiary)
can be seen resting on the Chalk.

Return by cliff path to Swanage.

E8.21   Lulworth Cove, Durdle Door and Bat's Head

Upper Jurassic. Cretaceous.

<u>Access</u>:- By public bus to Lulworth Cove (summer only) and back. By private
transport. Coastal and road walking. Low tide an advantage for some exposures.

<u>Maps</u>:- <u>Topo</u>:- 194. <u>Geol</u>:- Ew 342.

<u>Walking distance</u>:- 5 miles (8 km).

<u>Itinerary</u>:- Lulworth Cove is famous for the spectacular folds on the Middle
limb of the Purbeck Monocline. The bastions at the entrance are of Portland
Stone succeeded by the limestones with shales of the Purbeck and the brightly
coloured vertical or overturned Wealden Beds (where the Cove is at its
widest). The junction with the fault is an unconformity; the N. side of
the Cove cuts into the Gault. It should be noted that the cliffs above the
Cove are unsafe.

W. of the Cove Stair Hole shows crumpling of the Purbeck Beds. Follow the
road to the W. and then a path down to St. Oswald's Bay, passing over
Wealden Beds with bituminous seeps. To the N. the Upper Greensand contains
<u>Holaster laevis</u> and other fossils. Faults near here are now hidden by scree.

The Upper Chalk, mostly of the <u>Micraster corangium</u> Zone, is inverted. The
Red Hole at the cliff top is a solution-pipe filled with Tertiary sediments.
To the W. at Durdle Cove, thrusting is seen and fossils may be collected
from the Upper Greensand. Durdle Door is an arch in the Upper Chalk. Fine
cliff scenery can be seen by continuing W. to Bat's Head, another arch in
the Upper Chalk with an offshore stack.

Return to Lulworth Cove.

<u>E8.22   Weymouth, Portland and Chesil Beach</u>

Upper Jurassic. Coastal bar.

<u>Access</u>:- Road and coastal walking. By public bus back from Portland.

<u>Maps</u>:- <u>Topo</u>:- 194. <u>Geol</u>:- EW 312.

<u>Walking distance</u>:- 7.5 miles (12.0 km).

<u>Itinerary</u>:- Walk over bridge across Weymouth Harbour then S.W. along coast.
At the Western Ledges, just S.W. of the breakwater Corallian limestone with
abundant <u>Trigonia</u>, as well as clay and grit are exposed, and further S.W.
Kimmeridge Clay.

Onto causeway to Portland. To the N.W. the great bar of Chesil Beach is an
impressive sight.

At the S.E. end of the causeway walk S. along the W. shore then climb over
slipped Kimmeridge Clay to the main path below the West Weare Cliffs where
the various divisions of the Portland Beds can be seen, although vast
screes of quarry debris partly cover the slopes. The lower beds (Portland
Sand) consist mainly of sandstones and clays, the upper (Portland Stone)
mainly of limestones. At the base of the latter the Basal Shell Bed provides the best collecting in the Portland Beds in England. <u>Glaucolithites</u>
and <u>Kerberites</u> are the commonest ammonites, and <u>Trigonia</u>,<u>Chamys</u>, <u>Corbula</u>
and <u>Exogyra</u> among the most abundant lamellibranchs. The Cherty Series can
be examined in fallen blocks. Walk up a path, by the Blacknore Battery,
past an exposure of the Freestone Series to the cliff top then along the
path to a hairpin bend in the main road. Follow the Easton Road to the
offices of Dorset Limestone Ltd. Here there is an excellent collection
set out as a garden rockery, of the larger fossils from the quarries,
including the giant ammonite <u>Titanites</u>.

Return from town to Weymouth by public bus.

<u>E8.23   Bridport and Lyme Regis</u>

Jurassic. Cretaceous. Landslips.

<u>Access</u>:- By public bus to Bridport and Lyme Regis (service infrequent in
winter). By private transport. Low tide necessary for shore exposures
at Lyme Regis.

<u>Maps</u>:- <u>Topo</u>:- 193. <u>Geol</u>:- EW 326, 340, 327.

<u>Walking distances</u>:- At Bridport 4 miles (6.4 km). At Lyme Regis 2.5 miles
(4 km).

<u>Itinerary</u>:- By public bus or private transport to Bridport. Walk to shore
and turn E. to fine cliff exposures of Bridport Sands (Upper Lias).

Return to Bridport and travel to Lyme Regis. To shore on E. side of town
where fossiliferous Lower Lias limestones are exposed. Along shore N.E.
beneath cliffs of Lower Lias clays (above the limestones) which are overlain by Upper Greensand. This major overlap contrasts with the presence of
the full Jurassic succession E. of Weymouth (E8.20, E8.21). The cliffs
show large slips which should be treated with caution.

Continue along shore to W. side of R. Char then return to Lyme Regis by
high-level cliff path and examine Upper Greensand. Care should be taken
as the path has been affected by the slipping.

Return from Lyme Regis to Weymouth.

References

Bisat, W. A. and others. 1962. Geology around the University Towns: Hull.
    Geol. Assoc. Guide No. 11.
Chatwin, C. P. 1960. British Regional Geology: The Hampshire Basin and
    Adjoining Areas. M.G.S. 3rd ed.
Chatwin, C. P. 1961. British Regional Geology: East Anglia and Adjoining
    areas. M.G.S. 4th ed.
Gibbons, W. 1981. The Weald. Unwin.
Greensmith, J. T. and others. 1973. The Estuarine Region of Suffolk and
    Essex. Geol. Assoc. Guide No. 12.
Hemingway, J. E. and others. 1968. Geology of the Yorkshire Coast. Geol.
    Assoc. Guide No. 34. (rev. ed.)
Kent, P. 1981. British Regional Geology: Eastern England from the Tees to
    the Wash (formerly East Yorkshire and Lincolnshire). M.G.S.
Kirkaldy, J. F. 1967. Geology of the Weald. Geol. Assoc. Guide No. 29.
McKerrow, W. S. and Kennedy, W. J. 1973. The Oxford District. Geol. Assoc.
    Guide No. 3. (rev. ed.).
Pitcher, W. S. and others. 1967. The London Region. Geol. Assoc.
    Guide No. 30.

CHAPTER 9

Excursions to Tertiary Volcanoes

9.1 Western Scotland (Fig. 9.1)

Excursions to the Tertiary igneous centres of Scotland provide opportunities
to study deeply-eroded Eocene volcanoes and to travel through spectacular
mountain and coastal scenery. Five of the six major Scottish centres are
on islands and one is at the end of the long Ardnamurchan Peninsula. Access
from most of Britain is therefore lengthy. However, there are boat services
(which carry cars) from rail and road heads at Kyle of Lochalsh for Skye, at
Oban for Mull and at Ardrossan for Arran. Rhum, less accessible, and St.
Kilda, almost inaccessible, are not described.

Some other itineraries by land and sea are suggested in Ch. 10.

The Scottish Tertiary complexes penetrate the Northern Highlands, the
Caledonian Front, the Grampian Highlands and the Midland Valley. For the
stratigraphical successions in these regions see the relevant chapters.

Broadly, the sequence of igneous events in the complexes was as follows:-

(a) the extrusion of great spreads of mainly alkali-olivine-basalts;

(b) establishment of central vents giving rise to vent agglomerates
     with small flows of acid lavas and porphyritic basalts;

(c) intrusion of widely varied igneous rocks, many of which form ring-
     complexes containing ring-dykes and cone-sheets;

(d) the formation of great linear dyke-swarms (many dykes were, however,
     intruded during earlier stages).

For further details see Richey (1961) and Anderson and Owen (1980, Ch. 9).

Many striking aspects of the scenery are due to glacial sculpture.

Comparison with other regions:- Considerable Mesozoic successions are
preserved beneath the lavas. The Triassic is terrestrial, as in most parts
of the British Isles. The Jurassic, however, is of a distinctive "Hebridean"

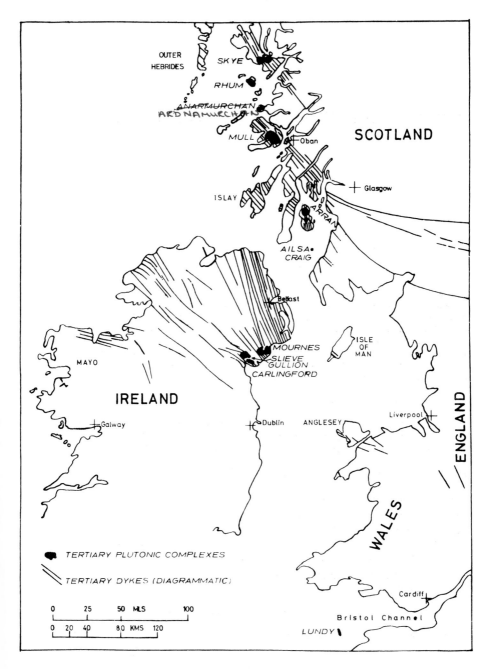

Fig. 9.1 Map of Tertiary igneous complexes.

facies with many sandy and shaley beds and several breaks in the sequence.
The Lower Cretaceous is missing. The Cenomanian contains white sandstones,
in places of high purity. The Chalk, harder than in Southern England, is
conspicuous in Northern Ireland and occurs to a limited extent in West Mull.

Skye (Fig. 9.2)

The best excursion centre is Broadford. For those without private transport
there are public buses to several parts of the island.

Very little guidance is needed to study the extensive Tertiary lavas of
Northern Skye; excursions to this and other areas are detailed by Brown and
others (1969).

Tertiary gabbro forms the spectacular Cuillins but traverses of these
3000 ft. peaks involve rough to difficult climbing. Mist is frequent,
and the compass in places unreliable. The gabbro can also be well seen
on Blaven, E. of the Cuillins.

Maps for Skye excursions:- Topo:- 32 Geol:- S 70, S 71.

E9.1   The Blaven area

Ordovician limestone. Mesozoic sediments. Tertiary lavas. Tertiary
gabbro, cone-sheets and dykes.

Access:- By private transport to W. side of Loch Slapin then rough walking
and some rock scrambling.

Walking distance:- 9 miles (14½ km).

Itinerary:- A stop should be made at Slapin School, where the road reaches
the E. shore of Loch Slapin. In crags above a small quarry 200 yds. N. by
W. of the school the Ben Suardal Group (L. Ordovician) of the Durness
Limestone yields an abundant gastropod/cephalopod/lamellibranch fauna.
Continue by road to a point on the W. side of Loch Slapin 2 miles (3.2 km)
S. of its head. Here calcareous sandstone of Inferior Oolite age contain
Pecten and Rhynchonella.

Park transport near Strathaird House and follow the path W. towards
Camasunary. On the lower ground the path crosses sandstones, shales and
limestones of the Great Estuarine Series and shales and sandstones of Oxford
Clay age before reaching Tertiary lavas; N.W. basic dykes are abundant.
Leave the path and circle the valley above Camasunary to gain the S. ridge
of Blaven. Tertiary gabbro is seen in contact with hornfelsed Liassic
sandstone and Tertiary lava.

On the climb up the rough and steepening ridge the gabbro is observed to
be cut by numerous basic cone-sheets and N.W. dykes. Between the S. top of
Blaven, 3031 ft. (924 m) and the N. or main top, 3042 ft. (924 m) there is
an awkward dip to a shallow col. From the summit in clear weather the view
is superb. Nearby the rugged gabbro peaks of the Cuillins contrast with
the granite cones of the Red Hills to the N. Further away the view ranges
from the Outer Hebrides to Ben Nevis, and to the S. the Tertiary centres of
Rhum, Ardnamurchan and Mull.

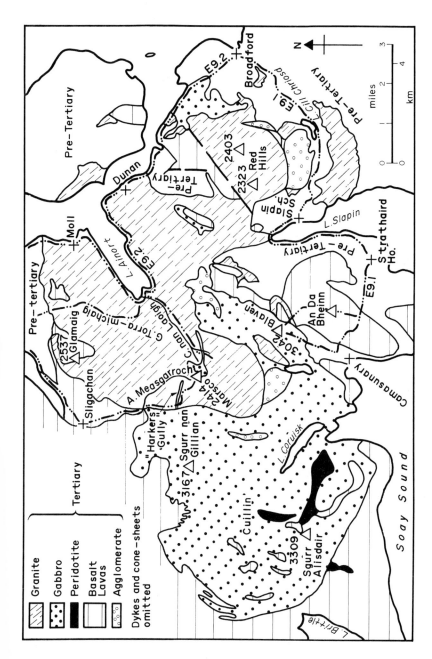

Fig. 9.2 Geological map, with routes, of Skye Tertiary Complexes.

From the col between the two tops a steep gully, determined by a fault-shattered dyke, should be descended S.E. The gabbro walls show abundant cone-sheets inclined W. at 45°. From the foot of the gully a traverse across scree S. leads to a col well below the S. Top. Here there is an intrusion of microgranite. From this col a route S. over the baked lavas of An Da Bheinn leads to the Camasunary path. The summit ridge can be avoided by traversing the lower part of the mountain where many of the features described above can still be seen.

E9.2   Western Redhills including Marsco area

Tertiary lavas. High-level granites and hybrid rocks. Glacial features.

Access:- By public bus or private transport to Sligachan, then rough walking.

Walking distance:- (from Sligachan over Marsco to head of Loch Ainort) 6 miles (9.6 km).

Itinerary:- The public bus goes through Glen Torra-Michaig, but with private transport the road N. of Loch Ainort should be followed to a small headland 550 yds. (0.5 km) N. of the Moll River. Coarse peralkaline arfedsonite-bearing granite outcrops and also a patchy hybrid rock, part of the Marscoite suite called glamaigite.

From Sligachan walk S. by E. up the glaciated Sligachan valley: exposures of Tertiary lavas give way to granite near the Allt na Measgarroch, and in this stream, above the track, the granite, with numerous dark xenoliths, is cut by thick dolerite dykes.

Further along the track, "Harker's Gully" is seen cutting the N.W. slope of Marsco. This mountain is made up of the following granites:-

Western Redhills or Glamaig Epigranite (earliest) - caps mountain and forms N.W. and N. slopes.

Southern Porphyritic Epigranite - partly encircles mountain as line of reddish-brown cliffs.

Marsco Epigranite - on lower S. and W. slopes, buff-weathering.

"Harker's Gully" follows a dyke-like mass of ferrodiorite (with the hybrid-rock marscoite on its N. margin) which has been interpreted as the remnants of a feeder-dyke of the Cuillins intrusion disrupted by the later granites..

Near the foot of the gully there is a large xenolith of Lewisian gneiss in the ferrodiorite. Opposite the huge overhanging shelter stone (protruding from the S. wall), at the top of the N. wall, marscoite is seen in a chilled contact with porphyritic felsite. Leave the steepening gully to the S. at the Second Terrace and continue up the W. slope of Marsco. In the bed of the first deeply-cut stream S. of "Harker's Gully" the roof of the Marsco Epigranite is seen beneath the Southern Porphyritic Epigranite.

Scramble up the steep but sound rock of the Southern Porphyritic Epigranite onto the N.W. ridge formed by the overlying Glamaig Epigranite. The broad N.W. ridge leads to the summit of Marsco, 2414 ft. (736 m), from which there is a magnificent view of the Cuillins.

From the summit a descent of 22 yds. (20 m) S.W. reveals exposures of the
Glamaig Epigranite/Southern Epigranite contact. From there walk about
160 yds. (150 m) S.E., just below the summit ridge, to study the complicated
junction between the Glamaig Epigranite and earlier gabbro.

Return to the summit then follow the S.E. ridge for 540 yds. (500 m) to a
grassy saddle, from which a steep grassy descent N. gains access to the
glaciated bowl of Corie nan Laogh, floored by Marsco Epigranite.

In bad weather it is advisable to leave "Harker's Gully" at the Second
Terrace via the N. wall and to walk round the northern slopes of Marsco
directly to Corie nan Laogh. The descent alongside the stream draining the
corrie to the road at the head of Loch Ainort provides another good section
through the Marscoite suite and a view of a bouldery lateral morain on the
S.E. slope.

Rejoin a bus or private transport at the road. Private transport can be
stopped at a roadside cliff at Dunan where an amphibolite-granite contains
druses filled with epidote, calcite, fluorite and zeolites.

Ardnamurchan (Fig. 9.3)

There are three intrusive centres, Centre 1 in the E., Centre 2 in the W.
and Centre 3 (youngest) about the middle of the complex.

Kilchoan, where there is accommodation, is the best place to stay to study
all three igneous centres.

Maps:- Topo:- 47 Geol:- S 51, S 52.

E9.3   Ben Hiant (Centre 1)

Moinian/Mesozoic/Tertiary lava succession. Vent agglomerates, pitchstone
lavas and intrusions of Centre 1.

Access:- By car to road cutting on S. bank of stream 0.4 miles (0.6 km) S.
of 13th milepost from Salen (room for only 2 cars).

Walking distance:- 2½ miles (4.0 km) over steep and rough ground. If
private transport is not available it may be possible to use public transport
at least one way on the road. Walking from Kilchoan and back by the road and
by a path directly E. from S. of Camphouse adds up to a walking distance of
12½ miles (20 km).

Itinerary:- From the road cutting traverse up the largest of three streams.
Starting about 300 ft. above the road, exposures of Moinian psammites with
micaceous layers are seen, cut by Tertiary dykes. Where the ground flattens
and becomes grassy, above the second waterfall, poorly exposed Triassic and
Liassic sediments occur. The Liassic is overlain by Tertiary plateau
basalts just below the confluence of two small streams. The more southerly
channel (separated from the northerly by an E.-W. dyke-controlled ridge)
should be followed to the well-defined contact between basalts and overlying
brecciated agglomerate filling an early vent of Centre 1; the contact dips
S.W. at 20°. Fresher and less brecciated agglomerate forms a steep slope
leading to an easily-observed, N.E.-trending, cone-sheet. From this point a
climb S.W. over undulating and boggy ground leads to outcrops of dolerite
with large porphyritic felspars. From here there is an ascent up dolerite

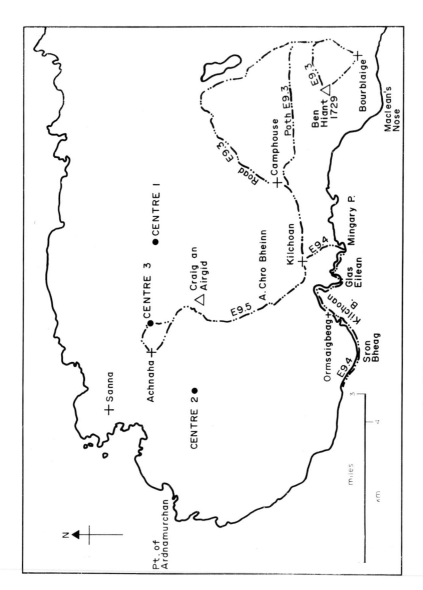

Fig. 9.3 Route map for Ardnamurchan.

crags to the summit of Ben Hiant, 1729 ft. (524 m). In clear weather there
is a fine view of the region and particularly of the trap-featured basalts
of Mull.

From the porphyritic dolerite a S.E. descent leads to andesitic pitchstone
lavas forming a hump above the abandoned village of Bourblaige. From the
hump a N.E. traverse, crossing spheroidal-weathering basalts, can be made
back to the road.

E9.4   Kilchoan shore sections (Centre 2)

Mesozoic sediments and Tertiary lavas. Vent-agglomerates and cone-sheets
of Centre 2 forming the best-developed cone-sheet complex in the British
Isles.

Access:- By walking from Kilchoan; the coastal exposures can be fully seen
only at low tide.

Walking distance:- 6 miles (9.5 km).

Itinerary:- S. from Kilchoan to Mingary Pier, about 65 yds. S. of which
the lower half of a composite sill is exposed, the upper half being cut off
by a massive basic sill. The composite sill consists of quartz-dolerite
showing both non-chilled, sharp, and gradational, contacts with overlying
granophyre. S. of the basic sill a composite cone-sheet occurs. From this
exposure walk E. then N. across a complex of cone-sheets, sills and dykes
cutting Moinian metasediments.

At the pebble beach of Port na Luing turn S. to the tidal island of Glas
Eilean where vent-agglomerates are well seen; the vent walls are preserved
to the S. and E. Northwards along the W. shore of the point basalt lavas
occur, faulted, near the inner part of Kilchoan Bay, against Liassic
sediments.

From Kilchoan bay the road and then a coast path leads to a small headland
about 300 yds. S.W. of Ormsaigbeag where there is access to the shore. The
headland consists of a massive quartz-dolerite dyke flanked by granophyre,
with hybridisation between the two components. From here to beyond the main
headland of Sron Bheag there is a magnificent display of basic cone-sheets
inclined N.W. to N. towards Centre 2. Significant exposures of the Mesozoic
country rocks include shales with Upper Liassic ammonites as lenticles within
a cone-sheet complex just S.W. of a marked ridge formed by a quartz-dolerite
sheet. About halfway between here and Sron Beag Inferior Oolite limestone
with belemnites and ammonites outcrops.

In the bay N.W. of Sron Beag there is a succession of Inferior Oolite sandstones and limestones/Tertiary white sandstone/mudstone with thin ironstones/
basalt lava. About ½-mile further N.W., just beyond an acid intrusion,
Upper Lias with about 3 ft. of Raasay Ironstone is seen; the ironstone is
largely altered to magnetite but belemnites are preserved.

There is a path up the cliff here but its continuation to Ormsaigbeag is
somewhat hazardous, and it is better to return along the shore, tide per-
mitting.

E9.5   Centre 3

Ring-dykes of the youngest complex. Glacial features.

Access:- By Kilchoan-Sanna road.

Walking distance:- 7 miles (11.2 km).

Itinerary:- Roadside exposures of the outermost and oldest ring occur near
the bridge over the Amhainn Chro Bheinn, where the road makes a sharp bend
N. Further N. the road crosses the Great Eucrite ring which is well exposed
on the slopes of Craig an Airgid where there are many features of glacial
erosion. It is worth climbing the hill in clear weather for the view over
Centre 3.

In the next mile along the road the following ring-dykes can be examined:

> Quartz-dolerite veined by granophyre (on W. side about 200 yds.
>   from road)
>
> Biotite-eucrite (biotite not seen in hand-specimen)
>
> Sithean Mor fluxion-gabbro (seen in stream)
>
> Inner Eucrite (up Allt Uaha Muice to N.W.)
>
> Quartz-gabbro

From the N. end of the houses at Achnaha (parking) a traverse E. reveals the
innermost and youngest intrusions. A climb over higher ground crosses the
Glendrian fluxion-gabbro. Beyond, tonalite is at first poorly revealed but
exposures improve eastwards and the rock can be seen to contain large,
platey biotites. Finally, quartz-biotite-monzonite is reached about ½-mile
(0.8 km) E. of Achnaha, on rising ground with a small cairn. The dramatic
effect of the geology on the topography is well seen; the Great Eucrite forms
a vast amphitheatre surrounding the lower-lying inner complex. From the
cairn a traverse can be made S. across the inner rings to the road N.W. of
Creag an Airgid, thence back to Kilchoan.

Mull

Mull can be reached by public boat service from Oban. There are limited
public bus services on the island. Cars can be ferried from Oban, and
generally be hired on Mull. The best centre for excursions is Salen where
accommodation is available.

Igneous activity migrated N.W. from an early caldera to the Beinn Chaisgidle
Centre and then to the Loch Ba Centre.

Three excursions are described. Those wishing to stay longer can refer to
the itineraries given by Skelhorn (1969).

Maps:- Topo:- 47, 48, 49 Geol:- S 43, S 44.

E9.6   Loch Don (Fig. 5.22)

Dalradian metasediments. Devonian lavas. Trias, Lias, Inferior Oolite and
Cretaceous sediments. Tertiary lavas. Circumferential folding around
Tertiary complex associated with early caldera. Cone-sheets. Dyke swarm.

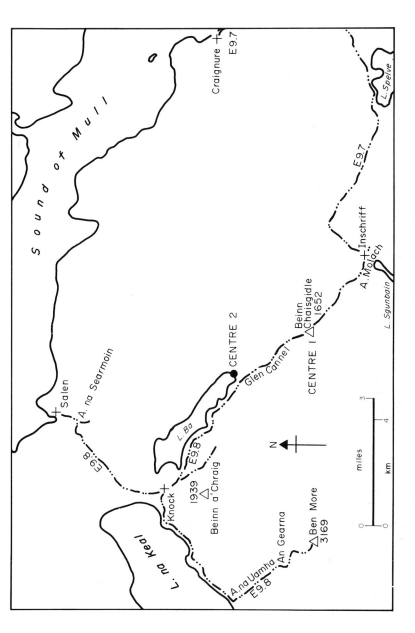

Fig. 9.4   Route map for Mull.

Raised beaches. Glacial features.

Access:- By car from Salen through Craignure to Lochdonhead. Also possible from Oban by using the boat service to Craignure. For a party it may be possible to arrange a crossing by motor-boat to Grass Point.

Walking distances:- From Lochdonhead back to Lochdonhead or from Grass Point back to Grass Point, $8\frac{1}{2}$ miles ($13\frac{1}{2}$ km). From Craignure $13\frac{1}{2}$ miles (21 km).

Itinerary:- From the Free Church on the S.W. outskirts of Lochdonhead a narrow road should be followed to Grass Point. The road first crosses Lower Old Red Sandstone basic lavas. At Ardnadrochet Farm a diversion should be made to the S. to Loch a' Ghleannain. Dalradian phyllites outcropping in the outflowing stream can be seen to be unconformably overlain by Lower Old Red Sandstone lavas 200 yds. (185 m) N. of the loch. These rocks lie on the axis of the Loch Don circumferential Anticline, which is asymmetrical, with a steeper E. limb. The lavas are overlain by Mesozoic sediments.

Return to the road and continue to Grass Point where Tertiary lavas are cut by acid cone-sheets and both by late basic dykes. The 100 ft. and 75 ft. raised beaches are present and there is a view to the N. of the Loch Don Re-advance Moraine. About $\frac{3}{4}$-mile (1.2 km) S. of Grass Point the lavas are underlain by a few feet of Tertiary mudstones and flint conglomerate resting on fossiliferous Upper Cretaceous sandstone overlying black shales probably belonging to the Great Estuarine Series. Continue along the coast to Port Donain where the lavas are underlain by fossiliferous Inferior Oolite overlying Lower Lias with Gryphaea and Pecten.

From Port Donain the line of a W.N.W. dextral strike-slip fault should be followed inland across Tertiary lavas in a syncline E. of the Loch Don Anticline. On this traverse the latter fold has Dalradian limestone and phyllite in its core overlain on its E. limb S. of the fault, by Trias/Lias/Inferior Oolite/fossiliferous Upper Cretaceous limestone. The Upper Cretaceous cuts across the Inferior Oolite to rest on Middle Lias.

To the W. Tertiary lavas re-appear in another syncline underlain at Rudha na Faing, by Mesozoic sediments on the E. limb of the Loch Spelve Anticline. Early basic cone-sheets are well-exposed, clearly of post-Anticline date.

Round head of Loch Spelve to road leading to Lochdonhead.

E9.7   Beinn Chaisgidle area (Fig. 9.4)

Intrusions associated with Beinn Chaisgidle area. If excursion is extended intrusions associated with Loch Ba centre. Glacial features.

Access:- By car from Salen through Craignure to Inshriff then by rough walking. By boat service from Oban to Craignure, thence by hired car.

Walking distance:- To top of Beinn Chaisgidle and back to Inshriff 6 miles ($9\frac{1}{2}$ km). Over Beinn Chaisgidle to Loch Ba and Knock 9 miles ($14\frac{1}{2}$ km).

Itinerary:- The car should be left near the bridge over the Allt Molach at Inshriff. Just downstream of the bridge a composite dyke with thin tholeiitic margins and a felsite centre cuts a quartz-gabbro ring-dyke. This is the

first of 11 ring-dykes, both basic and acid, exposed in the Allt Molach up
to its headstreams. However, it has been suggested that three of the
quartz-gabbro ring-dykes are in fact the same variable intrusion split by
acid ring-dykes. Numerous basic cone-sheets are also seen; the veining of
some of these by acid material would seem to be due to rheomorphism. Basic
material, mapped as a screen, between felsite (downstream) and quartz-gabbro
about 400 yds. upstream from the bridge is a Late Basic Cone-Sheet Complex.

From the head of the Allt Molach a traverse should be made (using compass in
mist) to the top of Beinn Chaisgidle, 1652 ft. (503 m). A quartz-dolerite
ring-dyke almost surrounds the summit area which is made of agglomerate in
annular vents cut by abundant Late Basic Cone-Sheets. The top gives a fine
view over many of the intrusions. Provided transport has been arranged at
Knock, those willing to make a fairly long walk can descend N.N.W. to Glen
Cannel Granophyre. The glen shows good glacial features. By keeping S.W.
of Loch Ba some of the geology described as part of the next excursion can
be seen.

E9.8   Loch Ba and Ben More (Fig. 9.4)

Tertiary basalt-lavas. Intrusions associated with Loch Ba Centre.

Access:- By walking along the road S.W. from Salen then on tracks and rough
mountainside. The walking distance can, however, be shortened if private
transport is available.

Walking distance;- Loch Ba (from Salen) 10 miles (16 km). Loch Ba (from
Knock) 3 miles (4.8 km). Ben More (from nearest point on road) 6 miles
(9.6 km). Loch Ba and Ben More (walking all the way) 18½ miles (29½ km).

Itinerary:- On the S. side of the road, just beyond the Allt na Searmoin,
a track leads to a quarry in two parts. The more northerly quarry is in
lavas, the more southerly in the Toll Doire Granophyre dyke, cut by thin
Late Basic Cone-Sheets. Continue to bridge over R. Ba at Knock, and turn
up track on S. side, then at a gate, branch off on a very rough path which
rises diagonally up Beinn a' Ghraig. Steam exposures show the Beinn a'
Ghraig Granophyre and, a little further on, diorite finely veined by the
granophyre. As the basic rocks are chilled against the granophyre it is
concluded that the veining is due to rheomorphism (see E9.16). To the E.
the Loch Ba Felsite Ring-dyke is magnificently exposed as a barrier standing
up above the granophyre, with contacts inclined outwards at 70°- 80°. To
the E., immediately S. of a lochan, four tuffisite dykes contain fragments
of Moinian psammites. North of the dykes, vent agglomerate is intruded by a
small rhyolite dome; these rocks belong to the Glen Cannel acid intrusion
vent complex.

Return to Knock. Those willing to make a very long day, or who have transport
available, should then go S. on the road to just beyond the Amhainn na Uamha,
the best point to start climbing Ben More. The gentle N.W. flank of An Gearna
leads to the steeper, but still easy, frost-shattered lava slopes which form
the top of the mountain. Ben More lies within the Zone of Pneumatolysis and
the lavas are altered. Trap-featuring can, however, be seen in several
faces. About 1800 ft. (549 m) the ordinary plateau-basalts are overlain by
the Pale Group of Ben More and from about 2000 to 2250 ft. (610-686 m)
mugearite occurs. From Ben More, 3169 ft. (965 m) the view is one of the
finest in Scotland. To the E. can be seen the peaks of the Western Grampians
and to the W. many of the Hebrides.

Return to road by same route.

Arran

Arran contains the two most accessible Tertiary complexes in Britain. The Northern Granite (the older) forms spectacular mountain scenery; to the S. there is the Central Ring Complex. The island is split by the Highland Boundary Fracture-zone so that in the N. the Upper Dalradian can be studied and in the S. a succession from Lower Old Red Sandstone to Triassic.

The island has thus been for long one of the most popular excursion areas in the British Isles. From the accommodation and transport aspects Brodick is the best centre; there is a daily boat, which carries cars, from Ardrossan, with train connection from Glasgow.

Four excursions are described. The excellent geological map and the excursion guide by Macgregor and others (1965) will suggest others to those who wish to make a longer stay.

Maps:- Topo:- 62, 68, 69 Geol:- Special sheet of Arran 1:50,000

E9.9   Corrygills and Clauchland (Fig. 9.5)

This is a useful first excursion as the whole or part of it can be after the arrival of the boat at Brodick.

Permian sediments. Numerous acid and basic Tertiary intrusions. Raised beach.

Access:- Road, path and rough shore walking.

Walking distance:- $8\frac{1}{2}$ miles ($13\frac{1}{2}$ km).

Itinerary:- The coarse breccia seen as the boat approaches Brodick Pier and along the shore is the Permian Brodick Breccia. This is penetrated by several Tertiary basic dykes and by a small explosion vent (opposite Gwyder Lodge). Towards the W. end of the village, behind the School House, a pitchstone sheet can be examined but care should be taken not to disturb the school.

A return should then be made through the village to the Lamlash road, which should be followed to the second turning off E., the Corrygills road. Follow this road, which becomes a track, S.E. towards the ridge formed by the Clauchland crinanite sill. On the way a pitchstone dyke, two crinanite dykes, an association of felsite and pitchstone and an irregular mass of porphyry (above South Clauchlands) can be examined. Above the porphyry the two leaves of the Dun Fionn pitchstone sill outcrop. The Clauchland sill should be followed down to Clauchland Point, near which both the upper and lower contacts of the southerly-dipping sill can be seen. Crinanite-pegmatite patches occur with large crystals of augite.

The Clauchland sill, along with two others to the S., has been interpreted as a thick cone-sheet related to a centre under Lamlash Bay.

From the point walk N.W. back to Brodick; although more can be seen at low tide, this is not essential as a path follows a well-marked 25 ft. raised beach. Numerous dykes cut the Brodick Breccia which contains a horizon with

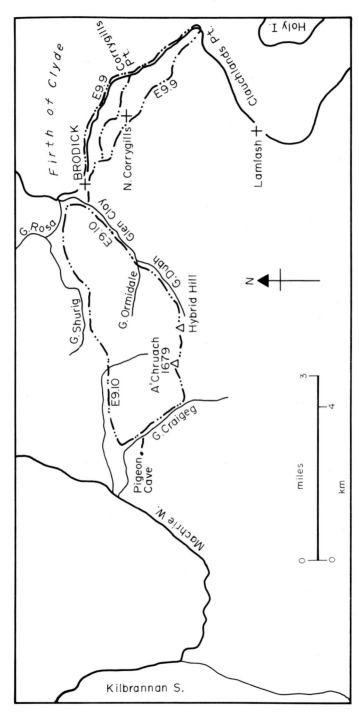

Fig. 9.5 Route map for Central Arran.

basalt fragments. There are also a number of felsite and pitchstone sheets.
The most famous, noted by Macculloch, outcrops about one third of a mile
S.S.E. of the mouth of the Corrygills Burn. At Corrygills Point a composite
felsite sheet, about 100 ft. thick and dipping steeply S., has flow-banded
upper and lower portions.

One and a half miles of raised beach lie between here and Brodick but a more
rapid way back is to return to Corrygills Bay and follow an uphill road to
the first farm at North Corrygills and then take a track to Strathwhillan
and Brodick.

E9.10   Gleann Dubh, Glen Craigag and Glen Shurig (Fig. 9.5)

Central Ring-Complex and its E. margin. Glacial features.

Access:- By Cloy Bridge in Brodick. Return by Glen Shurig (String Road).

Walking distance:- $11\frac{1}{2}$ miles ($18\frac{1}{2}$ km), 6 miles (9.6 km) if private transport
is available from foot of Glen Craigag.

Itinerary:- From Cloy Bridge follow the path on the N.W. bank of the stream
to opposite High Glencloy Farm. Here the Brodick Breccia (Permian) is intruded by a composite sill with spherulitic felsite margined by black pitchstone. The Brodick Breccia continues to the junction of Glen Dubh and Glen
Ormidale and for $\frac{3}{4}$-mile (1.2 km) up Glen Dubh. Both glens are noteworthy
for glacial corries, eroded in the rim of the Complex. In Glen Dubh a lower
terminal moraine is encountered, then the crescent of an inner terminal
moraine which once dammed a lake. Just upstream of the inner moraine a fault
brings up baked and epidotised Old Red Sandstone. A steep climb follows over
the outer intrusions of the Complex. Details are intricate, and the extent
to which they are worked out must depend on time and on the individual or
group interests. Broadly, up to the first long waterslide, gabbro is
separated by a screen of baked Lower Old Red quartzite from diorite and
microgranite. Where the main stream bends sharply W. a traverse should be
made to the next stream to the N. By following this stream to the foot of
steep slabs a gradual passage into gabbro can be seen and several irregular
felsite and microgranite bodies. The hillside to the E. shows reaction
products between the gabbro and acid rocks. From the top of the ridge both
the main stream and the northernmost burn can be seen to follow a gouge belt.
A felsite sheet is exposed in a long, steep gully. Near here a knoll known
as "Hybrid Hill" provides good exposures of hybridisation of basic and acid
rocks.

A traverse should then be made (mist is frequent and a compass may be
necessary) across peaty ground to A Chruach, 1679 ft. (515 m), the highest
point in the Complex. Poorly-exposed granite is part of an almost complete
ring. Continuation of the traverse W. across vent agglomerates leads to an
inner granite exposed in Glen Craigag. Descend this glen, reaching vent
agglomerates again about $\frac{1}{2}$-mile (0.8 km) above the road. Quarter of a mile
further downstream a diversion should be made $\frac{3}{8}$-mile (0.6 km) to the S.W.
Here the "Pigeon Cave" occurs in altered limestone believed to be a large
block of Cretaceous chalk in the vent agglomerate.

Descend Glen Shurig (String Road) to Brodick. In the first $1\frac{1}{2}$ miles (2.4 km)
exposures of granite, diorite and hybrid rocks are seen.

E9.11   Glen Rosa and Glen Sannox (Fig. 5.24)

Tertiary Northern Granite and part of its envelope. Tertiary dykes. Lower
and Upper Old Red Sandstone. Dalradian metasediments. Glacial features.

Access:- By roads and rough paths.

Walking distance:- 9 miles (14½ km) if transport is available on coast
road from Glen Sannox back to Brodick. Separate excursions can, however,
be made to the two glens.

Itinerary:- From Brodick follow the String Road and turn right just past
the ruins of a church where there is an exposure of Carboniferous sandstone.
Along the road to Glen Rosa Farm bright-red Upper Old Red Sandstone is seen
and at the farm darker sandstone of the Lower division. From the farm
follow the path up Glen Rosa. Hummocky terminal moraines probably mark a
late re-advance as they overlie the 100-ft. raised beach deposits.

Where the glen bends N.W. a sheet of metadolerite occurs flanked by steeply-
dipping to vertical Lower Old Red Sandstone. Quarter of a mile further on
Upper Dalradian turbidites outcrop, separated from the Old Red Sandstone by
the Highland Boundary Fracture-zone, although thermal metamorphism has
annealed the contact. The Dalradian continues to where Glen Rosa bends N.
just below the confluence of the Garbh Allt. The actual contact is not seen
in the main stream but can be studied in the Garbh Allt near the 700 ft.
(213 m) contour. Half a mile (0.8 km) above the granite margin the main
stream enters a gorge along a N.-S. tholeiite dyke.

Upper Glen Rosa is a spectacular glaciated valley flanked by granite with
dykes marked by gullies and ridge notches. A rough path continues over the
Saddle, a conspicuous col determined by a dyke, then drops very steeply to
Glen Sannox. About 1¼ miles from Sannox Bay the granite is seen close to
baked Lower Old Sandstone. About ½-mile (0.8 km) further on a barytes vein
in this formation was worked; specimens can be collected from the tip.

If private transport is used on the coast road back to Brodick a stop should
be made at Corrie to traverse a Carboniferous succession (low tide needed)
from the Calciferous Sandstone Series to the Coal Measures. The succession
is generally similar to that of the W. part of the Midland Valley but reduced to about 1500 ft. (460 m) occupying ¾-mile (1.2 km) of shore section.

E9.12   Ascent of Goat Fell (Fig. 5.24)

Tertiary Northern Granite and its S.E. envelope. Dykes.

Access:- By path.

Walking distance:- 9 miles (14½ km).

Itinerary:- N. from Brodick on the coast road, then branch off on a minor
road to the N. on the N. side of the Rosa Bridge. The minor road becomes a
path leading up the E. side of the Cnocan Burn. The path should be left
for the stream section 1.5 miles (2.4 km) N. of the Rosa Bridge. Lower Old
Red Sandstone becomes increasingly brecciated as the Highland Boundary
Fracture-zone is approached, beyond which are seen Dalradian metasediments,
also brecciated. Both formations are hornfelsed, which to some extent
obscures the structures due to faulting. The granite contact is reached

$\frac{3}{8}$-mile (0.6 km) further on. Here the path E. of the stream should be re-joined and followed over granite cut by W.N.W. dykes to the summit of Goat Fell, 2866 ft. (874 km). On a clear day the view is one of the finest in Scotland, embracing to the N. and N.W. the Highlands and part of the Inner Hebrides, to the E. and S.E. the Midland Valley and the Southern Uplands, while in the Lower Firth of Clyde the Tertiary microgranite island of Ailsa Craig leads the eye to the Tertiary basalts and underlying chalk of N.E. Ireland.

9.2   North-East Ireland

Antrim, in Northern Ireland, contains the largest spread of Tertiary lavas in the British Isles; they overlap the Highland Boundary Fracture-zone and therefore rest, generally with intervening Mesozoic, on the continuation of the Grampian Highlands and the Midland Valley (see also E10.29).

Further S. the three Tertiary complexes of the Mourne Mountains, Slieve Gullion and Carlingford (this third complex is in Eire) penetrate the continuation of the Scottish Southern Uplands. The broad sequence of igneous events is much as in Scotland (9.1).

The lavas in some districts can be divided into the Lower, Middle and Upper Basalts.

With private transport all the excursions can be done in long days from Belfast. Some may prefer local centres, suggested below.

Slieve Gullion and Carlingford are close to the Northern Ireland/Eire Border; local police should be consulted before undertaking excursions in this sensitive region.

E9.13   The Giant's Causeway (Fig. 5.26)

Portrush, with plentiful accommodation, can be reached by rail or bus from Belfast.

Columnar Tertiary lavas. Laterite. Dykes. Dolerite sill. Lias.

Access:- By public transport from Portrush to Bushmills or by car through Bushmills to Causeway Head.

Maps:- Topo:- $\frac{1}{2}$-inch Ireland 2 Geol:- N 17.

Walking distances:- From Bushmills and back 7 miles (11.2 km). From Causeway Head and back 3 miles (4.8 km).

Itinerary:- Before leaving Portrush a visit should be made to the shore to examine the large Portrush olivene-dolerite sill, the chilled top of which is in contact with hornfelsed Lias containing abundant ammonites.

From Causeway Head a path leads down to the world-famous Giant's Causeway. The columns occur in a lava belonging to the Middle Basalts; tilting of the columns is due to the lava of the Giant's Causeway having poured into a small valley in the Lower Basalts. There is a spectacular view across the bay to the E. of the columnar Middle Basalts, including the "Organ", of the bright red lateritic and bauxitic Interbasaltic Horizon and of the Lower Basalts.

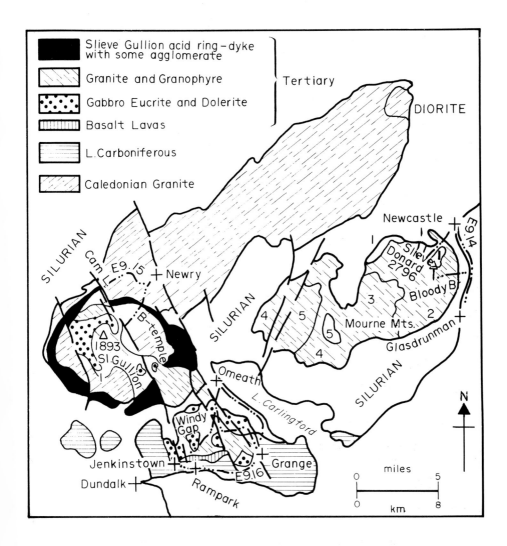

Fig. 9.6 Geological map, with routes, of Irish Tertiary Complexes.

Walk along the path past the "Organ" noting the dykes in the cliffs and particularly the intrusion which forms a great wall against the sky between the first and second amphitheatres.

Return by the same route.

E9.14   Mourne Mountains (Fig. 9.6)

The Mourne Mountains are best studied from Newcastle, Co. Down, where accommodation is available. The E. end can be visited without private transport.

Tertiary Granite Complex Cone-sheet and dykes. Altered Silurian. Glacial features.

Maps:- Topo:- ½-inch Ireland 9 Geol:- N 161.

Walking distance:- 12 miles (19 km). This can be reduced if transport can be obtained on coast road.

Itinerary:- From Newcastle the track up the U-shaped Glen River valley should be followed. A terminal moraine of a local glacier is seen. There are good sections in the river of hornfelsed Silurian which has a sharp contact with the quartzose G2 granite.

From the head of the Glen River there is a steep ascent to the summit of Slieve Donard, 2796 ft. (852 m), the highest point in Northern Ireland, which on a clear day gives a magnificent view which extends to the Isle of Man, the Lake District and North Wales. Around the summit a cap of the felspathic G1 granite rests on G2 and to the N.E. the cap includes altered Silurian.

From the summit a descent to the S. leads to the stream flowing down to Bloody Bridge. Further outcrops of G1 granite are seen, and around Bloody Bridge a huge moraine of large blocks deposited by Irish Sea ice.

The coast road, which follows the 25 ft. raised beach, should be followed S.W. to Glasdrumman. The shore-section shows Silurian cut by Tertiary dykes of the Mourne swarm and by a conspicuous cone-sheet. This has a central intrusion of quartz-felspar-porphyry bounded above and below by hornblende-olivene-basalt.

Return to Newcastle by coast road.

Note:- An extensive tour by car of the Mourne Complex is suggested by Charlesworth and Preston (1958 b).

E9.15   Slieve Gullion (Fig. 9.6)

The nearest centre is Newry.

Slieve Gullion Tertiary Ring-Complex penetrating S.W. end of Newry. Caledonian Complex.

Access:- By road from Newry.

Maps:- Topo:- ½-inch Ireland 9 Geol:- N 160.

Walking distance:- 11 miles (17½ km). Most of the excursion can, however, be done by car.

Itinerary:- From Newry follow the road W. to the N. end of Cam Lough, noting on the way outcrops of Caledonian granodiorite. Part of the almost complete Slieve Gullion ring of porphyritic granophyre can be studied in a small quarry just W. of the N. end of Cam Lough. The granophyre contains numerous xenoliths of granite, altered shale, basalt and dolerite. The N.W. Cam Lough transcurrent fault shifts the ring-dyke about a mile (1.6 km) to the S.E. on the E. side of Cam Lough. Here the dyke can be readily followed by eye up the side of Cam Lough Mountain and seen to incline outwards (N) at 65°-70°. Explosion breccias, affected by dynamic crushing, are seen on the same hillside. The breccia contains many fragments of Newry Granite.

By following the road on the S.W. side of the valley (and of the fault) it is possible to see the interior granophyre of the Tertiary Complex followed to the S. by gabbro.

At Ballintemple turn off on the road N.E. across the valley and continue in the same direction to Newry. About halfway to the town the ring-dyke is again crossed, cutting through Caledonian hornblende-granodiorite.

E9.16 Carlingford Complex (Fig. 9.6)

Tertiary lavas. Ring-dykes. Cone-sheets. Metamorphosed Carboniferous Limestone.

Access:- By road from Dundalk, Carlingford or Greenore.

Maps:- Topo:- ½-inch Ireland 9 Geol:- 1-inch Ireland 71.

Walking distance:- From Rampark, to which public transport should be available, the walking distance is 15 miles (24 km). Much of the excursion can, however, be done by private transport.

Itinerary:- From Rampark, on the coast road, a walk about ¼-mile (0.4 km) up the S.S.W. slope of Slieve Naglogh leads to exposures of Tertiary amygdaloidal basalt cut by basic cone-sheets inclined N.N.E. at 45°.

Return to road and continue to roadstone quarry in S. end of ridge S.W. of Grange. Here dolerite is veined by granophyre; the dolerite, however, seems to chill against the granophyre and this is probably a case of rheomorphism.

Continue to N. to an old quarry on the S.S.E. side of Barnavave where cone-sheets and later N.W. basalt dykes cut Carboniferous Limestone. A walk of about a mile (1.6 km) further N. leads to the E. slopes of Barnavave where the outer ring-dyke of eucrite is cut by granophyre and a skarn of Carboniferous Limestone contains various calc-silicate minerals.

Those on foot should then traverse W. to a road and walk to Windy Gap. The road follows a N.W. transcurrent fault, with a dextral displacement of about 0.7 mile (1.1 km), which cuts through the Complex. Those with private transport can drive through Carlingford to Omeath then S.W. to Windy Gap.

Near Windy Gap the eucrite of the outer ring forms craggy country with rough slabs reminiscent on a smaller scale of the Tertiary gabbro of Skye. Just S. of Windy Gap fork S. and follow the road across smoother granophyre terrain to Jenkinstown where the eucrite is seen on the S.W. margin of the Complex.

From Jenkinstown S.E. to coastal road near Rampark.

References

Anderson, J. C. G. and Owen, T. R. 1980. The Structure of the British Isles. 2nd ed. Pergamon.

Brown, G. M. and others. 1969. The Tertiary Igneous Geology of the Isle of Skye. Excursion Guide No. 13. Geol.Assoc.

Gribble, C. D. and others. 1976. Ardnamurchan. A Guide to geological excursions. Edinburgh Geological Society.

Macgregor, M. and others. 1965. Excursion Guide to the Geology of Arran. Geol.Soc. Glasgow.

Richey, J. E. 1961. British Regional Geology: Scotland: The Tertiary Volcanic Districts.

Skelhorn, R. R. and others. 1969. The Tertiary Igneous Geology of the Isle of Mull. Excursion Guide No. 20. Geol.Assoc.

CHAPTER 10

Geology of Some Transport Routes

Geologists who have to travel fairly long distances, particularly from overseas, often find it of interest to recognise the formations and structures they pass and to relate these to topography and economic development. This chapter is to fulfil this need for some at any rate of the routes likely to be used. The Geological Map of Great Britain on the 10 miles to 1 inch scale (3rd ed. 1979) should be available on these journeys. For Ireland there is the geological map on the scale of 16 miles to 1 inch.

The account starts with some major railway routes radiating from London, then goes on to a number of important roads and a few more railway routes in Scotland as well as local journeys to the islands.

For the road routes motorways are avoided as stopping to examine exposures is not permitted. It should be borne in mind that some types of public transport tickets allow a break of journey; a stopover can therefore be made to visit exposures along the route of particular interest.

Where routes described in this chapter overlap with excursions already described, to save space a cross-reference is given. In some mountain districts, roads and railways are more or less adjacent. Descriptions given for one type of transport can therefore be used for the other.

E10.1   London - Inverness (rail)

Every British formation from Eocene to Moinian (proterozoic) is crossed on this 568 mile (914 km) journey, the longest normally possible in one train in Britain. Topographical and economic differences mark many of the geological changes.

Sixteen miles (25½ km) out of London Euston station the line passes from the Tertiary onto the Chalk. The Tertiary is not seen but the Chalk accounts for the ascent from Watford to Tring Summit and is exposed in cuttings near the latter place. Between Tring and Rugby rich farmland is underlain by the Lower Cretaceous and nearly all the subdivisions of the Jurassic, dipping

291

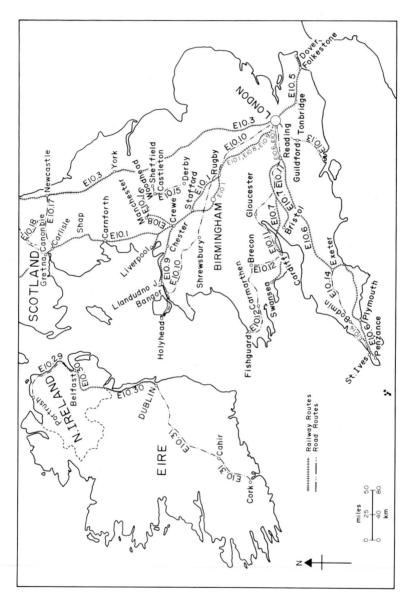

Fig. 10.1 Map of transport routes described for England, Wales and Ireland.

gently S.E. Near Leighton Buzzard the Lower Greensand, worked for industrial
sand, rests directly on the Kimmeridgian. The different formations are,
however, difficult to detect from the topography, apart from the Great
Oolite which makes something of a feature at Wolverton.

Just N. of Rugby the Triassic begins, and between Nuneaton and Atherstone
higher ground to the S.W. marks the outcrops of Cambrian quartzites and late
Precambrian volcanics (E7.29). Beyond Atherstone workings at the N. end of
the Warwickshire Coalfield are seen. Through Stafford and Crewe to Warrington flattish ground is underlain by Triassic, apart from a rise 18 miles
(29 km) N.W. of Stafford on the W. edge of the North Staffordshire Coalfield.

Between Rugby and Stafford some trains run via Birmingham. Triassic terrain
is interrupted W. of Coventry by the Warwickshire Coalfield and between
Birmingham and Wolverhampton by the industrialised "Black Country" over the
South Staffordshire Coalfield.

N. from Warrington the Pennines rise to the E., and from Wigan to Preston
some of the strata of the Lancashire Coalfield are seen in cuttings. Once
this coalfield is left heavy industry is not seen again until the Midland
Valley of Scotland. Between Lancaster and Carnforth the line is close to
Morecambe Bay across which rise the Lower Palaeozoic mountains of the Lake
District (Ch. 6.2). W. of the route at Carnforth quarried scarps of
Carboniferous Limestone dip S.E. off the Lake District Dome (Ch. 6.2). On
the long climb to Shap Summit the rugged scenery of the Ordovician Borrowdale
volcanics rises to over 3000 ft. (915 m) to the W., while to the E. the
Pennines close in. Between Oxenholme and Tebay Ludlow greywackes are seen
in cuttings above the glaciated Lune valley. Just over Shap Summit, 904 ft.
(276 m) there is a glimpse to the W. of the famous Shap Granite Quarry. E.
of Penrith red Permian sandstones make a scarp, and on the way down to
Carlisle the Permo-Triassic beds of the Vale of Eden are seen.

Seven miles N. of Carlisle the Scottish Border is crossed. High ground some
distance to the W. is formed of the Caledonian Criffel Granite (E6.22). The
route is by now entering the Southern Uplands (Ch. 6.3) but as far as
Beattock it follows for the most part Permian Basins within the Lower
Palaeozoic block. On the steep climb to Beattock Summit, 1014 ft. (309 m)
Llandovery greywackes with some shales are seen in cuttings; an interesting
feature is a quarry on the E. side in a far-flung basic dyke of the Mull
swarm. Rounded hills of Ordovician sediments flank the descent of the Upper
Clyde valley past Abington to the Southern Uplands Fault at Lamington. The
entry to the Midland Valley does not, however, bring about at first any
topographical change as Lower Old Red Sandstone lavas rise as hills N.W. of
the fault. To the W. the Lower Old Red Sandstone felsite laccolith of Tinto
is conspicuous. N. of Carstairs fluvioglacial deposits make striking ridges
(kames) on the E. side.

The Central Coalfield, and with it a major industrial zone, are entered at
Law Junction and continue past Motherwell, near which Coal Measures strata
are seen in cuttings almost to Stirling.

Important geological features are seen as Stirling is left towards the N.
To the E. the Ochil Fault (E7.48) has produced one of the finest fault
scarps in Britain with Lower Old Red Sandstone volcanics brought up to the
N. against Coal Measures. In front of the scarp the Wallace Monument is on
a Permo-Carboniferous quartz-dolerite sill. To the W. a faulted part of
this sill supports the Castle, and further W. the Clyde Plateau Lavas (Lower

Carboniferous) form the Fintry Hills. Lower Old Red Sandstone sediments, covered in places by thick fluvioglacial deposits, are seen near Dunblane. Outside Perth the Lower Old Red Sandstone lavas are penetrated in cuttings and in a tunnel.

Beyond Perth the Lower Old Red Sandstone sediments are hidden under thick fluvioglacial deposits and to the N.W. a sharp rise marks the beginning of the Grampian Highlands. The Highland Boundary Fracture-zone is crossed just before Dunkeld. There is a striking change in the nearby scenery; Upper Dalradian metagreywackes and slates of the Iltay Nappe form crags above the Tay valley. From Pitlochry many of the exposures described in E5.27 can be seen on the W. side, including the large working quarry in the Blair Atholl Limestone. On the opposite side here there is a view N.E. of Glen Tilt, eroded along a major fault. From Blair Atholl over Drumochter Summit, 1414 ft. (431 m), the highest point reached by a railway in the British Isles, numerous exposures are seen of Moinian flaggy psammites. Down the Spey valley thick fluvioglacial deposits predominate. On approaching Aviemore the granite mass of the Cairngorms, rising to over 4000 ft. (1219 m), is seen to the S.E., with northerly corries which are outlined in snow well into the summer.

The Moinian, with granite intrusions, is again visible between Aviemore and Inverness. At Slochd Mhor Summit, 1189 ft. (362 m), the route passes through a deep glacial overflow channel and another is seen at a lower level.

On the descent to Inverness there is a view across the Moray Firth and the flat ground underlain by Middle Old Red Sandstone to Ben Wyvis and other mountains of the Northern Highlands.

E10.2   London to Glasgow

For London to Motherwell see E10.1. From Motherwell to Glasgow Central station the route is over the Coal Measures; tips are evidence of widespread workings. Nearer Glasgow the faulted synclinal structure of the Glasgow district can be appreciated for not far S. of the line at Cambuslang the Clyde Plateau Lavas form high ground and much further to the N. the same lavas rise as the Campsie Fells, beyond the Campsie Fault (E7.45).

E10.3   London − Newcastle − Edinburgh − Aberdeen

The route from King's X station leaves the Tertiary of the London Basin 18 miles (29 km) out and climbs the dip-slope of the Chalk, seen in cuttings, to Stevenage. From here to Peterborough there are few features which enable the geology to be read as the Lower Cretaceous and Upper Jurassic are crossed. The Oolite outcrops account for the rise from Peterborough to Corby, but from there through York to Darlington the Lias and Trias form flat ground. Between York and Northallerton, however, the overlying Oolitic Series can be seen rising to the E. in the Yorkshire Moors and further away to the W. the high ground of the Pennine Arch. At Darlington the line passes onto the Permian Magnesian Limestone which is seen in a few cuttings and in a large quarry on the E. side of the line about 6 miles (9.6 km) N. of Darlington. The Coal Measures are entered near Durham where the Cathedral (E. side of line) is built on sandstone of this formation. A few further Coal Measures outcrops are seen onwards to Newcastle.

N. of that city the Coal Measures continue to Morpeth, marked by mines and
industrialisation. From Morpeth to Berwick the route is over the Lower
Carboniferous (E7.38, E7.39); the Great Whin Sill is seen in quarries.

On the ascent from Berwick folded Lower Carboniferous strata (E6.20) are
evident on the shore below. As the route climbs to Grantshouse Summit, the
Llandovery greywackes and shales of the Central Belt of the Southern Uplands
are seen. The Southern Uplands Fault is crossed just S. of Dunbar; Lower
Carboniferous strata occur on both sides of it. Beyond Dunbar the quarried
phonolite laccolite of Traprain Law (E7.52) makes a hill S. of the line and
Berwick Law a hill to the N. A few miles further W. the Lower Carboniferous
acid volcanics of the Garleton Hills (E7.52) are seen. The deep syncline
of the Midlothian Coalfield shows near Musselburgh by the presence of mines
and tips.

On the run into Edinburgh there is a striking view to the S. of the Arthur's
Seat Lower Carboniferous volcanic neck and associated trap-featured basalt
lavas and of the Salisbury Crags dolerite sill (E7.49).

By looking out to the S. just W. of Edinburgh Waverley station the Castle
Rock volcanic neck can be seen and after a long tunnel two features formed
by basic sills appear to the N., and to the S. the Lower Old Red Sandstone
volcanics of the Pentland Anticline (E7.49).

The presence of the oil-shales in the Lower Carboniferous is brought out by
the red tips of spent shale. As the Forth Bridge is crossed the sill of
quartz-dolerite (E7.51) which supports its N. end is clearly seen. Beyond
Burntisland Lower Carboniferous lavas appear in cuttings, and beyond Kinghorn it is possible to observe on the shore some of the Lower Carboniferous
strata described under E7.53. As the route turns N. the presence of the
Fife Coalfield is evident, and as the line passes onto the Lower Carboniferous
again the Lomond Hills to the W. mark a large quartz-dolerite sill.

On the N.E.-run through Cupar a strike outcrop of the Upper Old Red Sandstone
is followed and N.W. of Leuchars moundy fluvioglacial deposits are conspicuous.
At the S. end of the Tay Bridge Lower Old Red Sandstone basic lavas are
evident.

From Dundee to Arbroath the line mostly runs across a raised beach which
covers the Lower Old Red Sandstone.

For the main features of the geology from Arbroath to Stonehaven reference
should be made to E7.58.

Just N. of Stonehaven the Grampian Highlands are entered, and the orange-
weathering carbonated serpentine along the Fracture-zone (E7.58) can be seen
by looking down to the shore. Onwards to Aberdeen Upper Dalradian metasediments form fine cliffs.

E10.4   Edinburgh to Glasgow (rail)

For the first 4 miles (6.4 km) out of Edinburgh this line follows the
Aberdeen route (E10.3). W. of the divergence red tips indicate the former
working and processing of oil-shale in the Lower Carboniferous (Ch. 7.5).
At Ratho a feature on the S. side is due to a quartz-dolerite sill and near
Winchburgh a teschenite sill, in Upper Oil-Shale Group sediments, is seen in
a cutting. On the S. side at Lesmahagow rocky ground marks basalt lavas in

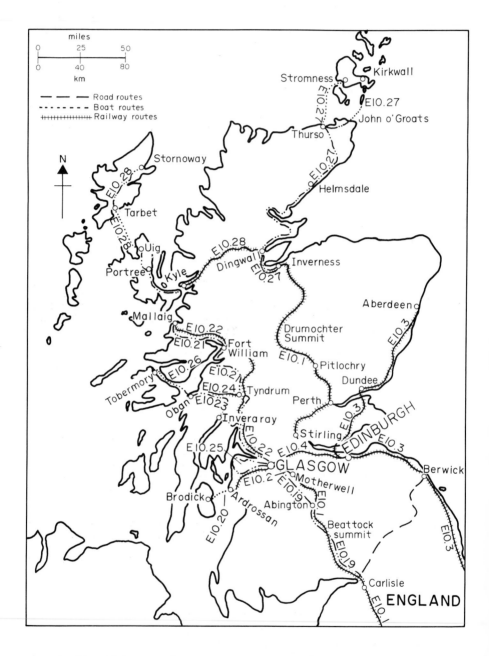

Fig. 10.2 Map of transport routes described for Scotland.

the Limestone Coal Group. Further W., before Polmont, ridges of fluvioglacial deposits (kames) are conspicuous on the N. side. The route is by
now crossing the Coal Measures in the centre of the Central Coalfield Basin.
West of Falkirk it passes onto the Millstone Grit on the W. flank of the
basin. Mines here are for fire clay in this formation, used for refractory
bricks.

Near Croy a thick quartz-dolerite sill is seen in a cutting. Mines further
W. are in the Lower Carboniferous Limestone Coal Group (Ch. 7.5). To the
N. the trap-featured Lower Carboniferous basalts of the Campsie Fells,
brought up by the Campsie Fault (E7.45) form high ground. A thick quartz-
dolerite dyke cutting Upper Limestone Group sediments is conspicuous in a
cutting 2.5 miles (4 km) beyond Lenzie.

Before the train enters the tunnel to Glasgow Queen St. station massive
sandstones in the Limestone Coal Group are seen.

E10.5   Folkestone and Dover to London (rail)

For visitors arriving from French ports this may be their first journey in
Britain. In fact the famous White Cliffs of Albion come into view long
before the boat reaches Folkestone or Dover. The Chalk is seen closer at
hand as the train climbs from Folkestone Harbour and turns W. towards
London. From Dover the same formation is traversed in cuttings and tunnels
before the junction with the Folkestone line is reached. Below the railway
the landslipped area known as the Warren consists of masses of Chalk which
have slipped over the underlying Gault (E8.14).

W. of Folkestone the Lower Greensand (Lower Cretaceous) is traversed in a
tunnel and cuttings and then for 25 miles (40 km) the line crosses flat
country, largely taken up by hop fields, underlain by the dominantly
clayey lowest Cretaceous beds in the heart of the Weald Anticlinorium (the
continuation of that of the Boulonnais in France).

Just before the line turns N.W. at Tonbridge a feature and cuttings mark
the Lower Cretaceous Tunbridge Wells Sandstone (E8.15). The route has now
turned towards the N. flank of the Anticlinorium and after crossing flat
ground corresponding to the Lower Cretaceous clayey beds again goes through
the Lower Greensand in a long tunnel before Sevenoaks. The scarp face of
the North Downs, formed of Chalk, is now ahead, and the railway passes under
the high ground in two tunnels; the Chalk can be seen in cuttings.

At Orpington the Chalk dips under the Tertiary of the London Basin, and the
heavily built-up ground all the way to the Thames and Victoria station is
mostly topographically subdued. However, there is a mile-long tunnel
between Penge East and Sydenham Hill through the London Clay (Eocene) which
is here capped by the sandy Claygate Beds.

E10.6   London to Exeter, Plymouth and Penzance

From London Paddington station to Reading the nearly flat route crosses
almost entirely the Tertiary. Thames terrace gravels have been worked in a
number of pits, some now ponds. The Tertiary continues to Newbury and then
the Chalk is entered, marked by a rising gradient. On leaving the Chalk the
line runs for a long way to Westbury over the Gault and Lower Greensand with
the Chalk, marked by the White Horse, rising to the S.

S.W. of Frome the Oolites can be seen but the topographical expression is
less than further N. Onwards to Taunton flattish country marks Liassic and
Triassic outcrops but sandstones in the latter formation S.W. of Taunton
account for the rise to Wellington tunnel. Down to Exeter the red beds of
the Permian are seen in places. The run along the coast S. of Exeter past
Dawlish is spectacular with red cliffs and stacks of Permian sandstone.

After Newton Abbot there are steep gradients as the line passes outcrops of
hard Devonian sediments. About 8 miles (13 km) E. of Plymouth a sharp rise
to the N. is due to the Dartmoor Hercynian Granite and its hornfels aureole.

W. of Plymouth ria type inlets of the sea can be seen. A number of exposures
of Devonian occur all the way to beyond Truro. N. of St. Austell huge white
tips (E7.6) mark the workings in china-clay derived from pneumatolysis of
the granite.

Between Redruth and Camborne numerous signs of mining for tin and other
metals are seen; the mineralisation is associated with the Carn Menellis
Hercynian Granite which rises to the S. As Penzance is approached the out-
standing feature is St. Michael's Mount, of granite and hornfelsed Devonian.

E10.7   London to Bristol and Swansea

For London to Reading see E10.6. W. of Reading the Thames has eroded the
Goring Gap, and here the Chalk is seen in cuttings. From Didcot to Swindon
the route follows Lower Cretaceous and Upper Jurassic along the Vale of
White Horse (E8.8) with the White Horse conspicuous on the Chalk Downs to
the S. Eroded down Upper Jurassic strata continue past Wootton Basset to
Chippenham. The obstacle of the Cotswold Hills is penetrated in the Box
Tunnel, through the Great Oolite which is seen in cuttings. The tunnel is
one of the classic early engineering works, planned, like the rest of the
London-Bristol railway by Brunel. Around Bath the hills consist of Jurassic
strata below the Great Oolite (E8.18).

Between Bath and Bristol there are cuttings in reddened Coal Measures sandstones. The reddening is due to iron oxide from the Triassic; Keuper Marls
are seen on the approach to Bristol.

On the Swansea route, which diverges W. of Swindon at Wootton Bassett, the
Great Oolite, seen in cuttings, is breached some 18 miles (29 km) further
W. by the long Badminton tunnel. Soon after leaving the tunnel Carboniferous
Limestone, dipping W. into the Bristol Coalfield Syncline, is seen and further
on, Coal Measures. The unconformably overlying Keuper Marls are cut through
at Stoke Gifford,then, after leaving a tunnel, a glimpse can be had of the
complicated Carboniferous Limestone and Coal Measures section described under
E7.15. The line descends through cuttings in Keuper Marls to the tunnel under
the Severn estuary, 4.35 miles (7 km), the longest in the British Isles. The
tunnel passes beneath the Triassic into Coal Measures overlying Carboniferous
Limestone, water from which caused great problems during construction, completed in 1886. Triassic sandstones are seen as the tunnel is left.

The continuation to Cardiff is mainly over coastal flats underlain by Flandrian
clays, but on the W. side of Newport station the Lower Old Red Sandstone is
penetrated in a cutting and in a tunnel.

W. from Cardiff faulted Liassic shales and limestones are seen on the E. side
of Bridgend station, part of the Mesozoic cover of the Palaeozoic in the

coastal region. Further W. very extensive blown sand deposits occur on the
S. side. As Port Talbot is approached the high ground marking the edge of
the South Wales Coalfield closes in from the N. Just beyond Neath station
there is a view upstream of the long, straight Vale of Neath eroded along a
N.E. fault-zone and containing a very deep buried valley (E7.22). The line
then passes through cuttings near Skewen in the Pennant Sandstone (Coal
Measures) before crossing another N.E. fault-guided valley, that of the
Tawe, just before Swansea.

E10.8   London to Manchester and Liverpool

For route to Crewe see E10.1.

Between Crewe and Manchester the line crosses Triassic. Outcrops of sandstone are seen at Alderley Edge.

The line to Liverpool diverges N. of Crewe and continues over Triassic.
Very little of the geology is seen.

E10.9   London to Holyhead

For route to Crewe see E10.1. Triassic strata underlie the continuation to
Chester. Sandstones form crags E. of Chester and on leaving the station
towards the W. red, dune-bedded sandstones are seen in cuttings. Around
Flint industrialisation marks the N. end of the North Wales Coalfield.
Inland from Rhyl the Vale of Clwyd is a Triassic-floored graben faulted down
into the Palaeozoic strata.

Carboniferous Limestone is seen S. of the line between Abergele and Colwyn
Bay and, as the picturesque Conway Estuary is crossed, a view to the N.
includes Great Orme's Head in the same formation. West of Conway there are
cuttings and tunnels through Ordovician volcanics, and high above the railway on the S. side a diorite intrusion is quarried on a large-scale at
Penmaenmawr.

Underlying late Precambrian volcanics are seen in cuttings near Bangor
station.

S.W. of Bangor the Menai Bridge links the mainland with Anglesey and from
there to Holyhead the route crosses the Precambrian Mona Complex (E6.3).
The surface is remarkably flat, and very few exposures show through the
Boulder Clay cover until near Holyhead, where quartzite forms Holyhead
Mountain.

E10.10   London to Holyhead (A5 road)

This is an historic route, with long straight stretches which follow the
Roman road to Cambria (Wales).

For over a hundred miles from London the road is nowhere far from the railway to Crewe and the same geological features can be seen (E10.1). The road
swings to the W. near Tamworth and for the next 30 miles (48 km) crosses
mainly an industrialised Triassic plain, although there is a noticeable
topographic change over the South Staffordshire (Cannock) Coalfield.

Near Wellington the conical Wrekin (type-locality for the late Precambrian
Uriconian) is seen to the S. and to the S.W. the Longmynd, also late Pre-
cambrian (E6.7). Coal Measures occur both near Wellington and near
Shrewsbury but are now of very little economic significance.

N. of Shrewsbury the route again crosses a Triassic plain, but where it
swings W. into the Vale of Llangollen industrialisation to the N. at Ruabon
is evidence for the presence of the North Wales Coalfield. Up the Dee
Valley to Corwen outcrops can be seen of Ludlow shales and sandstones, the
latter showing graded bedding in places. To the N., beyond Llangollen,
unconformably overlying Carboniferous Limestone forms conspicuous scarps.

Across the high ground N.W. of Corwen an anticlinorium which brings up
Ordovician sediments is crossed, but causes little topographical change.
The road then drops to the beautiful and deeply glaciated Conway Valley
beyond which there is a dramatic change in the scenery, due to the oncoming
of Ordovician igneous rocks. In fact, the road from Bettws-y-Coed to Bangor
is one of the most spectacular in the British Isles. It is possible to stop
at a number of places for closer examination.

Particularly striking are the Caradocian acid volcanics, the continuation of
those of the Snowdon Syncline (E6.1) which form the three-pronged peak of
Tryfan, 3010 ft. (917 m). Evidence of glaciation ranges from roches
moutonnées to corries and lake basins. The road then runs down the spectacular glaciated valley of Nant Ffrancon to Bethesda with its huge quarries
in Cambrian slate (E6.2). Between Bethesda and Bangor two ridges of late
Precambrian volcanics are crossed.

From Bangor to Holyhead the geology is the same as that described along the
railway (E10.9).

E10.11   London – Gloucester – Cardiff (road)

Of the several routes to Cardiff, the one described below offers the best
opportunity for geological observations.

From London via Reading to Didcot the geology has already been described
(E10.7). Westwards through Wantage and Faringdon the road (A417) follows
the Vale of White Horse along the outcrops of soft Lower Cretaceous and
Upper Jurassic sediments with the Chalk scarp to the S. (see also E8.8).
Just S. of Cirencester the route enters the Great Oolite and then crosses
this formation, forming the Cotswolds, by the dead straight Roman road,
which undulates over minor scarps and dip slopes. Before the descent to
Gloucester there is a fine view over the Severn Plain, floored mainly by
Lower Lias and Keuper Marl (Trias), and on the descent the Inferior Oolite
can be examined.

From Gloucester the A48 road should be followed to Cardiff. At Westbury,
Westbury Garden Cliff can be seen to the S. (and visited if time permits
and the tide is fairly low). The cliff shows a fine section of Keuper overlain by Rhaetic including the bone-bed. Hereabouts, at certain dates, the
Severn Bore can be well seen.

At Blakeney the Malvern-Bath Axis is crossed. Lower Old Red Sandstone outcrops on the shore to the S. and Silurian on the opposite coast. Rising
ground N. of Lydney marks the historic Forest of Dean Coalfield. The Coal
Measures rest directly on the Carboniferous Limestone or the Old Red Sandstone.

The limestone (D2) is seen to be folded by a beautiful anticline as the road winds down to the R. Wye. (Parking is possible nearby on the opposite side of the road.) As the road crosses the river there is a magnificent view of the classic incised meanders, in the limestone.

As Newport is approached red Keuper Marl is again seen. Between Newport and Cardiff the A48 mostly keeps just above the Flandrian clays of the coastal flats. Two miles (3.2 km) after crossing the R. Ebbw, Lower Old Red Sandstone marls and sandstone occur in cuttings. The descent to the R. Rhymney, as Cardiff is entered, is off the S. edge of the Rhymney Silurian Inlier.

E10.12   Cardiff – Brecon – Carmarthen – Fishguard

The direct way to Carmarthen is via Swansea, but a better picture of the geology of South Wales can be obtained along the route indicated above.

Leave Cardiff northwards on the old road (A470) to Merthyr. After passing under the M4 moundy sand and gravel deposits are seen which mark the limit of the Devensian ice. The road then passes through the fault-guided Taff Gorge in Carboniferous Limestone and a stop may be made to examine the geology described in E7.18.

The wide open part of the valley beyond is a glaciated basin. The route from here to Merthyr provides a cross-section of the South Welsh Coalfield. Exposures are mainly of the Pennant sandstone. Glaciated sections of the Taff valley are well seen beyond Pontypridd and, after passing through a post-glacial gorge at Quaker's Yard, below Merthyr.

1½ miles (2.4 km) N.W. of Merthyr, Millstone Grit quartzitic sandstone is seen, and further N.W. the Carboniferous Limestone. Gritty sandstones, dipping S., downstream from the reservoir, mark the rather thin development of the Upper Old Red Sandstone. The strong glaciation of the valley is evident. Upstream from the reservoir there are numerous exposures of the Lower Old Red Sandstone as the road climbs to the summit at the Storey Arms, 1414 ft. (439 m). To the E., on the descent to Brecon there is a view of the scarp face in Lower Old Red Sandstone, and with cwm (corrie) erosion of the Brecon Beacons rising to 2907 ft. (886 m). Half-a-mile (0.8 km) beyond the Storey Arms a roadside quarry in green Old Red Sandstone beds yields Goslingea breckonensis, an early vascular plant.

W. from Brecon past Trecastle, and on the A40 along the Cwydding valley, roadside exposures and the Capel Horeb Quarry show Downtonian (parking possible for cars only). A descending sequence (steep dips due to Carreg Cennen Disturbance) to the Ordovician (Bala) follows.

The Towy valley is entered at Llandovery and followed along the axis of the N.E.-plunging Towy Anticline to Llandeilo and Carmarthen.

At Carmarthen the river turns S. off the line of the anticline but the road continues over mainly Arenig strata on the axis to St. Clears.

Silurian occurs in and around Haverfordwest, a town where travel conditions may be difficult. N. for 6 miles (9.6 km) to Trefgarne over complicated terrain where there is Caledonian thrusting from the N. involving Precambrian, Cambrian and Ordovician strata. Volcanics of Precambrian and Ordovician age are present. Continue N. to Fishguard (E6.12) across mainly Arenig shales repeated by folds and an overthrust.

E10.13   London – Guildford – Portsmouth (road)

For about the first 30 miles (48 km) S.W. out of London the road crosses
the Tertiary and then passes through the Guildford Gap. This is due to the
narrow outcrop of the Chalk here, which dips very steeply N. and is also
faulted. There is little surface evidence of the geology as the route
traverses the Lower Cretaceous near the W. end of the Weald Anticlinorium
(cf. E10.5). 1.5 miles (2.4 km) S.W. of Petersfield the grey Lower Chalk
is seen in a quarry E. of the road, and a little further on the Middle
Chalk. Once over the higher ground formed by the Upper Chalk the road
descends slightly and passes onto the Tertiary of the Hampshire Basin.

The Upper Chalk is seen again, however, 1¼ miles (2 km) S.S.W. of Purbreck
Heath brought up by the Ports Down Anticline. Tertiary follows to Portsmouth
but is not seen.

E10.14   Exeter – Okehampton – Launceston – Redruth – St. Ives – Penzance
           (road)

On a southerly road route from Exeter to Penzance via Plymouth, St. Austell
and Truro, the geology seen is virtually the same as that along the railway
(E10.6). However, on a more northerly journey there are considerable
differences.

Westwards from Exeter the road crosses Upper Carboniferous sediments near
the centre of the South-West England Synclinorium; exposures are few. As
Okehampton is approached the high ground formed by the Dartmoor Hercynian
Granite is seen to the S. A diversion in this direction makes it possible
to examine at least some of the exposures described in E7.10. Beyond
Okehampton the route continues over the Upper Carboniferous but is gradually
edging towards the S. flank of the Synclinorium. S.W. of Launceston the
Upper Devonian outcrops, and spilitic lavas and metadolerites (greenstones)
form features.

Across Bodmin Moor picturesque granite scenery rises to nearly 1000 ft.
(305 m). On the S.W. side of the granite the road passes onto the Middle
Devonian and beyond Bodmin onto the Lower.

Between Redruth and Camborne (where there is a well-known School of Mines)
numerous signs of mining for tin and other metals are seen; the mineralisation
is associated with the Carn Menellis Hercynian Granite which rises to the S.

A short stop at much-painted St. Ives makes it possible to examine shore
outcrops of metadolerite (greenstone). Southwards from St. Ives through
Towednack exposures of the N.E. end of the Hercynian Land's End Granite are
seen. A short walk up the N.E. flank of Rosewall Hill, 2 miles (3.2 km)
S.W. of St. Ives, leads to disused workings for tin. Some minerals, particularly schorl, can be found on the dumps, but good specimens are difficult
to obtain. From above Penzance there is a fine view of Mount's Bay and
St. Michael's Mount, of granite and hornfelsed Devonian. At Gulval, just
before the coast is reached, greenstone (metadolerite) is seen.

E10.15   Derby – Matlock Bath – Bakewell – Castleton – Sheffield

Derby is built on the Trias but the road soon passes onto the Millstone
Grit on the S.E. side of the Peak District Dome, and for the next 11 miles

(17.6 km) the details of the topography are largely controlled by the hard sandstone beds in this formation. Near Matlock Bath the Carboniferous Limestone forms spectacular cliffs. From there through Bakewell to Grindleford Bridge the route crosses and recrosses the rather complicated junction of the Carboniferous Limestone and Millstone Grit on the E. side of the Dome. At Grindleford Bridge there is a direct route N.E. to Sheffield passing over Millstone Grit and Coal Measures, but a diversion N.W. to Castleton is much more profitable geologically.

At Castleton the road W. should be followed for $\frac{1}{2}$-mile (0.8 km) then a fork left for another $\frac{1}{2}$-mile (0.8 km). Here a footpath leads to the Blue John Mine which is open to the public (entrance fee) and where veins of Blue John (banded fluospar) can be seen.

In a gully 150 yds. (137 m) N.N.E. of the mine the unconformity can be seen between limestone (D$_1$) and Edale Shales (basal Namurian). On the way back to the road a stop can be made at Treak Cliff. Beyond a stile, on the N. side of a gully, reef limestones are exposed, with abundant brachiopods, polyzoa, lamellibranchs, gastropods, goniatites, trilobites etc. The nearby Treak Cliff Cavern (public cave, entrance fee) is worth a visit.

Return to Castleton and continue through Hope to Hathersage across lower ground underlain by Edale Shales. To the S. the Hope Cement Works use these shales and the limestone. Beyond Hathersage the road climbs to an upland plateau formed of the Chatsworth Grit near the top of the Millstone Grit. Spreads of boulders brought down from crags by solifluxion at the time of the "Newer Drift" are seen N. of the road. This goes up another scarp formed by a higher part of the Chatsworth Grit and reaches a summit at a height of 1250 ft. (381 m). It then descends the dip-slope of the Grit before passing onto the Lower Coal Measures forming Dore Moor. This formation is then crossed to the city centre of Sheffield with sandstone horizons showing in outcrops and quarries.

E10.16   Sheffield - Woodhead - Manchester

The road (A616) up the Don valley N.W. from Sheffield is over Lower Coal Measures with dip slopes of the Loxley Edge Rock coming down to the W., and scarps of the Greenmoor Rock and Greenoside Sandstone rising behind a power station to the E. At Oughtibridge the valley narrows, and the road crosses an anticline, striking E., bringing up the Rough Rock, $\frac{1}{4}$-mile (0.4 km) beyond Oughtibridge and the Rivelin Grit further on. To the E. of the road the wooded slopes of Wharncliffe rise steeply to 700 ft. (213 m) capped by the Greenmoor Rock and the Greenoside Sandstone. Old workings for coals and gannister are seen, and at More Hall old lead workings. A mile (0.8 km) beyond, the Rough Rock is crossed again on the N. limb of the anticline, and shortly after, at Deepcar, the road turns N.W. up the Little Don to Midhope-stones. Details of the Coal Measures onwards to Langsett are given in E7.35. Soon after Langsett the road passes onto the Millstone Grit forming the central part here of the Pennine Upfold and continuing across the Woodhead Pass, 1476 ft. (450 m) to the outskirts of Manchester. Across the high ground the topography is controlled by the scarps and dip slopes of the harder sandstone beds in the Millstone Grit. Most of Manchester is on the Coal Measures.

E10.17   Newcastle - Carlisle

This road (B6318) is of archaeological as well as geological interest as it
follows for the most part the Roman road close to Hadrian's Wall.

For the first 9 miles (14.4 km) from Newcastle the road passes over the Coal
Measures; the general dip is easterly. A fairly narrow outcrop of the Millstone Grit, which is much thinner than in the Pennines, follows and then the
Upper Limestone Group Carboniferous Limestone Series. Owing to the thick
cover of boulder clay, exposures are few. At Chollerford the road crosses
the North Tyne and continues over the Upper Limestone Group which contains
a considerable proportion of sandstone, seen in a number of outcrops.

8 miles (12.8 km) from Chollerford a stop should be made to visit the well-
preserved Roman Camp at Housesteads. $2\frac{1}{2}$ miles (4 km) further on a short
walk N. across moorland leads to the section of the Great Whin Sill with
the Roman Wall described in E7.40.

Where the road, generally straight, though undulating, for long stretches,
winds down to Greenhead, the thin S.W. end of the Great Whin Sill makes a
small exposure. For the next 7.5 miles (12 km) the route crosses the lower
part of the Carboniferous Limestone Series which contains numerous thin
limestones, some of which can be seen in disused quarries near the road.
$1\frac{1}{4}$ miles (2 km) E.N.E. of Brampton the Carboniferous Limestone is overlain
by the Triassic, but the unconformity is not seen and, in fact, only a few
exposures of these red rocks are seen near the rest of the route to Carlisle.

E10.18   Newcastle - Kielder - Beattock - Glasgow

For those travelling from Newcastle to the West of Scotland the route,
partially over secondary roads, across the Scottish Border near Kielder
is attractive both scenically and geologically. As far as Chollerford the
route is the same as that just described (E10.17). After crossing the
North Tyne the secondary road (B6320) should be followed from Chollerford to
Bellingham. At Barrasford the much quarried Great Whin dolerite sill is
seen. Otherwise there are few exposures owing to Boulder Clay cover.
Broadly, a descending sequence in the Carboniferous Limestone Series is
followed.

From Bellingham continue up the North Tyne valley passing onto the
Scremerston Coal Group (Lower Carboniferous). Outcrops are of sandstone
but the Group is very mixed lithologically. Just after passing the turn-
off for Falstone the road passes through a glacial overflow channel and a
short distance further on there is a good view of the dam which holds back
the Kielder Reservoir, the largest man-made lake in Europe. The dam is an
"earth-fill" structure built on thick Boulder Clay hiding a buried valley.

Old coal workings can be seen by walking up the hillside S.W. from the road
a short distance upstream from the dam. Three miles (4.8 km) further N.W.
the more extensive disused workings at Plashetts can be seen across the
reservoir. The Bakethin Dam, at its upper end, prevents the water from
flooding the flat ground towards Kielder village. The underlying Lower
Carboniferous rocks here are older than the Coal Group; only sandstone
outcrops are seen.

$2\frac{3}{4}$ miles (4.4 km) beyond Kielder the Scottish Border is crossed. Not far
beyond a good section in the Cementstone or Ballagan Group (E7.45) cut by

a basalt dyke is seen across the stream (easily forded except in flood).
Soon afterwards the road turns S.W. down Liddesdale. 2½ miles (4 km) beyond
Newcastleton strata of the Cementstone Group outcrop in a stream on the N.
side of the road. A mile (1.6 km) further on a quarry N. of the road shows
two Lower Carboniferous basalt lava flows with irregular columnar jointing.
These are approximately of the same age as the Clyde Plateau Lavas (E7.45).
Otherwise little is seen of the Lower Carboniferous rocks through which
Liddesdale runs.

At Canonbie red sandstones, dipping W.S.W. at 30°, are exposed in the left
bank of the R. Esk downstream of the bridge. These belong to the Barren Red
Measures (E10.2). The Productive Coal Measures outcrop to the N.; they are
not worked. To the S.W. Permian followed by Triassic (both poorly exposed)
comes on and continues to the junction with the main road (A74) near Gretna
where there is a view across the Solway Firth to the Lower Palaeozoic
mountains of the Lake District (Ch. 6.2).

For continuation to Glasgow see below (E10.19).

E10.19   Carlisle – Glasgow

As far as Abington the road closely follows the railway (for geology see
E.10.1). On the ascent to Beattock Summit, 1014 ft. (309 m) there are
several places where it is practicable to park and examine Lower Palaeozoic
sediments of the Southern Uplands, for example 6½ miles (10.4 km) from
Beattock, before the road passes from the N.E. side of the railway to the
S.W. Here Llandovery greywackes, dipping N. 15° W. at 50° are well exposed
on the N.E. side of the road.

At Abington the road turns N.W. and after passes over Caradoc greywackes
of the Northern Belt of the Southern Uplands for 3 miles (4.8 km) before
crossing the Southern Uplands Fault which is not exposed and makes no clear
feature.   There are few exposures, owing to drift cover, of the Lower Old
Red Sandstone sediments underlying the country beyond but basic lavas are
seen in cuttings where the road crosses moorland some 3 to 5 miles (4.4 to
8 km) from the Southern Uplands Fault. The lavas end at a major N.E. fault,
marked by a change of topography, between which and another N.E. fault before
Lesmahagow there is a 5 mile (8 km) wide syncline. The Lower Carboniferous
sediments on the flanks of the fold are not seen near the road, but Millstone Grit sandstones outcrop near Nether, and tips show the presence of
disused coal workings beyond the Douglas Water. The road passes over Lower
Old Red Sandstone at Lesmahagow and then crosses a fairly thin Lower
Carboniferous and Millstone Grit succession. Thick boulder clay mostly
obscures the solid geology. To the S.E. the conspicuous hill of Tinto marks
a Lower Old Red Sandstone felsite laccolith. At Stonehouse sandstone seen
in cuttings belongs to the Coal Measures of the Central Coalfield. These
continue to Glasgow and are marked by collieries (many disused) and tips.
When the road (A74, M74) was upgraded old workings caused construction
problems, notably at the big interchange near Baillieston.

On the run through the eastern suburbs to the centre of Glasgow the influence
of the drumlin topography on roads and buildings can be appreciated (see
also E7.44 and E10.20).

E10.20   Glasgow - Ardrossan - Brodick (Arran) (rail and boat)

For many geologists the journey from Glasgow via Ardrossan to Brodick is a
prelude to excursions in Arran (E9.7 - 12). For those who go by road one
of the routes to Ardrossan closely follows the rail route.

After crossing the R. Clyde the line turns W. to Paisley. Drumlins are well
seen (cf. E10.19), and the flat ground to the N. is the "100-ft. (30 m)"
raised beach floored with Flandrian clays. The underlying rocks are Lower
Carboniferous sediments above the Clyde Plateau Lavas. These are seen
across the Clyde to the N. with the trap-featuring evident, and to the S.
on the other side of the broad Glasgow Syncline. On the E. side of Paisley
a dolerite sill in the sediments is seen in a cutting.

Between Paisley and Dalry the railway breaks through the lavas which en-
circle the Glasgow district by the Lochwinnoch Gap, eroded along a N.E.
fault-system.

Near Dalry coals and ironstones used to be worked in the Limestone Coal
Group (see Ch. 7.5), and the area is still industrialised. Near Kilwinning
the Millstone Grit consists largely of basalt lavas altered to the Ayrshire
Bauxitic Clay (E7.54). Workings for refractory material in the clay can be
glimpsed.

As the line runs along the coast at Saltcoats the Coal Measures and intrusions described in E7.54 can be seen if the tide is not too high.

As the train swings round to Ardrossan Pier a volcanic neck can be seen to
the N. and a teschenite sill with ultrabasic floor on the shore; both are
in the Lower Carboniferous.

As the boat leaves the harbour a view to the N. shows the high ground formed
by the Clyde Plateau Lavas which are here directly underlain by the Upper
Old Red Sandstone exposed along the coast. In the same direction, a few
miles further out, the same lavas are seen forming the island of Little
Cumbrae (trap-featuring displayed) and at the S. end of the island of Bute.
Further N. and N.W. rise the Dalradian rocks of the Grampian Highlands
(Ch. 5.4). To the S. Tertiary microgranite forms the conspicuous island of
Ailsa Craig.

As Arran is approached (in good weather) the spectacularly glaciated
mountains of the Northern Granite (E9.11,E9.12) are displayed. To the S.
lower but rugged hills mark the Central Ring Complex (E9.10). The Permian
section seen as the boat approaches Brodick Pier is described in E9.9.

E10.21   Glasgow - Tyndrum - Glen Coe - Fort William - Mallaig (road)

The route from Glasgow to Cardross is described in E7.46 and from Cardross
to the Loch Sloy hydroelectric power station at Inveruglas on Loch Lomond
in E5.41.

An alternative route via Balloch is dealt with in the return section of
E5.41.

From Inveruglas to Ardlui the fjord character of upper Loch Lomond (it was
an arm of the sea until post-glacial times) is well seen as it cuts through
craggy mountains of Upper Dalradian metasediments. From Ardlui to

Crianlarich there is a spectacular climb up Glen Falloch, strongly over-deepened with hanging side valleys and a fine bridal veil waterfall on the E. side. Stops can be made to examine the metasediments which show increasing metamorphism as shown by the appearance of garnet and, in places, of albite.

N.W. from Crianlarich the Middle Dalradian occurs, and at Tyndrum a major N.E. fault is crossed with Moinian on its N.W. side. The fault is mineralised, and disused lead mines are seen on the S. side of the valley. Flaggy Moinian metapsammites can be examined on the way to Bridge of Orchy. Further N. the high mountains to the W. of Loch Tulla are in the Lower Devonian Etive Ring-Complex. Around the loch, in certain lights, the shore line of a glacial lake can be seen. The route next crosses the desolate moraine and peat covered waste of the Moor of Rannoch underlain by Caledonian quartz-diorite. Near Kingshouse, where a road down deeply glaciated Glen Etive branches off, there is a view of the great cliff of Lower Devonian volcanics, within the Cauldron Subsidence of Glen Coe, on Buchaille Etive Mor (the Great Shepherd of Etive). The road to Ballachulish, through historic Glen Coe, is one of the most spectacular in Scotland. Geological details of the route all the way to Fort William are given in E5.37 and E5.36.

As the road after leaving Fort William swings round the head of Loch Linnhe there are fine views in both directions of the great trench eroded along the Great Glen Fault and of the Ben Nevis Lower Devonian Ring Complex (E5.35). Westwards along Loch Eil metapsammites of the Loch Eil Division (thought to be the youngest in the Northern Highlands) can be examined. From the head of Glenfinnan, where there is a memorial to the raising of the standard by Bonnie Prince Charlie, to Lochailort the route is described in E5.20. The rocky scenery further W. above Loch nan Uamh (where Prince Charles landed before the 1745 Rebellion) is in Lewisian in the centre of the Morar Antiform (E5.18). As the road turns N. at Arisaig it passes onto the Upper Morar Psammite on the W. limb of the Antiform. There is a seaward view here (from N. to S.) of the Tertiary gabbro peaks of Rhum, the acid Tertiary lava of the Sguirr of Eigg and the Tertiary basalt lavas of Muck. As the route crosses the rock barrier at the W. end of Loch Morar it should be recollected that further inland this glaciated rock basin reaches a depth of 1017 ft. (310 m). On the seaward side of the approach to Mallaig station (E5.18) vertical Upper Morar Psammites cut by Tertiary dykes of the Skye Swarm are seen.

E10.22   Glasgow - Fort William - Mallaig (rail)

The West Highland Railway line is regarded by many as scenically the finest in the British Isles. From Glasgow to Bridge of Orchy it closely follows the road, although mostly at a higher level. For the geology to be seen refer to E7.46, E5.41 and E10.20.

From N. of Bridge of Orchy to Corrour the line crosses the Moor of Rannoch Caledonian quartz-diorite further E. than on the road. After passing over Corrour summit at 1384 ft. (422 m) the line descends above the E. shore of Loch Treig with a view to the W. of the high mountains eroded in the metasediments of the Transition Group between Moinian and Dalradian.

The dam at the N. end of Loch Treig has converted it into a reservoir from which water is carried in a 15 mile (24 km) tunnel to aluminium works at Fort William. As the line swings W. into Glen Spean the Laggan Dam, another

part of the power scheme, is seen to the E., sited on Caledonian quartz-
diorite which outcrops in a cutting near Tulloch. Between here and Roy
Bridge the line passes through the Monessie Gorge in the Leven Schists of
the Transition Group. On the valley side a shore line can be seen, corresponding to the lowest "Parallel Road" of Glen Roy (E5.34). The geology
from Roy Bridge to Spean Bridge is described in E5.34.

In clear weather the view to the S. as Fort William is approached is one of
the most striking mountain panoramas in Britain, with three peaks reaching
over 4000 ft. (1220 m).

Aonach Beag, 4060 ft. (1237 m), the most easterly, is in the metasediments
on the edge of the Lower Devonian Ben Nevis Complex (E5.35) the grey-
weathering Outer Granite of which rises as a high ridge immediately to the
W. Further W. the red Inner Granite forms Beinn Mor Dearg, 4012 ft. (1223 m)
and beyond deep Corrie Leis the great east cliff of Ben Nevis, 4406 ft.
(1345 m) consists of Lower Devonian volcanics (Fig. 5.18).

From Fort William to Mallaig the geology is the same as that seen along the
road (E10.20).

E10.23   Glasgow - Inveraray - Dalmally - Oban (road)

From Glasgow to Arrochar the route is the same as that of E5.41. The road
then swings round the fjord head of Loch Long beneath the strangely eroded
mountain known as the Cobbler, 2891 ft. (881 m) in Upper Dalradian metapsammites and mica-schists before climbing glaciated Glen Croe to a summit
at 860 ft. (262 m). A descent of beautifully U-shaped Glen Kinglass follows
to Loch Fyne, another fjord which is rounded to Inveraray. On its N.W. side
felsite sheets are seen in Middle Dalradian metasediments.

From Inveraray a long climb N. leads first over the Ardrishaig Phyllites
(Middle Dalradian) with felsite sheets and then over the Crinan Grits
(Middle Dalradian) with metadolerites and metabasalts, some intrusions,
some volcanics. From the summit there is a striking view N. of the Etive
Ring Complex (Lower Devonian (see E5.39)). Altered basic rocks can be
examined at the roadside on the descent to Loch Awe and towards Dalmally
they are seen in the Ardrishaig Phyllites. From Dalmally the road runs
N.W. to fault-determined Glen Strae. From here to Oban the geology is
described in E5.39.

E10.24   Glasgow - Oban (rail)

From Glasgow to Crianlarich the West Highland Railway (E10.22) is used.
From Crianlarich to Tyndrum the Oban line is lower, and striking morainic
features are seen S. of the line. On the same side at Tyndrum disused lead
mines along a major N.E. fault occur in the hillside. From Tyndrum to
Dalmally the railway runs along a glaciated valley. The complex Dalradian
geology cannot be made out from the train.

From Dalmally to Oban the line closely follows the road (E10.23 and E5.39).

E10.25   Glasgow - Paisley - Lochwinnoch - Largs - Gourock - Greenock -
          Paisley - Glasgow (road)

This circular tour gives a good impression of the geology of the western
part of the Midland valley and provides some fine views over the Firth of
Clyde.

From Glasgow past Paisley through the fault-determined Lochwinnoch Gap to
Dalry the geology seen is virtually the same as that described under E10.20.
From Dalry the Largs road should be followed N. This rises to 700 ft.
(213 m) across the Clyde Plateau Lavas showing trap-featuring of the basalt
flows in places. To the N.E. higher ground marks the Misty Law acid
volcanic centre.

Descend to Largs and turn N. along the coastal road which follows the
"30-ft. (9 m)" raised beach backed by a cliff. Along the shore there are
excellent exposures of the Upper Old Red Sandstone consisting of conglomerates
and sandstones showing thick cyclothems. One mile (1.6 km) N. of Knock
Castle an agglomerate neck is seen.

At Wemyss Bay Pier a thick N.W. Tertiary dolerite dyke of the Mull swarm
outcrops.

Just before the road turns N.E. to run temporarily inland at Inverkip the
Upper Old Red Sandstone is again well seen cut by two further dykes of the
Mull swarm.

The road reaches the coast again at the Cloch Lighthouse marking the entrance
to the inner Clyde. Here the basalt lavas reach the coast for a short distance to be followed by upfaulted sandy sediments of the lowest part of the
Carboniferous, intruded, at the outskirts of Gourock, by a trachyte sill.

Little is seen of the geology through the industrialised area of Greenock
and Port Glasgow. If time is short a fast run back can be made on the
motorway from Greenock, but the old road gives an opportunity for closer
observation and photography. The "30-ft. (9 m)" raised beach is followed
for a considerable distance, and the Clyde Plateau Lavas outcrop on both
sides of Langbank. Here there is a fine view to the N. across the Clyde
of the same, strongly trap-featured, lavas forming the Kilpatrick Hills
with Dumbarton Rock and other necks to the W. (E7.46). Beyond, there is a
glimpse of Ben Lomond in the Highlands. At Bishopton a major fault brings
down Lower Carboniferous sediments, not exposed near the road. The flat
ground is part of the "100-ft. (30 m)" raised beach.

Return to Glasgow.

E10.26   Oban - Tobermory (Mull) (boat)

The voyage from Oban to Tobermory is one of the most attractive coastal
sailings in the British Isles, both scenically and geologically.

As Oban is left the unconformity between Lower Old Red Sandstone conglomerates
and black Easdale Shales is clearly seen as is the "30-ft. (9 m)" raised
beach. The Dalradian metalimestone island of Lismore is a conspicuous feature
to the N.E. in Loch Linnhe. The Great Glen Fault runs between the island
and the Caledonian Strontian Granite Complex forming rugged mountains on the
mainland to the N.W. The Strontian Complex is the faulted continuation of the

Foyers Complex on the other side of the Fault 65 miles (104 km) to the
N.E. (E5.17).

To the S.W. the Tertiary ring intrusions of Mull (E9.7) rise as steeply-
sided mountains, but so complex is the geology that it is difficult to make
out the various units from a distance. In a good light, however, the cone-
sheets can be clearly seen. On the opposite side of the Sound of Mull the
spectacularly displayed Inninmore Fault brings a succession of Coal
Measures/fairly thin Mesozoic/Tertiary lavas to the W. down at least 1000 ft.
(305 m) against Moinian to the E.

At Loch Aline, on the N. shore of the Sound, where a stop (some accommodation
available) is well worth while, the lavas overlie very pure Cretaceous sandstone (mined for glass making) above Liassic limestones and shales with
Gryphaea and other fossils. Further down the Sound the lava pile of Ben
More (E9.8) becomes visible. The lavas, showing trap-featuring, now flank
the Sound all the way to Tobermory.

E10.27   Inverness – Helmsdale – Thurso – Orkney (road)

On the northern outskirts of Inverness the road passes over the hidden
Great Glen Fault, to enter the Northern Highlands, then crosses the bridge
over the Moray Firth narrows. The extent of glacial scour along the Fault
was shown by a bore in Inverness which went over 300 ft. (92 m) into sand
and gravel without reaching rock. As far as Edderton, in fact, the surface
geology is mainly glacial and post-glacial and little is seen of the underlying Middle Old Red Sandstone. The glacial deposits include both Boulder
Clay and outwash gravels; the 25-ft. (8 m) raised beach is conspicuous along
the coast.

Around the head of the Dornoch Firth the scenery changes and outcrops of
Moinian metasediments can be examined near the road. The route passes into
the Middle Old Red Sandstone again near Clashmore, and conglomerates can be
studied by parking further N. at the head of Loch Fleet and a short distance
further N.E. at the conspicuous Mound Rock. By stopping near the bridge at
Brora, Oxfordian white sandstone can be examined along the river gorge. The
sandstone forms part of a Mesozoic succession brought down near the coast by
the Helmsdale Fault (E5.14). By making another stop near Kintradwell Farm,
2.5 miles (4 km) beyond Brora, and crossing to the shore, Kimmeridgian
Boulder Beds can be studied; Rasenia walensis and other ammonites can be
found (E5.15). 6.5 miles (10.4 km) further on, 800 yds. (752 m) before the
hamlet of Portgower, the "Fallen Stack", the largest boulder, is seen on the
shore section (E5.15). The 25-ft. (8 m) raised beach is still seen here but
ceases a few miles further N. On the N. outskirts of Helmsdale there is a
quarry in the red Caledonian Helmsdale Granite. 9 miles (14.4 km) up the
R. Helmsdale, at Kinbrace, there was a rush in 1869-1870 to work alluvial
gold. The "Mother Lode" has never been found.

The irregular contact between the granite and overlying Middle Old Red
Sandstone (although the unconformity is not exposed) can be appreciated by
looking at quarries and cuttings along the road over the Ord of Caithness
to Ousdale (E5.14). At Berriedale the granite is again seen, and a short
distance inland Moinian granulite outcrops. To the W. the conical peak of
Morven is formed of one of the few quartzites in the Moinian.

Onwards to Thurso across moorland there are exposures of the Caithness Flags
(Middle Old Red Sandstone).

If the journey is continued to John o' Groats the exposures described in
E5.11 can be seen. A ferry (vehicles not carried) crosses to the S. end of
South Ronaldsay from which there is a public bus over this island and Burray
to Kirkwall (connected by causeways built during the 1939-1945 War) on the
Orkney "Mainland". Middle Old Red Sandstone forms the rocky coast.

The vehicle ferry leaves from the port of Scrabster. On the way from Thurso
the first part of excursion E5.12 can be undertaken. As the boat passes
the island of Hoy, hills of Upper Old Red Sandstone are seen. This formation
makes the famous stack known as the Old Man of Hoy and 1000-ft. (305 m)
vertical cliff of St. John's Head.

At Stromness, on the "Mainland", the ferry terminal, granite and Moinian
migmatites show beneath the Middle Old Red Sandstone.

The Sandwick Fish Bed in the latter formation has yielded many fine fossil
fish but specimens are now difficult to obtain. Permission must be sought
from the farmers on whose land the quarries occur.

E10.28   Inverness – Kyle of Lochalsh – Skye – Harris – Lewis (rail, boat and public bus)

From Inverness to Dingwall the route is mainly over Middle Old Red Sandstone
hidden by glacial deposits. The 25-ft. (8 m) raised beach is well seen
along the shore of the Beauly Firth. (Note comment on Great Glen Fault in
E10.27). Westwards from Dingwall some outcrops of the Middle Old Red Sandstone occur, then unconformably underlying Moinian psammites appear near
Auchterneed and rise in the spectacular, quarried, Raven's Rock S. of the
line. Moinian psammites and pelites outcrop through 3000 ft. (915 m)
mountain country to Achnasheen. At the W. end of Loch Luichart the penstocks and power station of part of Fannich Hydro-electric Scheme are seen
on the N. side. At Achnasheen there is a spectacular succession of lake and
river terraces.

After the long descent through Moinian of Strathcarron, the geology and
topography change dramatically to the N. at Achnashellach. Along the
Caledonian Front, Torridonian sandstones and Cambrian quartzites, forming
striking bare mountains over 3000 ft. (915 m) high, emerge from beneath the
Moinian along the Moine Thrust. Owing to the effect of the N.E. Strathcarron
Fault, however, the railway, as it runs along the S.E. shore of Loch Carron,
remains within the Moine Nappe as far as Stromeferry. The slope above is
exceedingly steep, and a shelter provides protection against rockfalls.
N.W. of the loch the Lewisian of the Kishorn Nappe (E5.8) is seen, and
further W. the bare Torridonian mountains of the Applecross Peninsula. As
the line runs round the coast, with its 25-ft. (8 m) raised beach, to Kyle
of Lochalsh, there are numerous exposures seen of Torridonian of the Kishorn
Nappe (E5.8).

From Kyle there is a ferry to Kyleakin in Skye, and from there on to
Broadford part of the route described in E5.9 is followed. The scenery
from Broadford to Sligachan is spectacular, the Tertiary granite of the Red
Hills contrasting with the black Tertiary gabbro of the jagged Cuillin.
(For details see E9.2). N. to Portree and N.N.W. to Uig the road crosses
the great spread of Tertiary basalts of North Skye.

From Uig a boat crosses to Tarbert, passing S. of the Shiant Islands consisting of a Tertiary dolerite sill in Liassic. As Tarbert in Harris is

approached the bare Lewisian Hills of the Outer Hebrides becomes more and more clear. At Tarbert (accommodation available) grey gneisses (probably Laxfordian) with basic lenses are easily studied.

There is a public bus to Stornoway (plentiful accommodation) in Lewis; the run gives many views of Lewisian scenery. Gneisses (probably Laxfordian) can be examined round the harbour, and a walk E. leads to exposures of conglomerates formerly thought to be Torridonian but now considered to be Permo-Triassic.

E10.29   Antrim Coast Road (Belfast – Portrush)

Portrush provides some of the best and most varied geology of any coastal route in the British Isles.

The 1:250,000 Geological Map of Northern Ireland will be found very useful.

Leave Belfast by the Shore Road. Just before the right-angled bend in the road S. of Whitehead, pipe amygdales are seen in a basalt face left of the road, below which columnar lava rests on Chalk with infilled swallow holes.

Through Whitehead then make an anticlockwise tour of Island Magee, examining the amygdaloidal lavas with reddened tops of the Gobbins; a Tertiary dyke at Portmuck in contact with altered Chalk partly replaced by magnetite and containing chalcedony and garnet; the interbasaltic bed (E9.13) on the E. side of Braun's Bay; and highly fossiliferous Lias (Lower Jurassic) at Barney's Point (opposite Magheramorne).

Leave Island Magee by the Ballycary Causeway and go on N. to the Magheramorne quarries where the Chalk, overlain by basalt lavas, is extensively worked down to the Greensand and Lias which are occasionally exposed at the bottom of the quarry. The Chalk contains <u>Ventriculites radiatus</u>, <u>Rhaladomya decussata</u>, <u>Belemnitella mucronata</u>, <u>Cideris sp.</u> and other fossils.

At Larne note should be taken of the raised beach spit of the Curran. Almost one mile (1.6 km) N. of Larne on the beach at Waterloo House Triassic marls are overlain by Rhaetic, Lias, Chalk and, above the road, by basalt lavas. At Ballygalley Head a dolerite neck shows columnar structure.

A diversion should then be made on a road which climbs W. from Carncastle. From the summit of the road walk N. about a mile (1.6 km) across the moor to Scawt Hill. Here a dolerite plug has pierced the Chalk; a number of rare calc-silicate minerals have been produced in the Chalk, and contamination of the magma has resulted in a hybrid zone containing titanaugite, nepheline, wollastonite, zeolites and other minerals.

Return to the Coast Road and continue past Garron Point, noting large land-slips. The road then crosses the mouth of Glenariff and the Highland Boundary. The Fracture-zone is probably under the Triassic outcrop here; a N.E. fault cutting strata up to the Tertiary lavas may mark a renewal of movement. At Waterfoot massive Triassic conglomerate forms the Red Arch. Beyond the Arch Upper Old Red Sandstone conglomerate is seen and the 25-ft. (8 m) raised beach. To the N. the Cushendall "porphyry" (dacite) outcrops; this is probably a Lower Old Red Sandstone lava. At Cushendun the spectacular conglomerate of the same age should be examined in 25-ft. (8 m) raised beach caves. The Lower Old Red Sandstone is an outlier N.W. of the Highland Boundary as it rests unconformably on Dalradian.

From the N. end of Cushendun follow the road leading W. to the main Cushendall-Ballycastle road and at the upper of two hairpin bends examine the Cushendun microgranite.

Continue down the Loughaveema overflow of glacial Lake Glendun. Loughaveema (the vanishing lake) lies at the junction of Dalradian schist and Chalk and drains very slowly through the latter to reappear in a spring about 1½ miles (2.4 km) away. By the roadside just above the lake Upper Dalradian "Green Beds" (cf. E5.27) with tourmaline can be seen. If time permits a diversion may be made E. to Tor Head to see the Dalradian described in E5.43.

Continue to Ballycastle. Schists of probable Middle Dalradian age can be seen in cuttings just outside the disused railway station.

The Ballycastle Coalfield (now not worked) lies N.E. of the town. Some ten seams occur, and the Lower Carboniferous strata in which they occur (cf. Ch. 7.5) can be seen in the coastal cliffs. Between these strata and the underlying Calciferous Sandstone a highly fossiliferous limestone is probably equivalent to the Hurlet Limestone of the Midland Valley of Scotland (cf. Ch. 7.5).

W. from Ballycastle to White Park Bay where fossiliferous Lias (Lower Jurassic) is overlain by Chalk, and several Tertiary dykes are seen.

Continue W. to the Giant's Causeway and Portrush (E9.13).

E10.30   Belfast - Dublin (rail)

The 1:250,000 Geological Map will be found useful for the N. part of this route.

For the first 12 miles (19 km) out of Belfast the line runs over the Triassic; to the N.W. the overlying Chalk and Tertiary lavas can be well seen. At Lurgan and Portadown the railway is on the lavas, but S. of Portadown it passes onto the directly underlying Lower Palaeozoic of the Irish "Southern Uplands" (Ch. 6.4) but very few exposures are seen owing to the glacial drift cover. More rugged scenery occurs across the Newry Caledonian Granite and where the line, near Hurrybridge, crosses granophyre of the Tertiary Slieve Gullion Complex (E9.15).

N. of Dundalk (close to the Eire border) the route passes onto the drift-covered Lower Palaeozoic (Silurian) again. North-east across Dundalk Bay the rugged scenery of the Tertiary Carlingford Complex (E9.16) can be seen. Near Drogheda the Carboniferous Limestone of Central Ireland is reached. Around Balbriggan Silurian shows through, and at Donabate a quarry can be seen in Ordovician andesite (E6.28). Boulder clay covers the Carboniferous Limestone from there to Dublin. To the E. Cambrian rises in the Nose of Howth (E6.29).

E10.31   Dublin - Cork (road)

From Dublin to Portlaoighise the route is over the drift-covered Carboniferous Limestone of Central Ireland. To the S.E. can be seen the high ground of the Leinster Massif (Ch. 6.5). As the road turns S.S.W. from Portlaoighise it passes over an ascending sequence (exposures are still few) on the W. edge of the Castlecomer Syncline. Ground rising to over 1000 ft. (305 m)

marking the Coal Measures outcrop is visible to the E. At Cahir the Galtee
Mountains to the W. reach 3015 ft. (919 m). They are formed of Old Red
Sandstone (probably Upper) within one of the major anticlines which rise
through the Lower Carboniferous of Central Ireland.

From Cahir to Mitchelstown the road follows an E.-W. syncline of Carboniferous Limestone. It then turns S. across higher ground at the W. end of
another anticline; a few outcrops of Old Red Sandstone can be examined.
A narrow syncline at Fermoy is overridden from the S. along the Hercynian
Front by Old Red Sandstone.

Sharp folding on E.-W. axes of this formation and of Lower Carboniferous
can be seen on the way to Cork. Near the city the exposures described in
E7.65 can be studied.

APPENDIX 1

Stratigraphical Grouping of Excursions

Sedimentary, metasedimentary and volcanic rocks

Post-glacial features and deposits, including raised beaches:- E5.2, 5.8, 5.14, 5.15, 5.29, 5.36, 5.38, 5.39, 5.40, 6.24, 6.25, 7.23, 7.44, 7.46, 7.48, 7.56, 7.57, 8.6, 8.10, 8.22, 9.9, 10.3, 10.7, 10.11, 10.20, 10.25, 10.27, 10.28, 10.29.

Glacial features and deposits:- E4.4, 5.6, 5.12, 5.16, 5.17, 5.20, 5.26, 5.27, 5.31, 5.32, 5.33, 5.34, 5.35, 5.37, 5.41, 5.42, 5.46, 5.47, 5.49, 6.1, 6.2, 6.4, 6.6, 6.15, 6.16, 6.17, 6.18, 6.20, 6.22, 6.23, 6.26, 6.30, 6.31, 7.17, 7.22, 7.40, 7.41, 7.42, 7.44, 7.46, 7.48, 7.49, 7.62, 7.63, 8.1, 8.4, 8.5, 9.2, 9.5, 9.7, 9.10, 9.11, 9.14, 10.1, 10.3, 10.4, 10.10, 10.12, 10.18, 10.19, 10.20, 10.21, 10.22, 10.23, 10.24, 10.27, 10.28, 10.29.

Pleistocene, non-glacial:- E7.5, 7.23, 8.6, 8.10.

Pliocene:- E8.6.

Oligocene:- E8.6, 8.17.

Eocene and Palaeogene (sedimentary):- E8.6, 8.9, 8.10, 8.11, 8.12, 8.16, 8.17, 8.20, 10.5.

Tertiary volcanic:- E5.43, 9.1, 9.2, 9.3, 9.4, 9.6, 9.8, 9.13, 9.16, 10.26, 10.28, 10.29, 10.30.

Cretaceous:- E5.43, 8.3, 8.4, 8.5, 8.7, 8.8, 8.9, 8.11, 8.13, 8.14, 8.15, 8.16, 8.20, 8.21, 8.23, 9.6, 10.1, 10.3, 10.5, 10.6, 10.7, 10.11, 10.13, 10.26, 10.29, 10.30.

Jurassic:- E5.9, 5.14, 5.15, 7.19, 7.20, 7.21, 8.1, 8.2, 8.4, 8.5, 8.7, 8.8, 8.18, 8.19, 8.20, 8.21, 8.22, 8.23, 9.1, 9.3, 9.4, 9.6, 9.13, 10.1, 10.3, 10.6, 10.7, 10.11, 10.26, 10.27, 10.29.

Triassic:- E6.26, 7.13, 7.14, 7.15, 7.19, 7.20, 7.21, 7.28, 7.29, 7.32, 9.3, 9.6, 10.1, 10.3, 10.6, 10.7, 10.8, 10.9, 10.10, 10.11, 10.15, 10.17, 10.18, 10.29, 10.30.

Permian:- E6.20, 7.11, 7.28, 7.36, 7.42, 7.47, 7.56, 9.9, 10.1, 10.3, 10.6, 10.18, 10.20.

Carboniferous:- E5.39, 5.45, 6.3, 6.14, 6.18, 6.19, 6.20, 7.9, 7.10, 7.13, 7.14, 7.15, 7.17, 7.18, 7.19, 7.20, 7.21, 7.22, 7.23, 7.24, 7.25, 7.26, 7.27, 7.28, 7.33, 7.34, 7.35, 7.36, 7.38, 7.39, 7.40, 7.41, 7.43, 7.44, 7.45, 7.46, 7.47, 7.49, 7.50, 7.51, 7.52, 7.53, 7.54, 7.55, 7.59, 7.60, 7.61, 7.63, 7.64, 7.65, 9.16, 10.1, 10.2, 10.3, 10.4, 10.7, 10.9, 10.10, 10.14, 10.15, 10.16, 10.17, 10.18, 10.19, 10.20, 10.25, 10.26, 10.29, 10.30, 10.31.

Devonian and Old Red Sandstone:- E5.11, 5.12, 5.13, 5.14, 5.22, 5.23, 5.24, 5.30, 5.35, 5.37, 5.38, 5.39, 5.40, 5.41, 5.42, 6.3, 6.18, 6.19, 6.23, 6.27, 6.33, 7.3, 7.4, 7.5, 7.7, 7.8, 7.11, 7.14, 7.16, 7.17, 7.18, 7.24, 7.26, 7.27, 7.31, 7.37, 7.46, 7.48, 7.49, 7.55, 7.57, 7.58, 7.60, 7.62, 7.63, 7.64, 7.65, 9.6, 9.11, 9.12, 10.1, 10.3, 10.6, 10.7, 10.11, 10.12, 10.14, 10.19, 10.20, 10.21, 10.25, 10.26, 10.27, 10.28, 10.29, 10.31.

Silurian:- E5.47, 5.48, 6.3, 6.7, 6.9, 6.10, 6.11, 6.16, 6.18, 6.19, 6.21, 6.22, 6.25, 6.26, 6.27, 7.16, 7.27, 7.28, 7.31, 7.56, 7.60, 9.14, 10.1, 10.3, 10.10, 10.11, 10.12, 10.19, 10.30.

Ordovician:- E5.2, 5.42, 5.47, 5.48, 6.1, 6.2, 6.3, 6.4, 6.6, 6.9, 6.10, 6.12, 6.13, 6.14, 6.15, 6.16, 6.20, 6.22, 6.23, 6.24, 6.25, 6.26, 6.27, 6.31, 6.32, 6.33, 6.34, 7.7, 7.41, 7.56, 7.58, 9.1, 10.1, 10.9, 10.10, 10.12, 10.19, 10.30.

Cambrian:- E4.4, 5.1, 5.2, 5.3, 5.4, 5.5, 5.6, 5.7, 5.8, 5.9, 5.42, 6.1, 6.2, 6.4, 6.5, 6.8, 6.12, 6.28, 6.29, 6.30, 7.7, 7.28, 7.29, 7.31, 7.58, 10.1, 10.10, 10.12, 10.28, 10.30.

Dalradian (partly Cambrian):- E5.26, 5.27, 5.28, 5.29, 5.30, 5.31, 5.32, 5.34, 5.35, 5.36, 5.37, 5.38, 5.39, 5.40, 5.41, 5.42, 5.43, 5.45, 5.46, 5.47, 5.49, 7.58, 9.6, 9.11, 9.12, 10.1, 10.3, 10.21, 10.22, 10.23, 10.24, 10.26, 10.29.

Shetland Metamorphic Rocks:- E5.21, 5.22.

Precambrian, other than below:- E6.1, 6.3, 6.7, 6.8, 6.9, 6.12, 6.34, 7.28, 7.29, 7.31, 7.32, 10.1, 10.9, 10.10, 10.12.

Torridonian:- E4.3, 4.4, 4.5, 4.6, 4.7, 5.4, 5.6, 5.7, 5.8, 5.9, 5.10, 10.28.

Moinian:- E5.1, 5.2, 5.3, 5.5, 5.10, 5.12, 5.16, 5.17, 5.18, 5.19, 5.20, 5.25, 5.26, 5.28, 5.29, 5.33, 5.34, 5.35, 5.36, 5.37, 5.39, 5.44, 9.3, 10.1, 10.21, 10.26, 10.27, 10.28.

Lewisian:- E4.1, 4.2, 4.3, 4.4, 4.5, 4.6, 4.7, 5.1, 5.3, 5.4, 5.5, 5.7, 5.8, 5.9, 5.10, 5.17, 5.18, 10.21, 10.28.

Intrusive igneous rocks

Tertiary:- E5.9, 5.10, 5.18, 5.19, 5.20, 5.35, 5.38, 5.40, 5.42, 6.26, 7.54, 7.55, 7.56, 9.1, 9.2, 9.3, 9.4, 9.5, 9.6, 9.7, 9.8, 9.9, 9.10, 9.11, 9.12, 9.13, 9.14, 9.15, 9.16, 10.1, 10.20, 10.21, 10.25, 10.26, 10.28, 10.29, 10.30.

Permian:- E7.30, 7.47, 7.54.

Hercynian and Permo-Carboniferous:- E5.11, 5.19, 5.20, 7.1, 7.2, 7.4, 7.5, 7.6, 7.10, 7.38, 7.40, 7.41, 7.43, 7.46, 7.48, 7.51, 7.53, 10.1, 10.3, 10.4, 10.6, 10.14, 10.17, 10.18.

Carboniferous:- E7.34, 7.44, 7.45, 7.49, 7.51, 7.52, 7.60, 10.3, 10.4, 10.12, 10.20, 10.25.

Devonian, South-West England:- E7.1, 7.7, 7.8, 10.14.

Devonian and Silurian (Caledonian):- E5.3, 5.4, 5.5, 5.12, 5.14, 5.17, 5.19, 5.20, 5.22, 5.25, 5.26, 5.31, 5.32, 5.33, 5.34, 5.35, 5.36, 5.37, 5.39, 5.40, 5.41, 5.46, 5.47, 5.48, 5.50, 6.13, 6.14, 6.15, 6.17, 6.21, 6.23, 6.25, 6.29, 6.31, 7.32, 7.37, 9.15, 10.1, 10.21, 10.22, 10.23, 10.26, 10.27, 10.29, 10.30.

Ordovician:- E5.47, 5.50, 6.1, 6.2, 6.4, 6.6, 6.10, 6.12, 6.17, 6.27, 6.31, 6.32, 7.56.

Precambrian:- E4.1, 4.2, 4.3, 4.4, 4.5, 4.6, 4.7, 5.16, 5.43, 5.44, 5.45, 5.46, 6.3, 7.27, 10.23.

Metamorphic rocks of doubtful age in South-West England:- E7.3, 7.12.

Centres in Alphabetical Order
with references to excursions from each centre

Aberdeen:- E5.31, 5.32.
Alnmouth:- E7.38, 7.39.
Arbroath:- E7.57, 7.58.
Arklow:- E6.31, 6.32.
Ayr:- E6.23, 7.54, 7.55, 7.56.
Bangor:- E6.1, 6.2, 6.3, 6.4.
Barmouth:- E6.5, 6.6.
Belfast:- E6.26, 9.13, 9.14, 9.15, 9.16
Birmingham:- E7.28, 7.29, 7.30, 7.31.
Bray:- E6.29, 6.30.
Bristol:- E7.13, 7.14, 7.15, 8.18, 8.19.
Broadford:- E5.9, 5.10, 9.1, 9.2.
Brodick:- E5.42, 9.9, 9.10, 9.11, 9.12.
Bude:- E7.9, 7.10.
Builth Wells:- E6.10, 6.11.
Buxton:- E7.33, 7.34.
Cardiff:- E7.16, 7.17, 7.18, 7.19, 7.20, 7.21.
Church Stretton:- E6.7, 6.8, 6.9.
Cork:- E7.64, 7.65.
Creeslough:- E5.46.
Donegal:- E5.44, 5.45.
Dublin:- E6.27, 6.28, 7.59.
Dumfries:- E6.20, 6.21, 6.22.
Dunbar:- E6.18, 6.19.
Durham:- E7.40, 7.41, 7.42, 7.43.
Durness:- E5.1, 5.2.
Edinburgh:- E7.49, 7.50, 7.51, 7.52, 7.53.
Fishguard:- E6.12.
Folkestone:- E8.13, 8.14, 8.15.
Fort William:- E5.33, 5.34, 5.35, 5.36, 5.37.
Gairloch:- E4.5, 4.6.
Galway:- E5.49, 5.50.
Glasgow:- E5.41, 7.44, 7.45, 7.46, 7.47, 7.48.
Helmsdale:- E5.14, 5.15.

Routes in Alphabetical Order
with references to geological
descriptions

Belfast to:- Dublin (rail) E10.30, Portrush (road) E10.29.
Cardiff to:- Brecon and Fishguard E10.12.
Carlisle to:- Glasgow (road) E10.19.
Derby to:- Castleton and Sheffield (road) E10.15.
Dover to:- London (rail) E10.5.
Dublin to:- Cork (road) E10.31.
Edinburgh to:- Glasgow (rail) E10.4.
Exeter to:- Launceston and Penzance (road) E10.14.
Folkestone to:- London (rail) E10.5.
Glasgow to:- Ardrossan and Brodick (rail and boat) E10.20, Fort William
and Mallaig (road) E10.21, Fort William and Mallaig (rail) E10.22,
Inverary and Oban (road) E10.23, Lochwinnoch-Largs-Gourock-Glasgow (road)
E10.25, Oban (rail) E10.24.
Inverness to:- Thurso and Orkney (road and boat) E10.27, Kyle of Lochalsh,
Skye, Harris and Lewis (rail, boat and road) E10.28.
London to:- Bristol and Swansea (rail) E10.7, Edinburgh and Aberdeen (rail)
E10.3, Glasgow (rail) E10.2, Gloucester and Cardiff (road) E10.11, Holyhead
(rail) E10.9, Holyhead (road) E10.10, Inverness (rail) E10.1, Manchester and
Liverpool (rail) E10.8, Plymouth and Penzance (rail) E10.6, Portsmouth (road)
E10.13.
Newcastle to:- Carlisle (road) E10.17, Kielder and Glasgow (road) E10.18.
Oban to:- Tobermory (boat) E10.26.
Sheffield to:- Manchester (road) E10.16.

Index